Chemical Demonstrations

Volume 5

Volume 5 Coauthors

RODNEY SHREINER, PH.D.
    Senior Scientist, Associate Director of the Wisconsin Initiative for Science Literacy,
    University of Wisconsin–Madison

JERRY A. BELL, PH.D.
    Professor Emeritus, Simmons College, Fellow of the Wisconsin Initiative for Science
    Literacy, University of Wisconsin–Madison

Chemical Demonstrations

A Handbook for Teachers of Chemistry

Bassam Z. Shakhashiri

VOLUME **5**

The University of Wisconsin Press

The University of Wisconsin Press
1930 Monroe Street, 3rd Floor
Madison, Wisconsin 53711-2059
uwpress.wisc.edu

3 Henrietta Street
London WC2E 8LU, England
eurospanbookstore.com

5   4   3   2   1

Printed in Canada

Library of Congress Cataloging-in-Publication Data
(Revised for volume 5)
Shakhashiri, Bassam Z.
Chemical Demonstrations.
Includes bibliographical references and index.
I. Chemistry—Experiments. II. Title.
QD43.S51983540'.7'881-70016
ISBN 978-0-299-08890-3 (v. 1)
ISBN 978-0-299-10130-5 (v. 2)
ISBN 978-0-299-11950-8 (v. 3)
ISBN 978-0-299-12860-9 (v. 4)
ISBN 978-0-299-22650-3 (v. 5)

To
Ron Perkins,
for his educational innovations in and out of the
high school classroom—a teacher for us all,

and to

Oliver Sacks,
a chemist at heart,
for making me aware of the complexity of memory,
emotions, and the brain, and helping me to
understand myself and to explain our beautiful chemical world.

B.Z.S

The demonstrations and other descriptions of use of chemicals and equipment contained in this book have been compiled from sources believed to be reliable and to represent the best opinion on the subject as of 2010. However, no warranty, guarantee, or representation is made by the authors or the University of Wisconsin Press as to the correctness or sufficiency of any information herein. Neither the authors nor the publisher assumes any responsibility or liability for the use of the information herein, nor can it be assumed that all necessary warnings and precautionary measures are contained in this publication. Other or additional information or measures may be required or desirable because of particular or exceptional conditions or circumstances, or because of new or changed legislation. Teachers and demonstrators must develop and follow procedures for safe handling, use, and disposal of chemicals in accordance with local regulations and requirements.

Contents

Preface   xi

Shining Light, Shedding Light   xiii
    Roald Hoffmann

Communicating Science via Demonstrations   xxi
    Bassam Z. Shakhashiri

Sources Containing Descriptions of Lecture Demonstrations   xxvii

Sources of Information on Hazards and Disposal   xxxi

Displaying Small Phenomena to a Large Audience   xxxiii

12 COLOR, LIGHT, VISION, PERCEPTION   1
    Rodney Schreiner, Jerry A. Bell, and Bassam Z. Shakhashiri

    Color   3
    More Properties of Light   7
    The Nature of Light   11
    Electromagnetic Radiation   22
    Interactions of Light and Matter   31
    Vision   66
    Perception   78
    The Demonstrations   84

The Production of Light   85

12.1 The Emission Spectrum from a Candle Flame   87
12.2 The Temperature Dependence of the Emission Spectrum
     from an Incandescent Lamp   93
12.3 Incandescence from the Combustion of Iron and of Zirconium 98
12.4 Chemical Reactions That Produce Light   102
12.5 Emission Spectra from Gas-Discharge Lamps   103
12.6 Colored Flames from Metal Ions   108
12.7 Light-Emitting Diodes: Voltage and Temperature Effects   111
12.8 Electrogenerated Chemiluminescence   118
12.9 Chemiluminescence   123
12.10 Chemiluminescence from the Explosive Reaction of Nitrous Oxide
     and Carbon Disulfide   124

Properties of Light   127

12.11 The Conversion of Light Energy to Thermal Energy   129
12.12 Refraction and Diffraction: The Separation of White Light
       into Colors   131
12.13 Disappearing Glass: Index of Refraction   135
12.14 Disappearing Gel: Index of Refraction   141
12.15 Observing the Transmission Spectra of Dyes   146
12.16 Dichroism: Transmission versus Reflection   151
12.17 Iridescence from a Polymer Film   154
12.18 The Photoelectric Effect   156
12.19 The Tyndall Effect: Scattered Light Is Polarized   160
12.20 Rainbow Spiral in an Optically Active Solution   163
12.21 A Sugar Solution Between Polarizers   166
12.22 The Birefringence of Calcite   167
12.23 A Liquid Crystal Display through a Polarizer   171
12.24 Laser Light Is Polarized   175

Perception and Vision   179

12.25 Additive Color Mixing   181
12.26 Subtractive Primary Colors   189
12.27 The Perception of Brightness Is Relative   192
12.28 The Hermann-Grid Illusion   196
12.29 Finding the Blind Spot   201
12.30 The Land Effect   204
12.31 Saturation of the Retina: Afterimage   208
12.32 The Persistence of Vision   212
12.33 The Imprecision of Peripheral Vision   215
12.34 The Pulfrich Phenomenon: Perception of Motion   218

Photemission: Fluorescence and Phosphorescence   221

12.35 Photoluminescence   223
12.36 The Halide Quenching of Quinine Fluorescence   227
12.37 Differentiation of Fluorescence and Phosphorescence   232
12.38 Phosphorescence Excitation: Energy and Color Relationship   235
12.39 Quenching Phosphorescence with Light   238
12.40 Quenching Phosphorescence with Thermal Energy   241
12.41 The Fluorescence of Molecular Iodine Vapor   243

Photochemistry   247

12.42 The Reversible Photochemical Bleaching of Thionine   249
12.43 Photochromic Methylene Blue Solution   256
12.44 The Photochemical Reaction of Chlorine and Hydrogen   260
12.45 The Effects of Solvents on Spiropyran Photochromism and Equilibria   261
12.46 A Copper Oxide Photocell   274
12.47 The Photobleaching of Carotene   277

12.48 Making a Cyanotype   283
12.49 An Iron(III)-Oxalate Actinometer   287
12.50 The Photoreduction of Silver Halide   291
12.51 Photochemistry in Nitroprusside-Thiourea Solutions   293
12.52 Photochromism in Ultraviolet-Sensitive Beads   300
12.53 The Photodissociation of Bromine and the Bromination
          of Hydrocarbons   304
12.54 The Photochemical Formation and Reaction of Ozone   308

Index to Volumes 1–5   311
Illustration Credits   323

Preface

When I was growing up in my native Lebanon, my mother knitted a yellow sweater for me. It was warm and comfortable, but what struck me the most was its bright color. I was fascinated by its beauty. I was curious, and I asked lots of questions. How long will the color last? Will it look the same after it is washed? Will the color clash with my favorite pair of pants? What makes it yellow? What is yellow? I was told the material was wool. I knew that wool comes from sheep and that their fleece is white. What made it yellow? I was told that wool was colored using different dyes and then made into yarn of various hues. I was fascinated and started thinking about color in my surroundings. I loved color combinations and wanted to know why some combinations are more harmonious than others. The sky is blue and clouds are white, but sometimes near sunset they have different colors. Trees have green leaves and some stay green year round, notably the Cedars of Lebanon. The leaves of other trees changed color and fell off in the fall—only to reappear, green, in the spring. Some lemons are green, while others are yellow. Watermelons are mostly green on the outside, but red on the inside. Looking in the mirror, I observed my blue eyes. At night, streetlights shed white light, and many stores had large illuminated signs displaying their names in red. I noticed the blue waves of the Mediterranean Sea and wondered why the tops of breaking waves were the same color as clouds. Color was everywhere—and ever changing.

Later, during my formal education, I began to understand more about color and was satisfied with the explanations to my youthful curiosities. However, this understanding led to further questions that directed my scholarly investigations to the properties of light and its interactions with matter. The contents of this volume reflect some of the fruits of these investigations.

Color, light, vision, and perception are the topics of this fifth volume in the series of handbooks aimed at providing teachers of science at all educational levels with detailed instructions and background information for using chemical demonstrations in the classroom and in public presentations. This volume deals with sense of sight and consists of one large chapter, with 54 demonstrations and 83 different procedures. The extensive introduction to this volume includes material aimed at reinforcing and expanding the background knowledge of the user. I believe firmly that whenever demonstrations are presented, the phenomena should be discussed and explained at a level suitable to the audience. A number of demonstrations included in this volume involve quite complex chemical concepts. The intricate details of electronic transitions in colorful (and all) chemical transformations are important, and I hope our explanations will enhance the teacher's ability to use the demonstrations effectively. Furthermore, I believe that by sharing fascinating aspects of the physiology of vision and of the psychology of perception, the teacher can succeed in triggering deeper interest in chemical biology, cognitive science, and neuroscience. I urge teachers in elementary and secondary schools, as well as teachers in colleges and universities, to use the material in the introduction to this volume, as well as in the demonstrations, to display intriguing chemical behavior and scientific concepts.

This volume is part of a continuing project whose purposes are to create, collect, develop, test, and publish demonstrations that will help teachers to connect chemistry with the sensations we receive via our senses of sight, hearing, smell, taste, and touch. Most classroom and public science demonstrations engage the brain through the eye and the ear, but very few involve olfaction, gustation, and touch. These latter topics are the subjects of forthcoming volumes.

I am fortunate that my longtime colleagues and personal friends Rodney Schreiner and Jerry Bell share the same interests. Their collaboration as principal coauthors has greatly enriched this project and made it possible. Without their contributions this volume would still be a figment of my imagination.

Rod Schreiner, Jerry Bell, and I are grateful for the suggestions and comments made by numerous colleagues around the country. In particular, we thank our good friend Ron Perkins, who often describes himself as "*just* a high-school teacher" but who is much more to us, for his careful review of the entire manuscript, for making detailed suggestions to help clarify concepts and improve procedures, and for contributing several "Wow! Neat! Super!" demonstrations. We deeply thank our friend Professor Roald Hoffmann for his captivating essay, the crowning jewel of this book, and for his gifts to science, the arts, and the humanities. We express deep appreciation for the perceptive suggestions made by my friends and colleagues Professor Richard N. Zare (Stanford), Professor John C. Wright (UW–Madison), Professor Nansi Colley (UW–Madison), Professor Emeritus Morton Z. Hoffman (Boston University), Professor John F. Berry (UW–Madison), Dr. Kenneth P. Fivizzani (ret. Ondeo Nalco), Dr. Rob McClain (UW–Madison), Professor George C. Lisensky (Beloit College), Professor Emeritus James H. Espenson (Iowa State), Professor Emeritus Frank A. Weinhold (UW–Madison), and Professor Emeritus Hyuk Yu (UW–Madison).

Invaluable testing of demonstration procedures was performed by Stacy Wittkopp (outreach specialist at WISL, the Wisconsin Initiative for Science Literacy), Andrew Aring (now in graduate school at Idaho), Eric Victor (now in graduate school at MIT), Patrick Meloy (Madison East High School), Jeremiah Walsh (Madison West High School), and Jennifer Wroblewski (Madison West High School). They helped assure the high standards of excellence for this project. Also, I thank my former graduate student Mary Ellen Testen (now teaching in Plainfield, Illinois) and my former undergraduate student Jim Maynard (UW–Madison lecture demonstrator).

Others at the University of Wisconsin–Madison provided invaluable assistance: Cayce Osborne in meticulously looking into every detail of correspondence with our collaborators and reviewers and for her design artistry in all aspects of WISL programs beginning in 2009, Patti Puccio in proofreading and for her association with our work beginning in 1970, and the entire University of Wisconsin Press staff, in particular Adam Mehring, the managing editor, and Gail Schmitt, the copy editor, for enhancing the clarity of the text, and Terry Emmrich, the production manager, for the timely production of this colorful volume.

With deep appreciation, I acknowledge the generous support from private donors, and at the University of Wisconsin–Madison, from the Department of Chemistry, the College of Letters and Science, the Graduate School, and the Office of the Chancellor. The University of Wisconsin–Madison makes it all possible.

Ultimately, I am blessed with the supporting love of my wife June and daughter Elizabeth.

Madison, Wisconsin   Bassam Z. Shakhashiri
December 2010   Professor of Chemistry

                           The William T. Evjue Distinguished Chair for the Wisconsin Idea
                           Director, Wisconsin Initiative for Science Literacy
                           University of Wisconsin–Madison

                           2011 President-elect, American Chemical Society

Shining Light, Shedding Light

Roald Hoffmann

FIREFLIES

At first, you're not sure it's real—it could be a passing flash on the retina, a car far off. But then you see a few lights flitting, the road bends into a spruce grove, your eyes adjust. And night strikes the screen for the pinpoint dance of fireflies chasing synchronicity.

In bright light, the beetle itself is unprepossessing. But maybe we shouldn't apply our standards of beauty to fireflies—we have enough trouble making sense of what attracts the male of our species to the female. But the phenomenon—chemiluminescence—is startling because it is relatively uncommon in the fauna and flora we see around us.

Fireflies have something to do with the appeal of color and light that is our subject. For not only do the tested demonstrations in this volume show the workings of light and color in chemistry. They also go beyond the utility of a physical phenomenon, in two ways. First they probe perception, for what the mind does with the wavelength of light that impinges on the retina is more than just register it. And second, these demonstrations touch directly on the beauty and wonder that makes chemistry more than just transforming molecules.

I will return to beauty and fireflies, but let me first delineate the special place of light absorption and emission in chemistry. Light and color in chemistry tell us of quantum mechanics at work. We use them as clues, signals from within. And light effects desired, occasionally dramatic, change.

THE QUANTUM IN CHEMISTRY

The strong, spatially expressed bonding propensities of carbon atoms imbue organic chemistry with a direct architectural sense. I have in mind the four bonds going off to the corners of a tetrahedron of a saturated C, the coplanarity of the six atoms in an ethylene.

Take taxol (paclitaxel), an important antitumor agent, a complex molecule with four rings and a number of substituents. There is much medicinal chemistry modifying the taxol structure, replacing a hydrogen by a succinate, removing an acetyl group.

taxol = paclitaxel

A simpler structure is found in the spiropyrans, molecules that figure in one lovely experiment in this volume. They are not only photochromic compounds with obvious applications in materials science, but their derivatives have been used in the treatment of hypertension and erectile dysfunction.

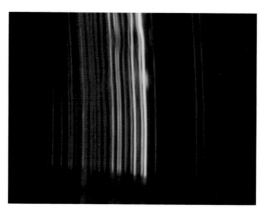

Benzospiropyran   Merocyanine

For a variety of purposes, not the least being simple curiosity, the spiropyran and taxol structures have been man- and woman-handled. In the fascinating range of modifications seen in the literature (and many more hidden in the notes of pharma-industry chemists), the three-dimensional structure of the molecules has played a key role—the oxetane ring of taxol positioned just this way, a reaction proceeding with inversion of configuration.

This is the classical mechanics side of chemistry. The motions of atoms in the course of reaction certainly figure in the thinking of chemists modifying these molecules. But the chemical imagination here is dominantly architectural in feeling—a ball and stick model of the molecule, and knowledge of the stereochemical outcome of chemical reactions will get you a long way in this wonderful game.

Consider now the color of a compound—for instance the change in color at the heart of the spiropyran experiment, from colorless to deep pink or blue, back to colorless. Or the orange of Vitamin B-2, riboflavin, the bright red glow of a neon lamp, the green of chlorophyll. There is no way to explain colors of compounds classically. Molecules have definite quantum states, they have filled and unfilled orbitals. The quantization of energy levels, at the heart of quantum mechanics, is directly displayed in those energy levels. Light is emitted (by those neon atoms in a discharge tube) or absorbed (by carotene in another experiment in this volume) at quite definite energy intervals. These are the colors of molecules as they impinge on the retina and—in much more fascinating detail—the spectrum of a molecule.

Reflection of a neon tube from the surface of a compact disc.

Any spectroscopic measurement or photochemical reaction is inherently quantum mechanical. Or shall we say, quantum chemical?

SIGNALS FROM WITHIN

The determination of molecular structure is a story of knowing without seeing. It is a thrilling detective story in which meager clues to the structure of a molecule, none definitive, are assembled by the molecular detective into an incontrovertible web of evidence for what atoms there are in a molecule, how they are connected to each other, and what their three-dimensional arrangement is in space.

We did not wait for microscopes to tell us the structure of carotene, benzospiropyran, or taxol. We did it with X-ray diffraction and spectroscopies of every sort, perturbing the molecule with light and measuring its response.

The light involved in our measurements is not just visible light. It is radiation that may range from the energetic extreme of X-rays to the slightest tickling of nuclear magnetic resonance. In every case we send in a beam of the radiation and measure its absorption, or lack of it, by an interposed sample of the compound (macroscopic) = molecule (microscopic) in question. Most of the time, the probing is nondestructive and can even be done on the material in situ, for instance reflection spectroscopy of the blue pigment in a Japanese scroll. Sometimes, one needs to take a physical sample. This is okay for Vitamin B-2 coming off an assembly line, less so for the Japanese scroll.

In X-ray crystallography, currently our major technique for getting precise molecular structures, one starts with a crystal or powder and measures (it used to be photographically, now by an electronic detector) the intensities and patterns of reflection of X-radiation from layers of atoms in a molecule in that crystal. With some clever programming and hale computers, one goes from diffraction pattern to structure on the atomistic level in a matter of minutes.

In infrared spectroscopy, one puts in infrared radiation of variable wavelength and studies how much of it is absorbed by a molecule. Absorbed in what way, and why? Well, a molecule has characteristic and well-defined energy levels corresponding to the motion of atoms around their equilibrium position in the ground state of a molecule. And it has excited vibrational states, different ones corresponding to the stretching of a CO bond in a ketone ($\sim$1700 cm^{-1}) to those in an alcohol ($\sim$1100 cm^{-1}). The infrared signature of each is distinct. And, as we will see below, those vibrations really matter in the atmosphere.

I cannot imagine the incredibly detailed microscopic knowledge we have of molecular structure without the intermediate, willed exploitation of light and color. And we are beginning to "see" molecules in the act of reaction, through femtosecond spectroscopies and state-selective reactions.

ESPECIALLY IN THE ATMOSPHERE

Light plays a special role in what goes on in the precious envelope that surrounds our planet. For it is the way that energy gets into molecules there, and—for better or for worse—does chemistry.

The atmosphere is tenuous, the pressure falling off exponentially with altitude. And much of the atmosphere is quite cold. Low pressure, low temperature means that there are fewer and fewer collisions per unit time as one goes to higher altitudes. And you can't have chemistry without collisions.

To put it another way, at the surface of the earth we may provide heat (from gas, coal, electricity) to some reagents in an enclosed vessel and thereby effect desired change. This happens in the Haber-Bosch process for making NH_3 from N_2 and H_2. Or in baking a meringue. In the upper atmosphere, the "vessel" is immense, with a low density of molecules in it. And there is no equivalent of a Bunsen burner or electric mantle to put energy into the atmosphere.

What the atmosphere has is light, a surfeit of it beamed to it from our sun. And the atmosphere cannot avoid reflected and emitted radiation from the earth. The way energy is traded in the atmosphere is mainly through the absorption and emission of light.

Let's take a topic of substantive current interest—global warming. No question the warming is there—I can show you a timeline of first and last frost in Atlantic City that goes back to 1874.

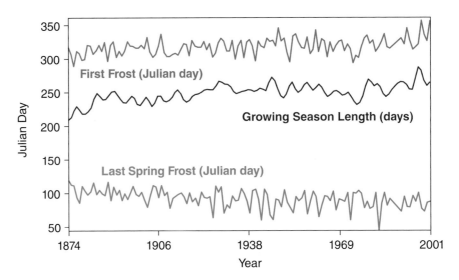

Growing season length and freeze dates at Atlantic City, New Jersey. The red line represents the annual length of the growing season, while the yellow area represents the growing season, the time between the last frost of spring (green line) and the first frost (blue line) of fall. From 1874 to 2001, the growing season has lengthened by 19 percent [1].

There be fluctuations, big ones. But there is also a trend that is clear—the growing season in Atlantic City has increased by over a month over 136 years. If this continues, there will be palms along the Boardwalk.

How does it happen? How does the atmosphere warm? Let's look at the energy of sunlight as it passes into the atmosphere. The atoms, ions, and molecules in the atmosphere absorb light. But not light of any wavelength. This is where quantum mechanics, as mentioned above, comes in, providing not only a set of quantized levels setting the color, but also a set of rules for what light will or will not be absorbed. Especially important for humans and for life as it has evolved is the absorption of the sun's light by a tenuous layer of ozone in the stratosphere. If concentrated at P = 1 atm, all the ozone in the atmosphere would be a layer a few millimeters thick. And you had better run—ozone, a lovely blue when concentrated—is very explosive.

The ozone molecules in the stratosphere absorb out of the sun's broad spectrum radiation a region in the ultraviolet that we have evolved to tolerate. I put it this way, for it is perfectly possible to envisage life without an ozone layer. But it would not be the life we know.

Of the sunlight impinging on the atmosphere 35 percent is reflected, 17.5 percent is absorbed by molecules in the atmosphere, 47.5 percent reaches the earth one way or another—directly or scattered by clouds and blue sky. What gets down here is absorbed—the greens, blues, and browns of the earth are testimony to that. Some of the energy of the absorbed sunlight is harnessed by photosynthetic systems, eventually leading to the local defeat of entropy represented by dandelion flowers and a poem.

Most of the energy that is absorbed by brown earth or blue water or green plant is degraded by a process called "intersystem crossing," by which an excited electronic state is transformed into a vibrationally excited ground state. The vibrationally hot molecules emit infrared radiation, and this high-wavelength, low-frequency light augments the so-called black-body radiation of the earth at its fluctuating surface temperature.

The earth shines in the infrared. Over 99 percent of the atmosphere, the recipient of this emission, couldn't care less. N_2, O_2, and Ar do not absorb infrared radiation to any significant amount. Why? Quantum mechanics again! The subject here is selection rules for vibrational spectroscopy: molecules whose dipole moment does not change in the course of a vibration will not absorb in the IR.

So H_2O, CO_2, CO, CH_4, NO do absorb infrared radiation, in different regions of the spectrum, some overlapping well, some badly, the spectrum of impinging IR radiation from the earth. They are "greenhouse gases." But wait, the story, a story of trading energies, is not quite over. How do the molecules that do absorb in the IR heat the atmosphere? Their own motions are enhanced by the energy of the infrared light they have absorbed—they vibrate like crazy. But these molecules are a tiny, trace part of the atmosphere. The temperature of the atmosphere is set by the velocities and collisions of the majority molecules—N_2, O_2, Ar. And these are not excited by IR radiation.

Not yet. The mechanism for heating the atmosphere is vibrational to translational energy transfer, from, say, CO_2 to N_2. The effect is quantum mechanical, but one can understand it semi-classically. While a CO_2 molecule is doing, say, an energetic asymmetric stretch, it collides with a relatively slow moving N_2 molecule.

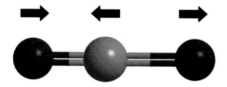

asymmetric stretch
2346.3 cm^{-1}

The vibration of the CO_2 delivers a kick to the N_2; the CO_2 vibrates less violently thereafter, the N_2 moves off with great velocity. Do it a billion times, you have made the N_2 molecules move faster, on average. The N_2 is hotter. And N_2 is 78 percent of the atmosphere.

The anthropomorphic language I use is imprecise. But it serves to make us less afraid of the quantum logic that governs the microworld. That energy is conserved does not mean that life (or the atmosphere) is static. Energy is traded continually. There is balance, dynamic balance. Or that balance is perturbed. By us.

BEAUTY

Now that was a slog through the intricacies of trading energy in the atmosphere. A necessary excursion, given the importance of the precious envelope of our planet. What does it have to do with the wonder of the clear, dark June night on which we first see fireflies? Or, in another season, seeing the aurora borealis? Or the rapidly moving clouds in clearest air the day after the hurricane spent itself?

What is the relationship of the facts and mechanisms, as best as we may discern them, to the emotion that comes over us when the sky, in its incredible range of presentations, presents such beauty?

One point of view is that knowing about things, here color and light, in the way chemists work diligently to gain such knowledge, detracts in some way from the gut feeling that beauty is before us. I don't think so. Philosophers have thought about beauty for over 2,500 years. Here, for instance, is what Immanuel Kant (who lived before and after Lavoisier) wrote about beauty:

> He who feels pleasure in the mere reflection upon the form of an object . . . justly claims the agreement of all men, because the ground of this pleasure is found in the universal, although subjective, condition of reflective judgments, viz. the purposive harmony of an object (whether a product of nature or of art) with the mutual relations of the cognitive faculties (the imagination and the understanding) [2].

A sentence Kantian in complexity, so let me try to unpack its meaning. Kant says that while beauty is in the eye of the beholder, there are also universals in the idea. That an object is beautiful derives from two sources—the immediate, emotional appeal of its harmony, coupled to our thinking about it, trying to understand, seeing the relations of an object to everything else in the world.

To see the fireflies, to love their light, links to our desire to understand—chemically and biologically—how and why they flash. That knowledge does not detract from but augments our appreciation of the fireflies. More of that anon. But let's return to doing things with lights.

PHOTODYNAMIC

It is one thing to measure, another to effect transformation. The absorption of light, using mechanisms honed by evolution, plays an essential role in biochemistry. And cultural evolution (science and technology) have introduced into our lives a variety of active photochemical devices—the phosphors on the display of your computer, the neon sign, the sensors in your camera, the silver halides in our old friend, photography.

The difference from spectroscopy, from using light to spy on molecular secrets, is that with photochemistry we aim to achieve change. And do it with large dollops of energy.

The photons of visible light absorbed by the photosynthetic assemblages of nature carry 40–80 kcal/mole. To introduce the same amount of energy into a cell through heating would necessitate heating to 200–1000°C, a process that would certainly destroy the cell. The lovely molecular machines in a leaf of a green plant take in that sunlight in one fell swoop. And then the energy is cleverly, rapidly (and efficiently, ~40 percent) shuttled by a cascade of fast reactions from biomolecule to biomolecule, effecting the transfer of protons across a cell membrane.

I want to tell you the story of another use of light, in healing. Though, in the nicely complicated way the world has of complicating our simplistic categorizations, that healing is accomplished by destruction. This is photodynamic therapy, and the particular use of porphyrins in it.

As it is, there are in our body destructive molecules—NO, and the singlet form of oxygen, an excited state of normal triplet state diatomic oxygen molecules. These destructive small molecules attack tissue components in a number of ways, and there have evolved ways for their controlled production in the body, for, if you like, housecleaning tasks. The generation and disposal of these internal reagents is nicely controlled. Singlet oxygen is essentially "biodisposable"—in water it has just a couple of microseconds to do its dastardly deeds. The idea of photodynamic therapy is simple: why not generate a multitude of such molecules, singlet oxygen in particular, at a desired place and time to destroy tumor cells and other tissue?

Outside the body, in the laboratory, we have come up with ways to "sensitize" oxygen, convert the ground triplet state of the molecule to the excited singlet state. The sensitizer is a usually more complex molecule that absorbs light, then undergoes another kind of "intersystem crossing" to an excited triplet state. With a proper match of energies and orbital shapes, that triplet state of the photosensitizer can interact with normal triplet oxygen and transform it to the highly reactive singlet form. Trading energies again, controlled by quantum mechanics.

David Dolphin of the University of British Columbia was and is an expert on the lovely class of molecules called porphyrins. They, or molecules related to them, play important biochemical roles—witness hemoglobin, coenzyme B-12, and chlorophyll. Some synthetic porphyrins are photosensitizers of singlet oxygen formation. Taking account of the biological constraints, Dolphin synthesized a set of porphyrin structures that are effective at the task. One of these, verteporfin = visudine^R, a mixture of two regioisomers and their mirror images, is widely used for treatment of one form of macular degeneration.

verteporfin

How nice that here a photochemical reaction is used for curing a disorder of our most prominent photochemical organ!

FIREFLIES

What makes them flash? It's luciferin, three rings of carbon, sulfur, and nitrogen, reacting with oxygen and ATP to a stressed molecule that gives off light with 100 percent efficiency.

luciferin

There's more to the story here, of reflecting uric acid crystals to make the light brighter. And how it's all for sex: male signals, the delay of the female flash back counts. I could tell you of the femme fatale of *Photuris*, imitating the flash pattern to poor male *Photinus* fireflies, promptly eaten to get a parcel of chemicals warding off spiders. But here and now I see the spruce's mass traced by fireflies, not falling like the dark angel, but in their criss-crossing outshining stars, imposing piecewise order.

Beautiful, physical, chemical, beautiful.

REFERENCES

1. Adam Markham and Cameron Wake, *Indicators of Climate Change in the Northeast*, Clean Air—Cool Planet: Durham, New Hampshire (2005).
2. Immanuel Kant, *Critique of Judgment*, trans. J.H. Bernard, p. 28, Hafner: New York (1951).

Communicating Science
via Demonstrations

Bassam Z. Shakhashiri

Science is one of the most important forces affecting human society. The differences between the way we live now, at the dawn of the twenty-first century, and the way our ancestors lived, say, two hundred years ago, are due mostly to science and its resultant technologies. It is impossible to understand contemporary society without some appreciation for science, how it works, and what it tells us about the physical world. Science is fundamentally a human endeavor, driven by the same impulses that motivate much of human activity: curiosity about the unknown, the thrill of discovery, delight in creativity, and the benefits derived from understanding. Fundamental, too, is the desire to share the curiosity, thrill, delight, and benefits. This desire to share is perhaps most acutely displayed by science teachers, whose deepest desire is to effectively communicate the beauty of science, both in and out of the classroom. One of the most effective means of communicating this beauty, of stimulating curiosity, and of sharing the thrill of discovery in science is through demonstrations of physical phenomena. Through my experience in presenting demonstrations, I have come to appreciate that there are characteristics common to effective demonstrations, and I wish to share some of these with you in what follows.

CLARITY OF PURPOSE

To assure that a demonstration communicates effectively, a teacher must have a clear idea about the purpose(s) and value of presenting the demonstration. This means knowing *what* the demonstration is about, *why* it is to be used, *how* it should be presented, and *where* and *when* it will be presented. The purposes might include displaying phenomena, illustrating principles, conveying attitude, sparking interest, nurturing curiosity, making connections, and so on. Engaging people of all ages through the senses in carefully crafted and well-orchestrated scientific adventures is my penultimate purpose. My ultimate purpose is to trigger cerebral and emotional discourse within the individual and among individuals to heighten the joy of learning.

Before selecting a demonstration, I always ask myself, What is the point? My purposes for using demonstrations in the classroom and in other public settings generally include displaying science phenomena and illustrating scientific principles. However, even more importantly, demonstrations convey the presenter's attitude toward science as a human endeavor, and in turn, learners' attitudes can be influenced by teachers doing experiments in classroom settings. Chemistry demonstrations help focus students' attention on chemical behavior and properties, and they increase students' awareness and knowledge of chemistry. The direct observation of a phenomenon as provided in a demonstration can stimulate the observer to develop an immediate concept of the phenomenon. A demonstration can also transform an abstract concept from the theoretical and imagined to the observed and real. Sometimes demonstrations are presented solely for the entertainment value of fire, smoke, and noise. However, to approach demonstrations simply as a chance to show off dramatic chemical changes or to entertain students is to fail to appreciate the opportunities they provide to communicate scientific concepts and to acquire knowledge of the properties of chemicals.

The demonstration should be a process, not a single event. The instructional purposes of a presentation dictate whether a phenomenon is demonstrated or whether a concept is developed and built by a series of experiments. In demonstrations, the teacher's knowledge of the behavior and properties of the chemical systems is key to successful instruction, and the way in which the teacher safely manipulates the chemical system serves as a model not only of technique but also of attitude. Demonstrations can involve student participation through responses to questions and suggestions, such as What will happen if you add more of . . . ? Even in a demonstration where the teacher directs the flow of events, the teacher can ask the same sort of "what if" questions and can proceed with further manipulation of the chemical system. In principle and in practice, every demonstration is a situation in which teachers convey their attitudes about the experimental basis of chemistry, motivate their students to conduct further experimentation, and lead them to understand the interplay between theory and experiment.

Demonstrations should not, of course, be considered a substitute for laboratory experiments. In the laboratory, students can work directly and at their own pace with the chemicals and equipment and make their own observations and discoveries. In the classroom, students witness chemical changes and chemical systems as manipulated by the teacher. The teacher controls the pace and explains the purposes of each step. Both kinds of instruction are integral parts of the education we offer students.

A HIERARCHY OF TEACHING AND LEARNING CHEMISTRY

In teaching and in learning chemistry, teachers and students engage in a complex series of intellectual activities. These activities can be arranged in a hierarchy that indicates their increasing complexity [1]:

(1)   observing phenomena and learning facts
(2)   understanding models and theories
(3)   developing reasoning skills
(4)   examining chemical epistemology.

This hierarchy provides a framework for including demonstrations in the teaching of chemistry.

At the first level, we observe chemical phenomena and learn chemical facts. For example, we can observe that at room temperature sodium chloride is a white solid composed of cube-shaped crystals. It dissolves in water to form a solution with different characteristic properties of its own. One such property, electrical conductivity, can be readily observed when two wire electrodes connected to a light bulb and a source of current are dipped first into a sample of the sodium chloride crystals and then into the solution. Additional phenomena and facts can be introduced: the white solid has a very high melting point, the substance is insoluble in hexane, its chemical formula is NaCl, etc.

At the second level, we explain observations and facts in terms of models and theories. For example, we teach that NaCl is an ionic, solid compound and that its aqueous solution contains mobile, hydrated ions: sodium cations, $Na^+(aq)$; and chloride anions, $Cl^-(aq)$. The mobility accounts for the electrical conductivity of the solution. The solid, which consists of Na^+ and Cl^- particles, is said to have ionic bonds; that is, there are electrostatic forces between the oppositely charged particles. The ions are fixed in place and arranged throughout the solid in a regular three-dimensional array called a face-centered cube, an arrangement that explains the cubical shape of the crystals. Here, the teacher can introduce a discussion of the ionic-bond model, bond energy, and bond distances. Similarly, a description of water as a molecular, covalent substance can be presented. The ionic- and covalent-bonding models can be compared and used to explain the observed properties demonstrated by a variety of compounds.

At the third level, we develop skills that involve both mathematical tools and logic. For example, we use equilibrium calculations in devising steps in a scheme for separating substances in aqueous solution. In these calculations, we combine the solubility-product, weak-acid dissociation, and complex-ion formation constants for competing equilibria in analyzing the behavior of a mixture of ions. The logical sequence of steps in the separation scheme is based on an understanding of the equilibrium aspects of solubility phenomena.

At the fourth level, we are concerned with chemical epistemology. We examine the basis of our chemical knowledge by asking questions such as, How do we know that the cation of sodium is monovalent rather than divalent? and How do we know that the crystal structure of sodium chloride can be determined from x-ray data? At this level we deal with the limits and validity of our fundamental chemical knowledge.

Across all four levels, the attitudes and motivations of both the teacher and the student are crucial. The attitude of the teacher is central to the success of interactions with students. Our motivation to teach is reflected in what we do and in what we do not do, both in and out of the classroom. Our modes of communicating with students affect their motivation to learn. All aspects of our behavior influence students' confidence and their trust in what we say. Our own attitudes toward chemicals and toward chemistry itself are reflected in such matters as how we handle chemicals, adhere to safety regulations, approach chemical problems, and explain and illustrate chemical principles. As I said at the outset, the most important purpose that lectures and demonstrations serve is to give teachers the opportunity to convey an attitude toward chemistry—to communicate to students an appreciation of chemistry and its usefulness, its cohesiveness, its value as the central science and as the science of the familiar, and its intellectual excitement and challenges.

PRESENTING EFFECTIVE DEMONSTRATIONS

In planning to use a demonstration, I always begin by analyzing the reasons for presenting it. Whether a demonstration is spectacular or quite ordinary, I undertake to use the chemical system to achieve specific teaching goals. I determine what I am going to say about the demonstration and at what stage I should say it. Prior to the lecture, I practice doing the demonstration. By doing the demonstration in advance, I often see aspects of the chemistry that help me formulate both statements and questions that I will use in class.

Because one of the purposes of demonstrations is to increase the students' ability to make observations, I try to avoid making statements like "Now I will demonstrate the insolubility of barium sulfate by mixing equal volumes of 0.1 M barium chloride and 0.1 M sodium sulfate solutions." Instead, I say, "Let us mix equal volumes of 0.1 M barium chloride and 0.1 M sodium sulfate solutions and observe what happens." Rather than announcing what should happen, I emphasize the importance of observing all changes. Often, I ask two or three students to state their observations to the entire class before I proceed with further manipulations. Occasionally pausing to point out an interesting observation or to pose a focusing question can increase student involvement by allowing them the opportunity to contribute to an explanation. In addition, I help students to sort out observations so that relevant ones can be used in formulating conclusions about the chemical system. Some valid observations may not be relevant to the main purpose of the demonstration. For example, when the above-mentioned solutions are mixed, students may observe that the volumes are additive. However, this observation is not germane to the main purpose of the demonstration, which is to show the insolubility of barium sulfate. However, this observation is relevant if the purposes include teaching about the additive properties of liquids.

Every demonstration that I present anywhere is aimed at enhancing the understanding of chemical behavior. The chemistry always speaks for itself more eloquently than anything

I can describe in words, write on a chalkboard, or show electronically on a screen. Modern technology now enables us to show an avalanche of text, a multitude of photos, and simulations of almost anything. The Internet provides access to mountains of information of widely varying validity. When technology or the Internet is used in an instructional program, the purpose for its use must be clear, and it should be the best way to achieve the purpose. Their convenience makes them attractive, but descriptions, photos, and simulations of phenomena are never as striking as the real thing.

In addition to following the advice given by Richard Ramette in his essay "Exocharmic Reactions" [2], here is a sampling of what to keep in mind when using demonstrations:

(1)   Be clear about the purpose(s) for doing the demonstration. What is the point you wish to make?

(2)   Plan and base your classroom lectures on one or several carefully selected demonstrations.

(3)   Study the details of the procedures to be followed, and practice more than once what you will be doing. Remember that you are presenting yourself as a skilled professional who clearly understands the science involved in the demonstrations.

(4)   Make sure to allow enough time for the demonstration. Be deliberate and do not appear rushed.

(5)   Prepare an outline or, even better, a "mini-script" to help your pacing.

(6)   Be confident in your presentation and always show enthusiasm.

(7)   Keep the area of the demonstration clean and clear. Ask the audience to focus their attention on the area of your work.

(8)   Be sure to stage your demonstration so everything is visible to your audience. Place an elevated platform above the table to improve visibility. Use a white table cover and white backgrounds for color changes. Use black backgrounds for white or cloudy phenomena and light emission. Scale the demonstration to fit the size of the audience, or if this is not feasible, use audio-visual devices as appropriate (see page xxxiii for information about such devices).

(9)   Practice, practice, practice!

(10) Demonstrations never "fail." A result may be unexpected or unanticipated by you, but this should be welcomed as an opportunity and a teachable moment, especially because it allows you to demonstrate the skill of analyzing a phenomenon that may be as puzzling for you as for your audience. Be prepared to try to explain whatever happens, but if you are not sure, admit the limits of your understanding. Tell the class that you will try to resolve the problem and will share what you learn and try again during the next class meeting.

Good teachers entice others to make connections between new observations and what they already know. Good teachers also encourage observers to share their observations. In the process, teachers often gain deeper insight into and appreciation of the science and beauty of a demonstration. The requirements for an effective demonstration go beyond the mere mechanics of mixing chemicals and manipulating equipment. In addition to the skillful handling of materials, an engaging stage presence is also essential to successful demonstrations and thus is conducive to better learning. Teaching is, I believe, the ultimate performing art.

USING THIS BOOK

This volume contains a single chapter entitled "Color, Light, Vision, Perception." A substantial introduction addresses the science background for the demonstrations that follow. The demonstrations are grouped into sections that deal with the production of light, the properties of light, perception and vision, photoemission (fluorescence and phosphorescence), and photochemistry. Each demonstration includes its own discussion, which employs terminology and concepts that are placed in broader context in the introduction. Accordingly, when teachers read the discussion section of any particular demonstration, they may find it helpful to refer to the introduction for background information. For additional information teachers may wish to consult other sources, including the references provided with the demonstrations.

Each demonstration has seven sections: a brief summary, a materials list, a step-by-step account of the procedure to be used, an explanation of the hazards involved, information on how to store or dispose of the chemicals used, a discussion of the phenomena displayed and principles illustrated by the demonstration, and a list of references. The brief summary provides a possible rationale for using the demonstration as well as a succinct description of the demonstration. The materials list for each procedure specifies the equipment and chemicals needed. Where solutions are to be used, the directions sometimes call for preparing stock amounts larger than those required for the procedure. The teacher should decide how much of each solution to prepare for use in practicing the demonstration and for the actual presentation. The availability and cost of chemicals may also affect decisions about the volumes to be prepared.

The procedure section often contains more than one method for presenting a demonstration. The alternative procedures sometimes offer different methods for displaying the same phenomenon, and in some cases they may demonstrate additional properties of the system of interest.

The hazards and disposal sections include information compiled from sources believed to be reliable. We have enumerated many potentially adverse health effects and have called attention to the fact that many of the chemicals should be used only in well-ventilated areas. In all instances teachers should inquire about and follow local disposal practices and should act responsibly in handling potentially hazardous material.

The purpose of the discussion section is to provide the teacher with information for explaining each demonstration. We include the discussion of chemical equations, relevant data, and properties of the materials involved, as well as a theoretical framework for understanding the phenomena demonstrated. Again, we remind teachers that they should refer to the chapter introduction for additional background information. Finally, each demonstration contains a list of references used in developing the procedures and providing information for the demonstration.

A WORD ABOUT SAFETY

Jearl Walker, who was a professor of physics at Cleveland State University and editor of the Amateur Scientist section in *Scientific American*, has been quoted as saying, "The way to capture a student's attention is with a demonstration where there is a possibility the teacher may die." Walker is said to have caught the attention of his students by dipping his hand in molten lead, by gulping a mouthful of liquid nitrogen, or by lying between two beds of nails and having an assistant with a sledgehammer break a cinder block on top of him. Walker reportedly has been injured twice, once when he used a small brick instead of a cinder block in the bed-of-nails demonstration and once when he walked on hot coals and was severely burned.

I disagree strongly with this kind of approach. Chemical demonstrations that result in injury are likely to confirm beliefs that chemicals are dangerous and that their effects are bad. In fact, every chemical is potentially harmful if not handled properly. That is why every person who does science demonstrations should be thoroughly knowledgeable about the safe handling of all chemicals used in a demonstration and should be prepared to handle any emergency. A first-aid kit, a fire extinguisher, a safety shower, and a telephone must be accessible in the immediate vicinity of the demonstration area.

Demonstrations involving volatile material, fumes, noxious gases, or smoke should be rehearsed and presented only in well-ventilated areas. Local procedures and ordinances for the disposal and storage of chemicals and equipment must be strictly followed. Wearing eye protection is mandatory everywhere, and shielding an audience from potential hazards, such as flying sparks, noxious fumes, ear-piercing sounds, etc., should be part of careful planning. Several of the demonstrations in this book can be hazardous. The procedures are written for experienced chemists, who fully understand the properties of the chemicals and the nature of their behavior. The authors take no responsibility or liability for the use of any chemical or procedure specified in this book.

I urge care and caution in handling chemicals and equipment. Remember to have clarity of purpose for every demonstration by answering the question What is the point?

REFERENCES

1. I have adapted many ideas from Paul Saltman's address at the Third Biennial Conference on Chemical Education, which was sponsored by the American Chemical Society, Division of Chemical Education, and held at Pennsylvania State University, State College, Pennsylvania (1974); see *J. Chem. Educ.*, 52, 25 (1975).
2. R. W. Ramette, "Exocharmic Reactions," in B. Z. Shakhashiri, *Chemical Demonstrations*, vol. 1, pp. xiii–xvi, University of Wisconsin Press: Madison (1983).

Sources Containing Descriptions of Lecture Demonstrations

We call attention to the following sources of information about lecture demonstrations. These lists are not intended to be comprehensive. Some of the books are out of print but may be available in libraries.

BOOKS

Alyea, H. N. *TOPS in General Chemistry*, 3d ed., Journal of Chemical Education: Easton, Pennsylvania (1967).

Alyea, H. N., and F. B. Dutton, eds. *Tested Demonstrations in Chemistry*, 6th ed., Journal of Chemical Education: Easton, Pennsylvania (1965).

Ammon, D., D. Clarke, F. Farrell, R. Schibeci, and J. Webb. *Interesting Chemistry Demonstrations*, 2d ed., Murdoch University: Murdoch, Western Australia (1982).

Arthur, P. *Lecture Demonstrations in General Chemistry*, McGraw-Hill: New York (1939).

Becker, B. *Twenty Demonstrations Guaranteed to Knock Your Socks Off*, Flinn Scientific, Inc.: Batavia, Illinois (1994).

Becker, B. *Twenty Demonstrations Guaranteed to Knock Your Socks Off*, vol. 2, Flinn Scientific, Inc.: Batavia, Illinois (1997).

Bilash, B. II, G. R. Gross, and J. K. Koob. *A Demo a Day Volume 2: Another Year of Chemical Demonstrations*, Flinn Scientific, Inc.: Batavia, Illinois (1998).

Blecha, M. T. "The Development of Instructional Aids for Teaching Organic Chemistry," Ph.D. Dissertation, Kansas State University, Manhattan, Kansas (1981).

Brown, R. J. *333 Science Tricks and Experiments*, Tab Books: Blue Ridge Summit, Pennsylvania (1984).

Chemical Demonstrations Proceedings, Western Illinois University and Quincy-Keokuk Section of the American Chemical Society, Macomb, Illinois, May 5–6, 1978.

Chemical Demonstrations Proceedings, Western Illinois University and Quincy-Keokuk Section of the American Chemical Society, Macomb, Illinois, May 4–5, 1979.

Chemical Demonstrations Proceedings, Western Illinois University and Quincy-Keokuk Section of the American Chemical Society, Macomb, Illinois, May 1–2, 1981.

Chemical Demonstrations Proceedings, Western Illinois University and Quincy-Keokuk Section of the American Chemical Society, Normal, Illinois, June 8, 1982.

Chen, P. S. *Entertaining and Educational Chemical Demonstrations*, Chemical Elements Publishing Co.: Camarillo, California (1974).

Davison, H. F. *A Collection of Chemical Lecture Experiments*, Chemical Catalog Co.: New York (1926).

Ehrlich, R. *Turning the World Inside Out and 174 Other Simple Physics Demonstrations*, Princeton University Press: Princeton, New Jersey (1990).

Faraday, M. *The Chemical History of a Candle: A Course of Lectures Delivered Before a Juvenile Audience at the Royal Institution*, The Viking Press: New York (1960).

Ford, L. A. *Chemical Magic*, T. S. Denison & Co.: Minneapolis, Minnesota (1959).

Fowles, G. *Lecture Experiments in Chemistry*, 5th ed., Basic Books, Inc.: New York (1959).

Frank, J. O., assisted by G. J. Barlow. *Mystery Experiments and Problems for Science Classes and Science Clubs*, 2d ed., J. O. Frank: Oshkosh, Wisconsin (1936).

Freier, G. D., and F. J. Anderson. *A Demonstration Handbook for Physics*, 2d ed., American Association of Physics Teachers: Stony Brook, New York (1981).

Gardner, M. *Entertaining Science Experiments with Everyday Objects*, Dover Publications, Inc.: New York (1981).

Gardner, R. *Magic Through Science,* Doubleday & Co., Inc.: Garden City, New York (1978).

Gilbert, G. L., et al. *Tested Demonstrations in Chemistry*, vol. 1, Denison University: Granville, Ohio (1994).

Gross, G. R., J. K. Koob, and B. Bilash II. *A Demo a Day: A Year of Chemical Demonstrations*, Flinn Scientific, Inc: Batavia, Illinois (1995).

Hartung, E. J. *The Screen Projection of Chemical Experiments*, Melbourne University Press: Carlton, Victoria (1953).

Herbert, D. *Mr. Wizard's Supermarket Science*, Random House: New York (1980).

Herbert, D., and H. Ruchlis. *Mr. Wizard's 400 Experiments in Science*, rev. ed., BookLab: North Bergen, New Jersey (1983).

Humphreys, D. A. *Demonstrating Chemistry: 160 Experiments to Show Your Students*, D. A. Humphreys: Hamilton, Ontario (1983).

Joseph, A., P. F. Brandwein, E. Morholt, H. Pollack, and J. Castka. *A Sourcebook for the Physical Sciences*, Harcourt, Brace, and World, Inc.: New York (1961).

Lanners, E. *Secrets of 123 Classic Science Tricks and Experiments*, Tab Books: Blue Ridge Summit, Pennsylvania (1987).

Liem, T. L. *Invitations to Science Inquiry*, 2nd ed., Science Inquiry Enterprise: Thornhill, Ontario (1990).

Lippy, J. D., Jr., and E. L. Palder. *Modern Chemical Magic*, The Stackpole Co.: Harrisburg, Pennsylvania (1959).

Lister, T. *Classic Chemistry Demonstrations*, Royal Society of Chemistry: London (1995).

Mattson, B., M. A. Kubovy, J. Hepburn, and J. Lannan. *Chemistry Demonstration Aids That You Can Build*, Flinn Scientific: Batavia, Illinois (1997).

McMillan, M. *A Demonstration-A-Day for High School Chemistry*, Science Source: Waldoboro, Maine (2000).

Mebane, R. C., and T. R. Rybolt. *Adventures with Atoms and Molecules: Chemistry Experiments for Young People*, Enslow Publishers: Hillside, New Jersey (1985).

Meiners, H. F., ed. *Physics Demonstration Experiments*, vols. 1 and 2, The Ronald Press Company: New York (1970).

Mullin, V. L. *Chemistry Experiments for Children*, Dover Publications, Inc.: New York (1968).

My Favorite Lecture Demonstrations, A Symposium at the Science Teachers Short Course, W. Hutton, Chairman; Iowa State University, Ames, Iowa, March 6–7, 1977.

Newth, G. S. *Chemical Lecture Experiments*, Longmans, Green and Co.: New York (1928).

Roesky, H. W., and K. Möckel. *Chemical Curiosities*, VCH Publishers, Inc.: New York (1996).

Sarquis, M., and J. Sarquis. *Fun with Chemistry: A Guidebook of K-12 Activities*, Institute for Chemical Education, University of Wisconsin–Madison: Madison, Wisconsin (1991).

Sharpe, S., ed. *The Alchemist's Cookbook: 80 Demonstrations*, Shell Canada Centre for Science Teachers, McMaster University: Hamilton, Ontario (undated).

Siggins, B. A. "A Survey of Lecture Demonstrations/Experiments in Organic Chemistry," M.S. Thesis, University of Wisconsin–Madison, (1978).

Summerlin, L. R., et al. *Chemical Demonstrations: A Sourcebook for Teachers*, vols. 1 and 2, American Chemical Society: Washington, D.C. (1985, 1987).

Sutton, R. M. *Demonstration Experiments in Physics*, McGraw-Hill Book Co.: New York (1938).

Talesnick, I. *Idea Bank Collation: A Handbook for Science Teachers*, S17 Science Supplies and Services Co.: Kingston, Ontario (1984).

Walker, J. *The Flying Circus of Physics—With Answers*, Interscience Publishers, John Wiley and Sons: New York (1977).

Weisbruch, F. T. *Lecture Demonstration Experiments for High School Chemistry*, St. Louis Education Publishers: St. Louis, Missouri (1951).

Wilson, J. W., J. W. Wilson, Jr., and T. F. Gardner. *Chemical Magic*, J. W. Wilson: Los Alamitos, California (1977).

Wray, T. K. *Chemical Demonstrations: The Chem Demo Book. A Guide to Fun, Safe and Exciting Chemical Demonstrations*, 2nd ed., Professional Environmental Trainer Association: Maumee, Ohio (1994).

ARTICLES

Bailey, P. S., C. A. Bailey, J. Anderson, P. G. Koski, and C. Rechsteiner. Producing a chemistry magic show. *J. Chem. Educ.* 52:524–25 (1975).

Castka, J. F. Demonstrations for high school chemistry. *J. Chem. Educ.* 52:394–95 (1975).

Chem 13 News No. 81. This November issue contained a collection of chemical demonstrations (1976).

Gilbert, G. L., ed. Tested demonstrations. Regular column in *J. Chem. Educ.*

Hanson, R. H. Chemistry is fun, not magic. *J. Chem. Educ.* 53:577–78 (1976).

Hughes, K. C. Some more intriguing demonstrations. *Chem. in Australia* 47:458–59 (1980).

Kolb, D. K., ed. Overhead projector demonstrations. Regular column in *J. Chem. Educ.*

McNaught, I. J., and C. M. McNaught. Stimulating students with colourful chemistry. *School Sci. Review* 62:655–66 (1981).

Rada Kovitz, R. The SSP syndrome. *J. Chem. Educ.* 52:426 (1975).

Schibeci, R. A., J. Webb, and F. Farrel. Some intriguing demonstrations. *Chem. in Australia* 47:246–47 (1980).

Schwartz, A. T., and G. B. Kauffman. Experiments in alchemy, Part I: Ancient arts. *J. Chem. Educ.* 53:136–38 (1976).

Schwartz, A. T., and G. B. Kauffman. Experiments in alchemy, Part II: Medieval discoveries and "transmutations." *J. Chem. Educ.* 53:235–39 (1976).

Shakhashiri, B. Z., G. E. Dirreen, and W. R. Cary. Lecture demonstrations, in *Sourcebook for Chemistry Teachers*, pp. 3–16, W. T. Lippincott, ed., American Chemical Society, Division of Chemical Education: Washington, D.C. (1981).

Steiner, R., ed. Chemistry for kids. Regular column in *J. Chem. Educ.*

Talesnick, I., ed. Idea bank. Regular column in *The Science Teacher.*

Wilson, J. D., ed. Favorite demonstrations. Regular column in *J. College Science Teaching.*

Sources of Information on Hazards and Disposal

In preparing the Hazards and Disposal sections of Volume 5, we have used the following sources.

BOOKS

Aldrich Chemical Company Catalog, Aldrich Chemical Co.: Milwaukee, Wisconsin (2009).

Bretherick's Handbook of Reactive Chemical Hazards, 7th ed., P. G. Urban, ed., Elsevier: Burlington, Massachusetts (2007).

Casarett and Doull's Toxicology: The Basic Science of Poisons, L. J. Casarett, C. D. Klassen, and J. Doull, eds., 6th ed., McGraw-Hill (2001).

Clinical Toxicology of Commercial Products, 5th ed., R. E. Gosselin, R. P. Smith, H. C. Hodge, and J. E. Braddock, eds., Williams and Wilkins Co.: Baltimore, Maryland (1984).

CRC Handbook of Laboratory Safety, 5th ed., A. Keith Furr, CRC Press: Boca Raton, Florida (2000).

Flinn Chemical Catalog and Reference Manual, Flinn Scientific, Inc.: Batavia, Illinois (2009).

Guide for Safety in the Chemical Laboratory, 2d ed., Van Nostrand Reinhold Co., Litton Educational Publishing, Inc.: New York (1972).

Handbook of Chemical Health and Safety, R. J. Alaimo, ed., American Chemical Society, Oxford University Press: New York (2001).

Hazardous and Toxic Materials: Safe Handling and Disposal, 2d ed., H. H. Fawcett, ed., Wiley Interscience: New York (1988).

Health and Safety Guidelines for Chemistry Teachers, American Chemical Society Deptartment of Educational Activities: Washington, D.C. (1979). The bibliography lists journal articles and books.

Laboratory Safety Guide, University of Wisconsin–Madison Safety Department: Madison, Wisconsin (2005).

The Merck Index, 14th ed., M. J. O'Neil, ed., Merck & Co., Inc.: Whitehouse Station, New Jersey (2006).

Prudent Practices in the Laboratory: Handling and Disposal of Chemicals, National Research Council, National Academy Press: Washington, D.C. (1995).

Registry of Toxic Effects of Chemical Substances, Department of Health, Education and Welfare (NIOSH): Washington, D.C., revised annually, available from Superintendent of Documents, U.S. Government Printing Office, Washington, D.C. 20402.

Safety in Academic Chemistry Laboratories, 5th ed., American Chemical Society Committee on Chemical Safety: Washington, D.C. (1990). The bibliography lists many journal articles and books.

Safety in the Chemical Laboratory, M. M. Renfrew, ed., Journal of Chemical Education: Easton, Pennsylvania, Vol. 4 (1981).

Safety in the Chemical Laboratory, N. V. Steere, ed., Journal of Chemical Education: Easton, Pennsylvania, Vol. 1 (1967), Vol. 2 (1971), Vol. 3 (1974).

Sax's Dangerous Properties of Industrial Materials, 11th ed., R. J. Lewis, Wiley-Interscience, (2000).

PERIODICALS

Journal of Chemical Education. See the Chemical Laboratory Information Profiles (CLIPs), published by the ACS Division of Chemical Education.

Chemical Health and Safety. Published by Elsevier Science Inc. for the ACS Division of Chemical Health and Safety.

Displaying Small Phenomena
to a Large Audience

Among the reasons for presenting science demonstrations is to provide the audience with direct experience of natural phenomena. Such experience moves concepts from the realm of the imagined to that of the observed. Chemical demonstrations that involve bulk changes in readily available material can usually be presented on a scale large enough to be viewed directly. Seeing the color of three liters of solution oscillate between colorless, orange, and blue (Demonstration 7.1, Briggs-Rauscher Reaction) or seeing two liters of solution glowing brightly blue in a darkened room (Demonstration 2.4, Oxidations of Luminol) produces an unforgettable impression.

Nevertheless, sometimes phenomena occur on a scale that is too small for an audience of more than a few observers to see. Such is the case when phenomena are inherently small scale, such as the glow of a light-emitting diode (Demonstration 12.7) or the birefringence of calcite (Demonstration 12.22). In other cases, the materials may be too costly to allow a large-scale presentation. Presenting these phenomena to a large audience can be facilitated by techniques that allow the entire audience to view the phenomena simultaneously. Simultaneous viewing can be provided by a system that enlarges and projects the visual effect onto a screen. A variety of such projections systems are available, and each has its own advantages and limitations.

A simple and common method for enlarging and projecting an image is the use of an overhead projector. Such projectors consist of a box with a glass top and a mirror mounted on an arm over the top. Inside the box is a very bright lamp, usually cooled by a fan. The light from this lamp is collimated by a Fresnel lens just below the glass top and directed onto the mirror, and the mirror reflects the light onto a screen for viewing. Objects placed on the glass top (stage) of the projector produce a shadow that can be viewed on the screen. Opaque objects produce dark shadows; colored transparent objects produce colored shadows. Thus, overhead projectors are useful to enable a large audience to see color changes in small amounts of transparent material, such as solutions, or shape changes in small opaque objects, such as growing metallic crystals. An overhead projector will not work for displaying color changes in opaque objects. A great many creative techniques for displaying chemical demonstrations using an overhead projector are described in a series of articles, "Overhead Projector Demonstrations," edited by Doris Kolb in the *Journal of Chemical Education.*

Another method for enlarging and projecting an image that is becoming ever more common is provided by a video-display system. Such a system has two major parts: a camera and a display. The camera can be a video camera that generates a video signal directly or a camera designed to be connected to a computer. The display can be either a panel on which the image can be viewed, such as an LCD or CRT, or a video projector. A video camera connects directly to the display, while the computer camera requires a computer to generate the signal for the display. Some computers have a built-in camera, a "web cam." Small objects of all sorts and colors can be accurately displayed to a large audience using a video-display system, as long as the camera can be focused on the objects and the panel or projection are large enough and bright enough to be viewed. In general, video projectors are not as bright as display panels, and projections need to be viewed in dim ambient light. This makes video projectors inappropriate for use in bright areas.

Video displays use a three-color system to produce what appears to be a full-color image. Only three discrete colors (red, green, and blue) are actually emitted by the panel or projector. (Some may add a fourth color; nevertheless, what is emitted are discrete colors, not a continuous spectrum.) This discrete-color system can produce visual artifacts, especially when color vision is the subject of the display. For example, some video systems, when displaying an emission line spectrum such as the reflection of a neon lamp from the surface of a compact disc, show emission lines in the correct relative positions, but with the wrong color. Therefore, each video-projection system should be checked for accuracy of image before it is used.

Document cameras are devices that employ a video-projection system in a configuration similar to an overhead projector. In place of a mirror mounted over an illuminated glass, a document camera uses a video camera to capture an image and transfer it to a video display. Many document cameras have a stage that can be lighted from below, as that of an overhead projector, or from above. When lighted from below, the objects on the stage of a document camera produce an image similar to that produced by an overhead projector, and when lighted from above, the image is similar to that produced by a video system (which the document camera is). Because the image is displayed by a three-color video system, the same precautions regarding color accuracy must be taken with a document camera as with any other video system.

Because video display systems are readily available and generally convenient, it may be tempting to prepare and display an electronic recording of a phenomenon, rather than displaying it directly. However, a prerecorded image is never as impressive as one created "live." Audiences are familiar with contemporary motion pictures and videos that contain all manner of "special effects," in which physically impossible phenomena appear to occur. For this reason, audiences have generally become skeptical of what they see on motion-picture or television screens. This skepticism makes it advisable to present demonstrations of physical phenomena so that the audience can view the phenomena directly, without the intermediation of video. Nevertheless, there are cases in which video can make visible that which would otherwise be unseen. Furthermore, audiences are adept at recognizing and distinguishing between "live" video and that which has been "canned." Live video can make otherwise invisible phenomena observable by a large audience, and give the group an experience similar to direct observation.

Chemical Demonstrations

Volume 5

12

Color, Light, Vision, Perception

Rodney Schreiner, Jerry A. Bell, and Bassam Z. Shakhashiri

Through our eyes we see light, and the way light interacts with the objects around us tells us a great deal about our surroundings. Color is one of the most striking characteristics of the objects we see, and color often helps us distinguish among them or detect changes in them. But this visible light is only a tiny fraction of the electromagnetic radiation produced by processes in the world and universe we live in. The electromagnetic spectrum extends from beyond radio waves at the long-wavelength end of the spectrum through microwaves, infrared (IR) and visible waves to ultraviolet (UV) radiation, x-rays, and gamma rays at the short-wavelength end (Figure 1). (Sometimes "radiation" is equated with the emissions from radioactive materials, but only gamma radiation is truly electromagnetic radiation. Alpha and beta emissions are helium nuclei and electrons, respectively.) The word "radiation" is derived from a Latin noun that means "ray" and its verb form "to emit rays." Electromagnetic radiation is a form of energy that is emitted by moving (accelerating) electric charges and is transmitted through space as electromagnetic waves (such as a ray of sunshine, for example). This is much like a water wave that transmits energy from the source (say a stone dropped into the water) to a floating object that the energy in the wave causes to bob up and down.

Many familiar words are derived from this same Latin source word and share its meaning. For example, a radiator that keeps a room warm emits invisible infrared electromagnetic radiation with wavelengths that are longer than visible radiation. Radios pick up signals carried by electromagnetic waves that have wavelengths much longer than visible or infrared radiation. The gamma radiation from radioactive sources (objects that spontaneously—actively—emit these rays) has wavelengths that are much shorter than visible radiation.

Electromagnetic waves, like all waves, are characterized by their wavelength and their frequency, which are inversely proportional to each other. In other words, waves having a long (large) wavelength have a low (small) frequency, and vice versa. Also, as is the case for all waves, the product of wavelength and frequency is equal to the speed of the wave. In the case of electromagnetic radiation, this product is the speed of light in the medium through which the wave is traveling. Also for electromagnetic radiation, its energy is proportional to its frequency and therefore inversely proportional to its wavelength.

Amazingly, the electromagnetic spectrum spans more than 24 orders of magnitude in energy, in frequency, and in wavelength. Extremely low-energy, low-frequency waves of electric circuits and power lines have wavelengths in the range of 10^6 m (1000 km). High-energy, high-frequency gamma waves that we have detected from unknown sources billions of light-years away in the universe have very short wavelengths, about 10^{-18} m (1×10^{-18} m = 1 attometer [am]). (There does not appear to be a limit to the longest and shortest electromagnetic waves, so the range is even larger than noted here.) Visible light, wavelengths from about 7×10^{-7} to 4×10^{-7} m (700 to 400 nanometers, 1 nanometer = 1×10^{-9} m; Figure 1), is

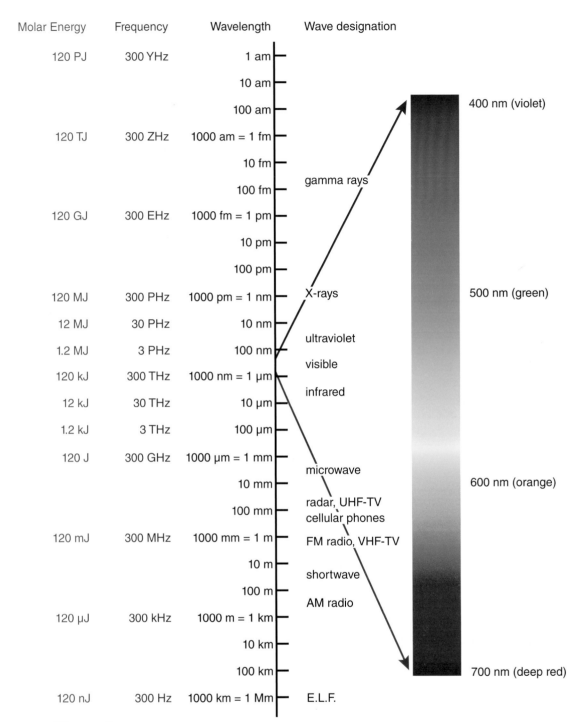

Molar Energy	Frequency	Wavelength	Wave designation
120 PJ	300 YHz	1 am	
		10 am	
		100 am	
120 TJ	300 ZHz	1000 am = 1 fm	
		10 fm	
		100 fm	gamma rays
120 GJ	300 EHz	1000 fm = 1 pm	
		10 pm	
		100 pm	
120 MJ	300 PHz	1000 pm = 1 nm	X-rays
12 MJ	30 PHz	10 nm	
1.2 MJ	3 PHz	100 nm	ultraviolet
120 kJ	300 THz	1000 nm = 1 µm	visible
12 kJ	30 THz	10 µm	infrared
1.2 kJ	3 THz	100 µm	
120 J	300 GHz	1000 µm = 1 mm	
		10 mm	microwave
		100 mm	radar, UHF-TV cellular phones
120 mJ	300 MHz	1000 mm = 1 m	FM radio, VHF-TV
		10 m	shortwave
		100 m	
			AM radio
120 µJ	300 kHz	1000 m = 1 km	
		10 km	
		100 km	
120 nJ	300 Hz	1000 km = 1 Mm	E.L.F.

400 nm (violet)

500 nm (green)

600 nm (orange)

700 nm (deep red)

Figure 1.   A portion of the electromagnetic spectrum.

a minuscule part of the electromagnetic spectrum. In our everyday speech, "light" is usually limited to the electromagnetic radiation we can see. However, you have probably heard of "black light" (invisible ultraviolet radiation with wavelengths just a bit shorter than visible light) and recognize that we sometimes use "light" to mean invisible radiation as well. In what follows, we will occasionally refer to invisible, as well as visible, electromagnetic radiation as "light." The context should make it clear whether we mean the visible or a wider range of electromagnetic radiation.

Is it chance that has limited the wavelengths of electromagnetic radiation we can see, or is there some fundamental reason for the limitation? In order to answer this question, we first have to answer another: What is light? That is, what does all you have just read mean and how do we know? And to make the story more interesting, light sometimes exhibits the properties of waves and sometimes the properties of particles, so we will have to examine this wave-particle duality.

The way light interacts with matter reveals much about the composition of both, even if we limit our observations to the light we can see. How are our eyes and brains constructed to detect and process the information provided by this light? Scientific instruments extend our abilities to detect light, both visible and invisible, and can provide more information than our eyes alone. To interpret what we see and what our instruments detect, we need a basic understanding of the properties of light and how it interacts with matter. Some of these properties are part of our everyday experience, but we do not always pay conscious attention to them. Other properties of light are observed only under special conditions that can be incorporated into experiments and demonstrations to make these properties apparent.

This introductory material includes an examination of the properties of light that are part of our everyday experience and others that require experimentation, the sources of light, the interactions of light with matter, especially as they are related to color and vision, and the processing required for seeing and interpreting what is seen. These topics are so intertwined that our examination will be more panoramic than linearly discursive.

COLOR

Color is perhaps the most familiar and aesthetically pleasing property of light. There is red light, green light, and blue light, and light of a great variety of tints and shades. What causes some light to be red and other light green has been an intriguing question to humans since ancient times. Why do you observe the colors of the rainbow when white light from the sun interacts with uncolored materials such as raindrops in the air after a storm, the shiny surface of a compact disc (CD), or the edges of a colorless piece of glass? (See Demonstration 12.12, Refraction and Diffraction: Separation of White Light into Colors.)

In 1666 Isaac Newton (1643–1727) performed a series of experiments that led him to advance the understanding of light and color. He put a small hole in a window shade, allowing a shaft of sunlight to enter a darkened room. He placed a clear, colorless glass prism in this beam of white sunlight. The beam of light was bent (refracted) as it passed through the prism and fell onto a white surface. On the white surface, the beam appeared not as a spot of white sunlight but as a band of colors—red, orange, yellow, green, cyan, blue, and violet (the names Newton associated with the different colors)—that blended one into the next. The colors appeared in the same order as they do in a rainbow (or on the surface of a CD). A colored image was produced from white light and a colorless glass prism. This colored image is called a *spectrum,* from the Latin word for "image."

Newton's production of colors from white light and colorless glass was not new. What was new was a further experiment and his explanation of the results. He replaced the white surface on which the colors were observed with a second prism held upside down with respect to the first. When he did this, the band of colors recombined into a spot of white light (Figure 2).

Newton explained his observations by stating that white light is a mixture of all colors of the rainbow. It appears white because that is the way our eyes and brain interpret the mixture of all colors when it enters our eyes. White light is composed of a range of different kinds of light, each of which is bent (refracted) by a different amount in passing through a prism. The kind that is bent least gives the sensation of red. The kind that is bent slightly more gives the sensation of orange. And finally, the kind that is bent most gives the sensation of violet.

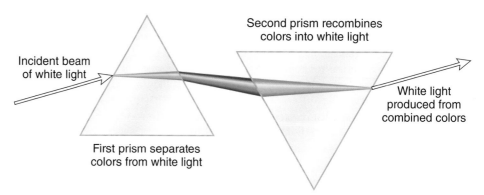

Figure 2.   Schematic representation of Newton's experiment showing that the white-light spectrum formed by one prism could be recombined to white light by a second prism.

This theory of light easily explains the color of objects. A colored object absorbs some colors and bounces back (reflects) other colors. For example, a ripe apple appears red in sunlight because it absorbs from the white sunlight all colors except red and reflects the red light to our eyes. Transparent substances can also be colored if they absorb some colors and transmit others. (See Demonstration 12.15, Observing Transmission Spectra of Dyes.)

Color Perception

When an object reflects only green light into our eyes, we perceive the object as green. However, an object that appears green may not necessarily be reflecting green light. We must distinguish between *physical color* and *physiological color*. A physical color is identified by the amount of refraction it undergoes in passing through a transparent prism, as in Newton's experiment. Physiological color is the color our brain interprets light to be when it is detected by our eyes; it is the sensation we have. The physiological system works in such a way that physical green light is sensed as green. However, the physiological system can also produce the sensation of green from a mixture of colors that does not contain physical green light. For example, a mixture of blue light and yellow light gives the sensation of green. (See Demonstration 12.6, Colored Flames from Metal Ions.) Light that is colored by reflection from a colored surface or by transmission through colored glass may not actually contain the physical color that corresponds to the physiological one we sense. We can determine the physical colors in light by passing the light through a prism and separating the physical colors.

Just three colors of light—red, green, and blue—when combined in the appropriate proportions can produce the sensation of any other color. (See Demonstrations 12.25, Additive Color Mixing; 12.26, Subtractive Primary Colors; and 12.30, The Land Effect.) This ability of red, green, and blue to create the entire spectrum of colors is a result of the way our eyes interact with light. In the retinas of our eyes, there are three different types of color-sensitive cells. One reacts most strongly to red-yellow light, another to green light, and the last to blue light. The degree to which a particular color of the spectrum or a particular mixture of colors activates each of these cell types produces the sensation that we perceive as a particular color. Light that is physical yellow activates the three cell types in a particular ratio. A combination of the physical colors red, green, and blue that activates the three cell types in the same ratio also gives the sensation of yellow.

This aspect of our color vision is exploited in the screens of color televisions and computer monitors. The screen is covered by an array of pixels ("picture elements"), each of which is divided into subpixels. In cathode ray tube (crt) or plasma screens, the subpixels are dots of materials that can emit red, blue, or green light. On a crt screen, beams of electrons, one for each color, from the back of the picture tube activate the dots in the subpixels to emit light. On a plasma screen, the dots are activated by high-energy emission from a plasma (ionized gas) in a cell directly behind the dot. In either case, the subpixels are activated so that the pixel glows in a particular red-green-blue ratio. When viewed from a distance, the light from each subpixel in a pixel falls on the same place on your eye's retina. Your eye responds to the particular ratio of red, green, and blue light, and your brain interprets the ratio as a particular color. If you look at a television tube up close, the light from the individual dots will fall onto different spots on the retina, and you can see the individual dots of red, green, and blue. The mechanism by which a liquid crystal display (LCD) screen produces the light from its red, green, and blue subpixels is different (see below), but control of the red-green-blue ratio is still responsible for the color you see from each pixel.

The Solar Spectrum

When a beam of sunlight passes through a prism, it is separated into a spectrum of colors. Although we refer to the spectrum as containing red, orange, yellow, green, blue, and violet light, it actually contains many more than just these six colors (and also extends into the infrared and ultraviolet we cannot see). These color names refer to regions of the spectrum (Figure 3), and each region contains a range of different colors. For example, in the orange region of the spectrum, the color is not all the same. It shades almost imperceptibly from red at one end to yellow at the other. Sunlight is a mixture of a vast number of colors.

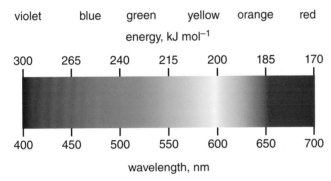

Figure 3.   Colors of the visible spectrum with corresponding wavelengths and energies.

When a very high resolution spectrum is examined closely, however, the spectrum of sunlight contains gaps. If sunlight passes through a very narrow slit, forming a very narrow beam of sunlight, and this beam passes through a prism, the spectrum that is produced contains many very narrow black gaps, called *spectral lines,* where a color seems to be missing. These lines are always in the same place in the spectrum of solar radiation, even if the light is reflected from the moon or from the planets. However, when light from another star (gathered by a telescope) is passed through a prism, the spectrum usually contains a different pattern of black lines.

Emission Spectra

When an element is vaporized and heated hot enough or given enough energy, often with an electrical discharge, it emits colored light. Sodium vapor emits a strong yellow light, which is seen in sodium vapor street lights; potassium vapor, a dim violet light; mercury vapor, a greenish-blue light; and neon gas, a red-orange light that is familiar in neon signs. When the emitted light is passed through a slit and a prism, it is separated into discrete colors (Figure 4a). The separated colors are called an emission spectrum (pl., *spectra*). Each element produces a characteristic emission spectrum, no matter which of its compounds is vaporized. Thus, we can determine the elements in a substance by vaporizing the substance and examining its emission spectrum. (See Demonstrations 12.5, Emission Spectra from Gas-Discharge Lamps; and 12.6, Colored Flames from Metal Ions.)

Several elements were discovered through their emission spectra. When certain minerals were heated to incandescence (hot enough to glow), their emission spectra contained patterns of colors that did not correspond to any known elements. In one case, a mineral produced a strong emission in the blue region of the spectrum, where no other element was known to emit. The unknown source of the emission was called *cesium,* from the Latin for "sky blue." The reactive alkali metal element was later extracted from the mineral and characterized chemically as well as spectroscopically.

Absorption Spectra

Shining white light through the vapor of a substance can produce gaps in the spectrum caused by absorption of some colors by the substance. For example, if white light passes through sodium vapor and then through a prism, the spectrum that is produced will contain gaps at exactly the same place in the spectrum at which sodium vapor emits light (Figure 4b). The gaps are the *absorption spectrum* of sodium.

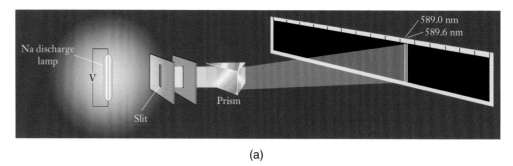

(a)

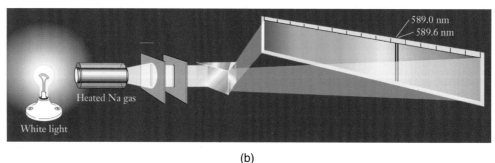

(b)

Figure 4.   Comparison of (a) an atomic emission spectrum and (b) an atomic absorption spectrum.

To account for the dark gaps in the solar spectrum, the German physicist Gustav Kirchhoff (1824–87) proposed that these are also an absorption spectrum. The vapor that forms the solar atmosphere absorbs light emitted from the sun. Sodium in the solar atmosphere absorbs at the same place in the spectrum that sodium absorbs on earth, and the same is true for the other elements; thus elements in the solar atmosphere can be identified by examining the gaps in the solar spectrum. In fact, the element helium was identified in the solar atmosphere before it was found on earth. The solar spectrum contained absorptions that did not correspond to those of any known element, so the element responsible was named *helium,* from the Greek for "sun." Only later was an element whose absorption spectrum matched that observed from the sun found on earth. (Note that helium, a noble gas, is named with the "ium" ending, as though it were a metal. Because almost all the other elements detected in the solar atmosphere were metals, the new element was also assumed to be a metal. It is the only nonmetal in the periodic table with the "ium" ending.)

MORE PROPERTIES OF LIGHT

Experience tells us, and we take for granted in our everyday life, that light rays travel in straight lines. When we look at an object, we are sure that the object we see exists in the direction in which we are looking. There are a few exceptions to this general rule, such as mirrors, but we soon learn to make adjustments for these.

Light travels unhindered only through a vacuum. All forms of matter absorb light to some degree. Many substances do so to such an extent that they absorb all the light that falls on them. Such substances are called *opaque,* from a Latin word meaning "dark." Materials through which light passes without being completely absorbed are called *transparent,* from the Latin word meaning "to show through." Even transparent materials become opaque when they are sufficiently thick: we can easily see through water that is a few meters deep, but light cannot penetrate water beyond about 30 meters.

Light is a form of energy and energy cannot be destroyed. When light is emitted (leaves a system), energy is lost from some part of the emitting system. When light is absorbed, it is converted to another form of energy. The absorption of light by an opaque material or by a sufficient thickness of a transparent material may seem to destroy the light, but it is actually converted into the increased motion of atoms and molecules, raising the temperature of the material. (See Demonstration 12.11, The Conversion of Light Energy to Thermal Energy.)

Reflection

Light travels in a straight line only when it travels through a uniform medium. A vacuum is a uniform medium, and so is air that has a constant temperature and density throughout. When the medium changes, light no longer necessarily travels in a straight line. When light strikes an object made of opaque material, some of the light is absorbed and the light that is not absorbed changes direction abruptly, bouncing back from the surface like a billiard ball from the edge of a billiard table. The bouncing back of light from an opaque body is called *reflection,* from Latin words meaning "to bend back."

Light reflection follows rules similar to those that govern the bouncing of a billiard ball. Imagine a flat surface capable of reflecting light, such as a mirror. A beam of light traveling perpendicular to the surface would be reflected directly back along the path from which it came (Figure 5). If the beam were traveling at an angle from the perpendicular, it would be reflected at the same angle on the other side of the perpendicular (Figure 6). A billiard ball striking the edge of a pool table behaves the same way.

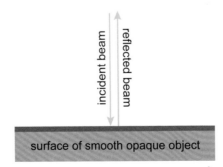

Figure 5.   Reflection of a light beam striking a plane surface perpendicularly.

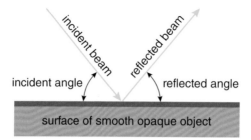

Figure 6.   Reflection of a light beam striking a plane surface at an angle.

Most surfaces, however, are rough, even those that appear flat. Very flat surfaces, such as those of mirrors, are rare. A beam of light falling on a rough surface does not strike every part of the surface at the same angle (Figure 7). Some parts of the beam strike the surface head on, while other parts strike at various other angles. Each part of the beam is reflected at an angle equal to the angle at which it strikes. Because the angle varies from one part of the beam to another, the reflection is broken up and scattered in all directions. This type of reflection is called *diffuse reflection*. Nearly all of the objects in the world around us have rough surfaces, and almost all reflection we see is diffuse reflection. This type of reflection allows us to see objects from all sides, because light is reflected in many directions.

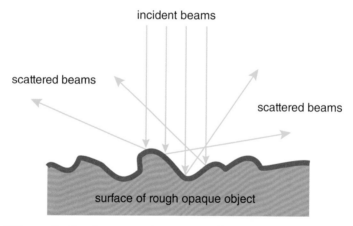

Figure 7.   Diffuse reflection from a rough opaque surface.

If a surface has only a tiny amount of roughness, only a portion of the reflection will be diffuse. Much of the light that strikes the surface will be reflected at the same angle at which it struck the surface (Figure 8).

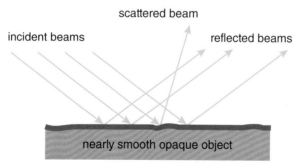

Figure 8.   Reflection from a nearly smooth opaque surface to produce highlights.

A large fraction of the reflected light is not scattered. When this happens, you can see the reflecting object from various angles by the diffuse reflection, but you will see far more light if you observe the object at the reflecting angle. At that angle you will see a "highlight." Such a surface is the polished skin of an apple in which you can see images of bright lights, such as a window, reflected from the surface (Figure 9).

Figure 9. Reflections in the polished skin of an apple: on the right, light from a window, and on the left, the cup.

If a surface is extremely flat, virtually none of the reflection is diffuse. Nearly all of a beam that strikes the surface will be reflected without scattering. When you view such a reflection, you interpret the reflected beam as you would the original beam. For instance, the light reflected diffusely from an apple makes a pattern that the eyes and brain interpret as an apple. When the reflected light from an apple first strikes an extremely flat surface, such as a mirror, and is then reflected to your eyes, you will still interpret the light as from an apple (Figure 10). You cannot determine the path followed by the light that reaches your

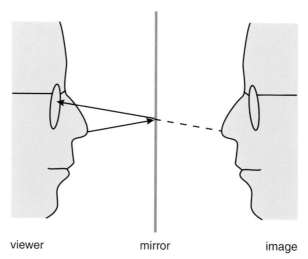

viewer   mirror   image

Figure 10. Light from the viewer's nose travels to the mirror, where it is reflected to the viewer's eye. The viewer interprets the reflected light as having traveled in an uninterrupted straight line from the image behind the mirror.

eyes; you cannot tell whether the light has been reflected or not. The apple seen reflected in a mirror is seen as if it were behind the mirror. You perceive it to be where it would be if the light had come straight to you without interruption, instead of striking the mirror and being reflected to you. (The illusions in a carnival house of mirrors are possible because we are so accustomed to interpreting light as traveling in straight, uninterrupted lines.)

Refraction

The path followed by a beam of light can change direction when it passes from one transparent medium into another. When a beam of light enters a flat slab of clear, transparent glass at an angle from the perpendicular, the beam of light changes direction as it enters the glass. The beam is bent so that the angle from the perpendicular is smaller in the glass than in the air. This change in direction of light when it passes from one transparent medium into another is called *refraction,* from Latin words meaning "to break back." You can easily observe refraction if you partially immerse a stick in water so some of the stick is in the water and some in the air (Figure 11). The stick seems to bend where it enters the water. As the stick is lowered or raised in the water, the place where it seems to bend stays at the surface of the water.

Figure 11.   A pencil propped between the bottom and side of aquarium partially filled with water.

Refraction is also the basis for an old parlor trick (Figure 12). A coin is placed at the bottom of an empty cup, and the cup is viewed from such an angle that the rim hides the coin. When the cup is filled with water, the coin becomes visible, even though neither the cup, coin, nor eye has moved. To explain this, one must assume that light reflected from the coin changes direction of travel on passing from the water into the air.

"Heat waves" rising from the highway on a hot day, desert mirages, or the shimmering you observe when one liquid is poured into a different liquid are other phenomena caused by refraction. In these cases the changes of medium are from warmer to cooler air or from one liquid to another. The waves, mirages, or shimmering you observe are due to the movement of the media with respect to one another and consequently the shifting of the directions from which the refracted light reaches your eyes. As we will see below, all these refraction phenomena are a result of the different speeds of light in the different media—air and water, cooler and warmer air, and different liquids, for example.

Figure 12.   Both photos show the same cup with a coin on the bottom. The only difference is that on the right the cup is filled with water.

THE NATURE OF LIGHT

The discovery that white light was a mixture of many colors created serious problems for physicists. How was light of one color different from light of another color? And more fundamentally, what was the nature of light in general?

There are two general ways we can explain the transmission of energy from one object to another without the objects actually touching. One is by way of particles streaming across the space separating the objects. The other is by way of waves propagating through the space. Both explanations were advanced for light in the latter half of the seventeenth century, and our modern understanding of light accommodates both viewpoints. In those early days, however, they were in conflict because each could explain many observed phenomena, although neither could explain all.

To choose between conflicting or overlapping theories, we generally look for and test conditions under which they predict different observable phenomena. For light, these phenomena include the properties of light described in the previous section. Unfortunately, determining the conditions under which the particle or wave model (or neither) was better was not possible with the available experimental techniques of the seventeenth and eighteenth centuries. In this section, we will look briefly at basic arguments for the particle model and then more extensively at the wave model, which, by the beginning of the twentieth century, had become the accepted model.

The Particle Model of Light

The particle model, which Newton supported, is the intuitively simpler of the two alternatives and can explain why light rays travel in straight lines. Suppose luminous objects are constantly firing tiny particles outward in all directions. If these particles have no mass, a luminous body would not lose mass simply by being luminous, and light particles would not be affected by gravity. When traveling in an unobstructed path, the particles, if unaffected by gravitational force, would travel in a straight line at a constant velocity. They would be stopped and absorbed by opaque barriers, but those speeding past the edge of the barrier would cast a sharp boundary between the illuminated area beyond and the barrier-shaded area.

Newton believed the alternative, wave, model was untenable. The waves that were familiar to scientists at the time were water waves and sound waves, which do not necessarily travel in straight lines or cast sharp shadows. Sound waves curve about objects, as we know whenever we hear a sound around a corner. Water waves visibly bend about an object such as a floating log of wood. It seemed reasonable to suppose that this behavior was characteristic of all waves.

The Wave Model of Light

But the particle theory had its difficulties. Beams of light could cross at any angle without affecting each other in direction or color, which meant that the light particles did not seem to collide and rebound as would be expected for particles of matter. Furthermore, despite ingenious suggestions, there was no satisfactory explanation why some light particles were sensed (seen) as red and others as green. The particles had to differ among themselves, of course, but *how* was a mystery.

The strongest supporter of the wave model of light in the seventeenth century was the Dutch mathematician, astronomer, and physicist Christiaan Huygens (1629–95). As an astronomer, Huygens constructed his own telescopes, improving on previous instruments with new designs based on the rules of geometric optics. (His study of the rings of Saturn, showing that they were composed of discrete masses—rocks—is honored in the name of the Cassini-Huygens space probe now investigating that planet more closely.) To explain these rules, he proposed a wave model of light that fit the known facts of geometric optics, even though there was no direct evidence to support a wave model.

Huygens pictured each point on a wave front acting as a source of circular waves, expanding outward indefinitely (Figure 13). These circular waves combine, adding to and canceling each other as they spread. Only along the wave front do they add together; at all other points, they cancel. So the wave front advances forward in a straight line (at least if only its middle portion is considered), forming a new source for circular waves.. Furthermore, the waves would have no mass and there would be an infinite supply of them, like water waves and sound waves.

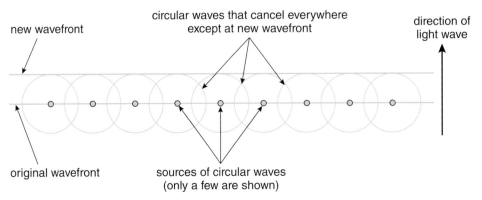

Figure 13.   Schematic diagram of Huygens' waves.

Huygens' wave theory could explain the reflection and refraction when, for example, the straight wave front of a beam of light in air strikes a plane surface of glass obliquely (Figure 14). In Figure 14a, an incident wave front (red) traveling from the upper left approaches the air-glass interface at a 45 degree angle. At the interface, there are an infinite number of points, only some of which are represented, and when the incident wave front reaches the interface, at each point a circular wave is created (Figure 14b). The reflected wave (green) created at each point propagates into the air at the same speed as the incident wave. Therefore, the distance traveled in any time period is the same for the incident and reflected waves, and their wavelengths are the same. The refracted wave (blue) created at each point along

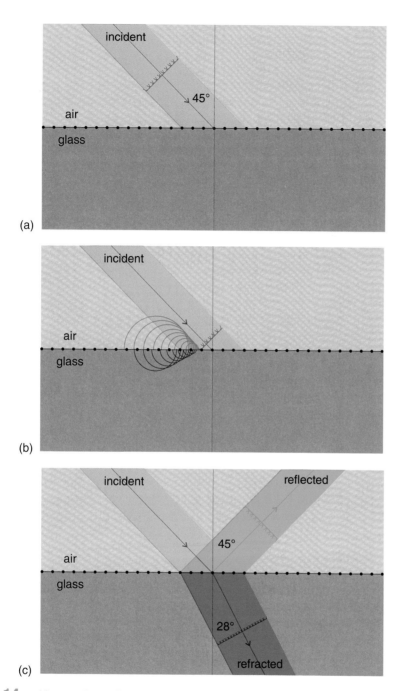

(a)

(b)

(c)

Figure 14. Huygens' wavefront (a) approaching (red) an air-glass interface, (b) reflected (green) and refracted (blue) at interface, and (c) propagating as reflected and refracted waves.

the interface travels more slowly in glass than the incident wave travels in air. Therefore, the wavelength in glass is shorter than that in air. In Figure 14c, the reflected wave front (green) travels away from the interface at the same angle as the incident beam approached. The refracted wave front (blue), because it travels more slowly and with a shorter wavelength (although its frequency is unchanged), travels through the glass at an angle closer to the perpendicular than that at which the incident beam entered.

As the refracted beam emerges from the other side of the glass, the circular waves created at the points where it emerges travel faster in the air and reverse the change in direction shown in the figure. If the sides of the glass are parallel, the emerging light beam takes on its original direction, as shown in Figure 15.

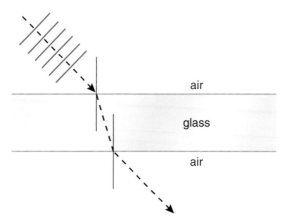

Figure 15.   The emergent beam that results from refraction of a light beam at the air-glass and glass-air interfaces of a thickness of glass with parallel surfaces travels in the same direction but is offset from the axis of the incident beam.

The Huygens' wave model explains refraction by assuming that the velocity of light is less in glass than in air. With additional assumptions, it can also explain the formation of a spectrum. If light is a form of wave, it must have a *wavelength.* Assume that the wavelength varies with color, being longest at the red end of the spectrum and shortest at the violet end. Further, suppose that short wavelengths are slowed more on entering glass from air than are long wavelengths. Thus, red light would be least refracted, orange next, and so on. In this way, light passing through a prism would be expected to form a spectrum. (See Demonstrations 12.16, Dichroism: Transmission versus Reflection; and 12.17, Iridescence from a Polymer Film.)

Newton could also explain refraction with his particle model, but he had to assume that the velocity of the light particles increased in passing from a medium of low optical density (air) to one of high optical density (glass or water). Here was a clear-cut difference between the two models. A measurement of the velocity of light in different media would tell whether Newton's or Huygens' assumption was correct. However, it was not until the middle of the nineteenth century that the measurement could be made showing that the velocity of light in water is about three-quarters the velocity in air.

Diffraction

There was another difference in the predictions of the two models. Newton's light particles traveled in straight lines in all portions of a light beam, so the beam was expected to cast absolutely sharp shadows. Not so with Huygens' waves. Each point in the wave front served as a focus for waves in all directions, but through most of the wave front, a wave to the right from one point was canceled by a wave to the left from the neighboring point on the right, and so on. After all cancellations were taken into account, only the forward motion was left. There was an exception, however, at the ends of the wave front. At the right end, a rightward wave was not canceled because there was no rightward neighbor to send out a leftward wave. At the left end, a leftward wave was not canceled. A beam of light, therefore, had to "leak" sideways if it behaved as a wave. In particular, if a beam of light passed through a gap in an opaque barrier, the light at the boundary of the beam, the light just skimming the edge of the gap, ought to leak sideways so that the illuminated portion of a surface farther on ought to be wider than one would expect from strictly straight-line travel (Figure 16).

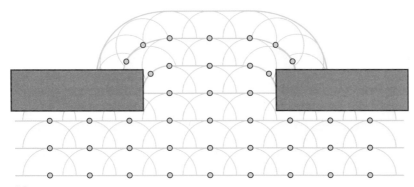

Figure 16.   Huygens' light waves "leaking" sideways when passing through a gap.

This phenomenon of a wave bending sideways at either end of a wave front is called *diffraction* and is easily observed in water waves and sound waves. Thus, the observation that light passing through a gap in a barrier did not seem to exhibit diffraction favored the particle model. However, what was not clearly understood in Newton's time was that the smaller the wavelength of any kind of wave, the smaller the diffraction effect. Therefore, if one made still another assumption—that the wavelength of light waves was very small—the diffraction effect would be expected to be very hard to observe and the wave model not so easily dismissed.

In fact, the diffraction of light was observed in the seventeenth century. An Italian physicist, Francesco Maria Grimaldi (1618–63), passed light through two apertures and found that the final band of light on the illuminated surface was a bit wider than it ought to have been if light had traveled through the two apertures in straight lines. In other words, diffraction had taken place. Even more important was the observation that the boundaries of the illuminated region showed color effects, with the outermost portions of the boundary red and the innermost violet. Eventually, it was understood that these color effects also fit the wave model. If red light had the longest wavelengths it would be most diffracted, while violet light, with the shortest wavelengths, would be least diffracted by the same aperture.

Indeed, this principle is used to form spectra by means of diffraction gratings. When fine parallel lines are scored on glass, they will represent opaque regions separated by transparent regions, or gaps, at the edges of which diffraction can take place. If the gaps are very narrow, the glass will consist entirely of gap edges, so to speak. If the scoring is very straight and the gaps are very narrow, the diffraction at each edge will take place in the same fashion, and the diffraction at any one edge will reinforce the diffraction at all the others. This will produce a spectrum as good as, or better than, any that can be formed by a prism. Lines can be scored more finely on polished metal than on glass. In this case, each line is an opaque region separated by a reflecting region, and this will also form a spectrum, although ordinary reflection from unbroken surfaces will not. Today, diffraction gratings are used much more commonly than prisms to produce spectra. (See Demonstration 12.15, Observing the Transmission Spectra of Dyes.)

Newton knew about Grimaldi's experiments and even repeated them, particularly noting the colored edges. However, the phenomenon seemed so minor to him that he disregarded its significance and continued to advocate the particle model. Newton's great prestige carried the day, and throughout the eighteenth century, light was considered by almost all physicists to be particulate in nature.

Interference

In 1801 Thomas Young (1773–1829) conducted an experiment that revived the wave model most forcefully. He let light from a slit fall upon a surface containing two closely adjacent slits (Figure 17). Each slit served as the source of a spreading beam of light and the two beams overlapped before falling on a screen. If light were composed of particles, the overlapping region should receive particles from both slits. With the particle concentration doubled, the overlapping region should be uniformly brighter than the regions on the outskirts, beyond the overlap, where light from only one beam would be received. Instead, the overlapping region consisted of stripes: bright bands alternating with dim ones.

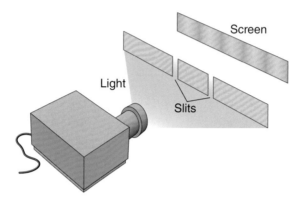

Figure 17.   Schematic representation of Young's two-slit experiment.

This result presented a serious difficulty for the particle theory of light, but for the wave theory there was no problem (Figure 18). At some points on the screen, light from both

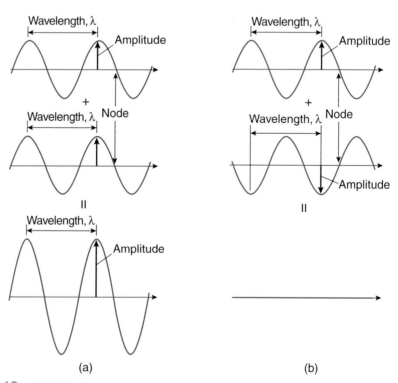

Figure 18.   (a) Reinforcement and (b) cancellation of superimposed waves.

the first and second beam would consist of waves that were *in phase,* that is, crest matching crest, trough matching trough. The two light beams would reinforce each other at those points so that there would be a resultant wave of twice the amplitude and, therefore, a region of doubled brightness. At other points on the screen, the two light beams would be *out of phase,* with crest matching trough and trough matching crest. The two beams would then cancel, at least in part, and the resultant wave would have a much smaller amplitude than either component; where canceling was perfect, there would be no wave at all. A region of dimness would result.

The type of light particle Newton envisaged could not interfere with and cancel another particle. A wave can and does interfere with and cancel another wave. *Interference patterns* can easily be demonstrated in water waves (Figure 19), and interference is responsible for the phenomenon of beats in sound waves, used, for example, in tuning a piano. Young showed that the wave model accounted for the observed interference pattern.

Figure 19.   Constructive and destructive interference of water waves.

Furthermore, from the spacing of the light and dark interference bands, Young could calculate the wavelength of light. If the ray of light from one beam is to reinforce the ray of light from the second beam, the rays must be in phase. That means the distances from the point of reinforcement on the screen to each slit must differ by an integral number of wavelengths. Young chose the interference bands requiring the smallest difference in distances and found the wavelength to be about a fifty-thousandth of an inch. This is certainly small enough to account for the difficulty of observing light diffraction effects. It was also possible to show that the wavelengths of red light are about twice the wavelengths of violet light, which fit the requirements for spectrum formation in the wave model.

In the metric system, it is convenient to measure the wavelengths of light in *nanometers* (nm). The spectrum extends from about 760 nm for the red light of longest wavelength we can see to 380 nm for the violet light of shortest wavelength. The position of any spectral line can be located in terms of its wavelength. The wavelength ranges for the different colors are roughly red, 760–630 nm; orange, 630–590 nm; yellow, 590–560 nm; green, 560–490 nm; blue, 490–450 nm; and violet 450–380 nm (Figure 3). Note that Young's fifty-thousandth of an inch is about 5×10^{-7} m = 500 nm.

The wave model of light was not accepted at once despite the conclusiveness (in hindsight) of Young's experiment. However, all through the nineteenth century, additional evidence in favor of light waves turned up, and additional phenomena that the particle model could not explain were readily and elegantly explained with the wave model. Consider, for example, the color of the sky.

Light Scattering

When a light beam meets an object, what happens to the light depends on the size of the object. If the object is greater than 1000 nm in diameter, visible light is absorbed. If the object is smaller than 1 nm in diameter, the light is likely to pass on undisturbed. If, however, the object lies between 1 nm and 1000 nm in diameter, it will be set to vibrating as it absorbs the light, and it may then emit a light ray equal in frequency (and wavelength) to the original but that travels in a different direction. This is called *light scattering*. The tiny water or ice particles in clouds or fog are of a size to scatter light in this fashion. Therefore, a cloud-covered sky is uniformly white or, if the clouds are thick enough to absorb a considerable fraction of the light, various shades of gray. This same kind of scattering is responsible for the snowy plume of spray at the base of a waterfall and for the whitecaps on a wind-whipped lake or breaking ocean waves.

As particles grow smaller, the amount of scattering becomes a function of the light's wavelength. Light of short wavelength is scattered more than light of long wavelength. The British physicist John Tyndall (1820–93) studied this phenomenon and found that light passing through pure water or a solution of small-molecule substances, such as sugar, underwent no scattering. The light beam, traveling only forward, cannot be seen from the side and the liquid is *optically clear*. If the solution contains particles large enough to scatter light, however (for example, molecules of proteins or small aggregates of ordinarily insoluble materials, such as gold or iron oxide), some of the light is scattered sideways, and the beam can then be seen from the side. This phenomenon is the *Tyndall effect*. (See Demonstration 12.19, The Tyndall Effect: Scattered Light Is Polarized.)

The English physicist John William Strutt, Lord Rayleigh (1842–1919), studied scattering in greater detail and derived an equation that showed how the amount of light scattered by molecules of gas varied with a number of factors, including the wavelength of the light. He showed the amount of scattering was inversely proportional to the fourth power of the wavelength. Since violet light had one-half the wavelength of red, violet light is scattered more, by a factor of 16 (= 2^4).

Why Is the Sky Blue?

Over short distances, the scattering by particles as small as the gas molecules of the atmosphere is insignificant. If, however, the miles of atmosphere stretching overhead are considered, scattering mounts up and, as Rayleigh showed, is almost entirely toward the violet end of the spectrum. Enough sunlight is scattered to drown out the feeble light of the stars, which are present in the sky during the day as well as at night. Furthermore, the scattered light that illuminates the sky is heavily represented in the shorter wavelength region and appears blue. Seen from space, the earth retains that distinctly bluish appearance that results from atmospheric light scattering. The sun itself, with that small quantity of shorter wavelength light missing from its spectrum, appears slightly redder than it would be if the atmosphere were absent.

This effect is enhanced when the sun is on the horizon. Its light then comes obliquely through the atmosphere and travels through a greater thickness of air (Figure 20). Enough light is scattered from even the middle portions of the spectrum to lend the sky a faintly greenish hue, while the sun itself, with a considerable proportion of its light scattered, takes on a very ruddy color. Reflected from broken clouds, this light can produce a beautiful effect. Since the evening sky, after the day's activities, is dustier than the morning sky, and since the dust contributes to scattering, sunsets tend to be more spectacular than sunrises. After gigantic volcanic eruptions, tons of fine dust are hurled into the upper atmosphere, and sunsets remain particularly beautiful for months afterward.

On the moon, which lacks an atmosphere, the sky is black, even when the sun is present in the sky. Shadows as well are pitch black on the moon. On earth, the dust normally present in the atmosphere scatters light, so shadows are not absolutely black. Although darker by far than areas in direct sunlight, shadows receive enough scattered light to make it possible to read newspapers in the shade of a building or even indoors on a cloudy day.

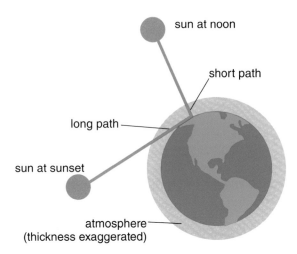

Figure 20.   At sunset, light from the sun travels a longer path through the atmosphere than at noon.

Wave Nomenclature

As we continue to develop the wave model for light, it will be useful to use the conventional nomenclature for wave descriptions. Wavelength is the distance between sequential maxima (or minima) of a wave. In the representation of waves in Figure 18, the wavelength of the waves is denoted by its usual symbol, the Greek letter λ (lower case lambda). If we focus attention on one of the maxima in a traveling wave, such as the water waves shown in Figure 19, we can measure the distance the maximum moves in a given amount of time (one second, for example), which is the velocity of the wave. There is no standard symbol for wave velocity; let us call it s, for speed. Now, if we refocus our attention on a particular location in the path of the wave, we can count the number of wave maxima that pass by in one second, which is the frequency of the wave. The usual symbol for frequency is the Greek letter ν (lower case nu).

To determine the relationship among λ, ν, and s, consider that if we know the distance the wave moves in one second, s, we can divide by the length between each maximum, λ, to find the number of waves required to cover the entire distance in one second, which would be the number of waves passing one point in one second, ν.

$$\nu = \frac{s}{\lambda}$$

$$s = \nu\lambda$$

For a particular kind of wave, if the wave velocity is constant, the product, $\nu\lambda$, is constant. Thus, the wavelength and frequency of a wave are inversely proportional—the shorter the wavelength, the higher the frequency. This makes sense, because the shorter the wavelength, the more waves are required to cover the distance the wave travels in one second.

Polarized Light

There are two important classes of waves that have different properties. Water waves are *transverse waves*, undulating up and down at right angles to the direction in which the wave as a whole is traveling. Sound waves are *longitudinal waves*, undulating back and forth in the same direction in which the wave as a whole is traveling. Which variety represents light waves?

In 1669 a Dutch physician, Erasmus Bartholinus (1625–98) discovered an interesting property of Iceland spar, which is a transparent, crystalline form of calcium carbonate. When the crystal was placed on a surface bearing a black dot, two dots were seen through the crystal. When the crystal was rotated in contact with the surface, one of the dots remained motionless while the other rotated about it. Apparently, light passing through the crystal split into two beams that were refracted by different amounts. (See Demonstration 12.22, The Birefringence of Calcite.)

Using the wave model, this phenomenon could be easily explained if light waves were transverse waves. Imagine a beam of light moving toward you with the light waves undulating in planes at right angles to the line of motion, as they would in water waves traveling toward you. Say the light waves are moving up and down. They might also, however, be moving right and left and still be at right angles to the line of motion (Figure 21). They might even be moving diagonally at any angle and still be at right angles to the line of motion. When the component waves of light are undulating in all possible directions at right angles to the line of motion and are evenly distributed, we have *unpolarized light.*

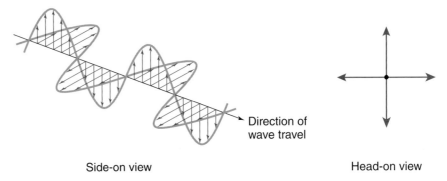

Side-on view   Head-on view

Figure 21. Side-on and head-on views of transverse waves at right angles to one another.

All undulations taking up diagonal positions can be divided into an up-down component and a left-right component. Therefore, for simplicity's sake, we can consider unpolarized light as consisting of an up-down component and a left-right component present in equal amounts (intensities). It is possible that the up-down component may be able to slip through a transparent medium where the left-right component might not. As an analogy, suppose you hold a rope that passes through a gap in a picket fence. If you make up-down waves in the rope, they will pass through the gap unhindered (Figure 22a). If you make left-right waves, those waves would collide with the pickets on either side of the gap and be damped out (Figure 22b).

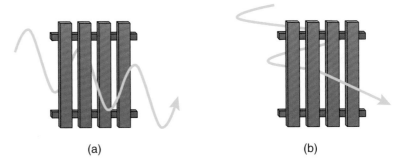

(a)   (b)

Figure 22. Picket-fence analogy for (a) transmission and (b) absorption of one component of unpolarized light by a polarizing medium.

The way light passes through a transparent substance depends on how the atoms making up the substance are arranged—that is, how the gaps between the atomic "pickets" are oriented. In most cases, the arrangement is such that light waves in any orientation pass through with equal ease. Light enters unpolarized and emerges unpolarized. In the case of Iceland spar, however, only up-down light waves and left-right light waves can pass through. One of these passes with greater difficulty, is slowed up more, and therefore is refracted more. The result is that at the other end of the crystal two rays emerge—one made up of up-down undulations only and one made up of left-right undulations only. Each of these is a ray of polarized light with half the intensity or brightness of the original unpolarized light. Because the undulations of the light waves in each of these rays exist in one plane only, such light is usually called *plane-polarized light.*

Polarizers

Materials that transmit light of only one polarization, up-down or left-right, are called *polarizers,* because the transmitted light is polarized. Inexpensive sheets of polymeric (plastic) polarizers are now readily available and used in many applications. One application is in antiglare sunglasses. Light that reflects from a road surface is partially polarized because the pavement material reflects more light of that polarization. The direction of polarization of the sunglass lenses is chosen to block the direction of polarization of the light reflected from the road, thus reducing the glare. Figure 23 uses the picket fence analogy to show how light polarized by one polarizer (or a phenomenon like road glare) can be transmitted or blocked by another.

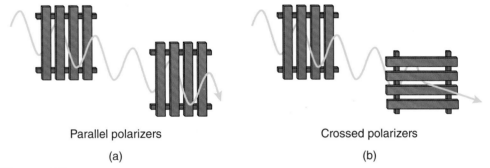

Parallel polarizers   Crossed polarizers

(a)   (b)

Figure 23.   Transmission and cancellation of light by (a) parallel and (b) crossed polarizers.

Liquid crystal displays (LCD) for color televisions and computer monitors make use of polarized light. Each subpixel in the LCD display is a many-layered sandwich, part of which is a pair of polarizers that are crossed so that light coming from the rear of the screen is blocked, as on the right in Figure 23. In front of the second polarizer is a red, green, or blue filter that transmits only this subpixel color. Between the polarizers is a liquid crystal layer held between two transparent electrodes (electric conductors). Liquid crystals are composed of long polar molecules that stack up next to one another in an orderly way, much like the molecules in a solid crystal, but are held together by relatively weak forces so that they can reorient easily under the influence of external forces like an electric field. (See below for more about fields.)

Long, oriented molecules can interact with polarized light and change the direction (angle) of polarization. (See Demonstration 12.23, A Liquid Crystal Display through a Polarizer.) If the direction of polarization of the light passing between the two polarizers is changed somewhat, then some of the light can get through the second polarizer and reach the observer as a red, green, or blue beam, depending on the filter color. When no electric potential is applied to the electrodes, the arrangement of the polarizers and the liquid crystal is such that no light gets through. As an increasing potential is applied to the electrodes, the orientation of the molecules in the liquid crystal changes and more and more light is transmitted through the sandwich and thus through the filter to our eyes. The balance of light from the red, green, and blue subpixels is controlled by the potential applied to each subpixel electrode and thus produces a full color image on the screen, as discussed in the color perception section above.

Other molecules can also interact with polarized light to rotate the angle of polarization. These molecules are said to be *optically active* or *chiral.* The requirement for a molecule to be optically active is that it not be identical to its mirror image. Such mirror-image molecules rotate polarized light in opposite directions by exactly the same amount. The amount and direction of rotation of polarized light by optically active molecules depends on the wavelength of the light. The angle of rotation for an optically active compound as a function of wavelength is called *optical rotatory dispersion* (ORD). The ORD is directly related to the molecular structure of a compound and has been extensively studied to develop the correlations necessary to use it as a tool for determining the three-dimensional structures of newly synthesized molecules and isolated natural products. (See Demonstrations 12.20, Rainbow Spiral in an Optically Active Solution; and 12.21, A Sugar Solution Between Polarizers.)

ELECTROMAGNETIC RADIATION

By the mid-nineteenth century, the evidence for the wave model of light was overwhelming, but there remained a problem. Other, familiar waves moved in a medium—water waves on water and sound waves in air, for example—and were easy to visualize. For light, the question was: What is (are) the component(s) that undulates to form a light wave?

To begin the answer we need to consider the experimental work that was being done in the fields of electricity and magnetism. Many scientists contributed to this work, but the English chemist and physicist Michael Faraday (1791–1867) played a major role. Two of the most important observations are easily visualized. When an electric current (electrons moving through an electrical conductor) was passed through a coil of wire, the coil acted like a magnet and could influence another nearby magnet—for example, a compass needle. If the ends of this coil of wire were attached to the input of a galvanometer (an instrument to measure electrical current) and a magnet was passed through the coil, the galvanometer detected a momentary electric current produced in the wire. Faraday used these phenomena to invent electric motors that generated work from electromagnetic interactions and an electric current and dynamos (generators) to do the reverse, produce an electric current from electromagnetic interactions and an input of work.

Fields

Both of the phenomena described above involve action at a distance. We are familiar with magnets attracting or repelling one another when they are brought close but not touching. And we know that electrically charged objects also attract or repel one another without touching. What is it that connects one magnet to another, for example, so that the motion of one influences the behavior of the other when they are not touching? Faraday proposed the idea of a *field* that surrounded a magnet or a wire through which electric current flowed and could interact with other such fields. Magnetic fields can be visualized by their influence on finely divided iron (filings) scattered on a surface with a magnet (or magnets) below, as shown in Figure 24.

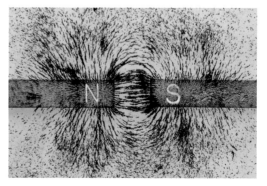

Figure 24.   Visualization of a magnetic field with iron filings.

Few of Faraday's contemporaries accepted the concept of electric and magnetic fields, and he lacked the necessary mathematical background to develop it further. That development was the work of the Scottish physicist James Clerk Maxwell (1831–79). Starting with the ideas that a changing electric field could create a magnetic field and that a changing magnetic field could create an electric field, Maxwell went a step further to combine the two ideas and proposed that an oscillating or undulating magnetic field could generate an undulating electric field that, in turn, could generate an undulating magnetic field. Thus the undulating magnetic and electric fields could coexist and generate each other forever.

Maxwell brought together the mathematics required to describe this electromagnetic model and used his equations to calculate the speed with which the fields would generate one another. The data required for this calculation were known from experiments on electrical and magnetic phenomena, and the result of the calculation was the speed of light in a vacuum, c (the standard symbol for the speed of light), 3.00×10^8 m s^{-1}, which had also been measured by this time. Maxwell recognized that this could not be coincidence and concluded that light is an electromagnetic wave, often represented schematically as in Figure 25.

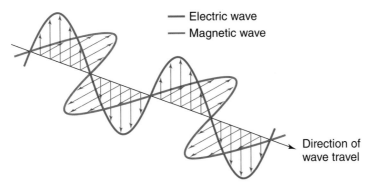

Figure 25.   Representation of an electromagnetic wave.

This was a satisfying conclusion that seemed finally to provide both theoretical and experimental verification of the wave model of light. Electromagnetic waves were produced by the motion of charges in matter and then were propagated through space as the magnetic and electric fields generated each other. A light wave could go on forever, unless it encountered an object with which it could interact. That is why we can see with our naked eye the light from stars that are millions of light years away and, with telescopes to gather more light, stars that are billions of light years away.

Black-Body Radiation

Kirchhoff's work on spectroscopy showed that a substance that absorbed certain wavelengths of light better than others would also emit those wavelengths better than others once it was heated to incandescence. Imagine a substance capable of absorbing all the light, of all wavelengths, that fell upon it. Such a substance would reflect no light of any frequency and would appear perfectly black. It is natural to call such a substance a *black body*. If a black body is brought to incandescence, its emission should, by Kirchhoff's rule, be as perfect as its absorption. It should radiate (emit light, electromagnetic radiation) in all frequencies, since it absorbs in all frequencies. Furthermore, since it absorbs light at each frequency more efficiently than a nonblack body, it would also radiate more efficiently at each frequency.

Kirchhoff's work increased the interest of physicists in the quantitative aspects of radiation and in how radiation from a body varied with its temperature. It was common knowledge that the total energy radiated by a body increased as the temperature increased, and in 1879 the Austrian physicist Josef Stefan (1835–93) made this quantitative. He showed that the total energy radiated by a body increased as the fourth power of the absolute temperature, a relationship that became known as *Stefan's law.*

Consider a body that is maintained at room temperature, 300 K, and is radiating a certain amount of energy. If the temperature is raised to 600 K, the melting point of lead, the absolute temperature has been doubled, and the total amount of energy radiated is increased by 2^4 (= 16) times. If the same body is raised to a temperature of 6000 K, the surface temperature of the sun, its absolute temperature is 20 times higher than it was at room temperature and it radiates 20^4 (= 160,000) times as much energy.

In 1884 the Austrian physicist Ludwig Boltzmann (1844–1906) gave this finding a firm mathematical foundation and showed that it applied, strictly, only to black bodies and that nonblack bodies always radiate less energy than Stefan's law would require. Because of his contribution, the fourth-power relationship is sometimes called the *Stefan-Boltzmann law.*

But it is not only the total quantity of energy that changes with rising temperature. The wavelength (and frequency) distribution of the light waves emitted also changes in a way that is part of our common experience. For instance, for an object at the temperature of a steam radiator (less than 400 K), the radiation emitted is in the long-wavelength infrared (electromagnetic radiation beyond the red end of the visible light spectrum). Your skin absorbs the infrared and you feel the radiation as heat, but you see nothing. A radiator in a dark room is invisible.

As the temperature of an object goes up further, it not only radiates more infrared, but the wavelength distribution of the radiation also changes. When it reaches a temperature of 950 K, the object appears a dull red in color and can be seen in a darkened room because it emits enough radiation that is at short enough wavelengths for us to see. As the temperature goes higher still, the red brightens and eventually turns first orange and then yellow as more and more of still shorter wavelengths of light are emitted. (See Demonstration 12.2, The Temperature Dependence of the Emission Spectrum from an Incandescent Lamp.) At a temperature of 2000 K, an object, although glowing brightly and easily visible in a lighted room, is still emitting radiation that is largely in the infrared. When the temperature reaches 6000 K, the temperature of the surface of the sun, the emitted radiation is chiefly in the visible light region of the spectrum. It is probably partly because the sun's surface is at that particular temperature that our eyes have evolved to be sensitive to that particular portion of the spectrum.

Measuring Black-Body Radiation

Toward the end of the nineteenth century, physicists attempted to determine quantitatively the distribution of wavelengths in the radiation emitted at different temperatures. To do this accurately they needed a black body to be sure that at each wavelength all the light possible (for that temperature) was being radiated. No actual body absorbs all the light falling upon it, so no actual body is a true black body. In the 1890s a German physicist, Wilhelm Wien (1864–1928), thought of an ingenious way of circumventing this difficulty. Imagine a furnace with a small hole in it. Any light of any wavelength entering that hole would strike a rough inner wall and be mostly absorbed. What was not absorbed would be scattered in diffuse reflections that would strike other walls and be absorbed there. At each contact with a wall, additional absorption would take place, and only a vanishingly small fraction of the light would manage to survive long enough to be reflected out the hole again. That hole, therefore, would act as a nearly perfect absorber and would, therefore, represent a black body. If the furnace were raised to a certain temperature and maintained there, then the radiation emitted from that hole is *black-body radiation* and its wavelength (or frequency) distribution could be studied.

In 1895 Wien made such studies and found that, at a given temperature, the energy radiated was a function of the frequency of the emission. As the frequency increased, the energy emitted increased, reached a peak, and then began to decrease at higher frequencies. When Wien raised the temperature, he found that more energy was radiated at every frequency and that, again, the energy emitted reached a peak. The new peak, however, was at a higher frequency than the one at the lower temperature. As he continued to raise the temperature, the peak of energy emitted moved continuously in the direction of higher and higher frequencies. The frequency at which the peak emission occurred (ν_{max}) varied directly with the absolute temperature. Stated in terms of the wavelength of the emitted radiation (which is inversely proportional to the frequency), the emitted energy increased as the wavelength got shorter, reached a peak at some wavelength that varied inversely with the absolute temperature, and then decreased at shorter wavelengths, as shown in Figure 26.

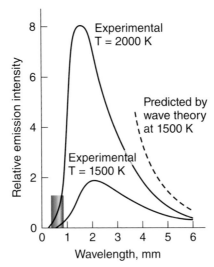

Figure 26.   Relative emission intensity as a function of wavelength from black bodies at two temperatures. The visible wavelength region is denoted.

Both Stefan's law and Wien's law are of importance in understanding more about the stars. From the nature of a star's spectrum, one can obtain a measure of its surface temperature. From this, one can obtain a notion of the rate at which it is radiating energy and, therefore, of its lifetime. The hotter a star, the more short-lived it may be expected to be. Wien's law explains the colors of the stars as a function of temperature. Reddish stars are comparatively cool, with surface temperatures of 2000–3000 K. Orange stars have surface temperatures of 3000–5000 K, and yellow stars (such as our sun), of 5000–8000 K. There are also white stars with surface temperatures of 8000–12,000 K and bluish stars that are hotter still.

At the other end of the black body temperature scale is radiation that is left over from the origin of the universe in the Big Bang. The expansion of the universe since its birth about 14 billion years ago has lowered its temperature, so black body radiation from this source now peaks in the microwave region of the spectrum at a wavelength of 1.9 mm, which corresponds to a temperature of about 2.7 K. This cosmic background microwave radiation was discovered in 1964 by two radio astronomers, Arno Penzias and Robert Wilson, who earned the Nobel Prize for this work that provides present-day evidence for the Big Bang.

Interpreting Black-Body Radiation

The shape of the black-body radiation curve was a puzzle for physicists at the end of the nineteenth century. The wave model predicted a result that was inconsistent with the experiments and that was physically unreasonable. In the wave model, the black-body furnace was filled with waves of all wavelengths that were produced by vibrating (oscillating) charges in the walls of the furnace. These charges were assumed to vibrate with different frequencies, which were the frequencies of the emitted waves, and all frequencies were possible. In order for a wave to "fit" in the furnace, an integral number of its wavelengths had to span the dimensions of the space inside. The shorter the wavelength of the wave, the easier it was to fit within these constraints. Thus, the furnace should be filled with many more short-wavelength (high frequency) waves than long-wavelength (low frequency) waves. The probability that any wave would escape from the hole was assumed to be the same, so emission would be proportional to the number of each kind inside, and more high-frequency than low-frequency waves would be emitted.

Lord Rayleigh worked out an equation based on this model and found that the amount of energy radiated over a particular range of frequencies should vary as the fourth power of the frequency. Sixteen times as much energy should be radiated in the form of violet light as in the form of red light, and far more still should be radiated in the ultraviolet. In fact, by Rayleigh's formula, virtually all the energy of a radiating body should be radiated very rapidly in the far ultraviolet. Some people referred to this as the violet (or ultraviolet) catastrophe.

The point about the violet catastrophe, however, was that it did not happen. To be sure, at very low frequencies the Rayleigh equation held, and the amount of radiation climbed rapidly as the frequency of the radiation increased. But soon the amount of radiation began to fall short of the prediction (see Figure 26). It reached a peak that was considerably below what the Rayleigh equation predicted for that frequency and, at still higher frequencies, rapidly decreased, although the Rayleigh formula predicted a continuing increase.

Planck's Quantum Hypothesis

In 1899 a German physicist, Max Planck (1858–1947), began to consider the problem. He found a mathematical relationship that described the curve and then searched for a model from which this relationship could be derived. Rayleigh's analysis, it seemed to Planck, was mathematically and logically correct, provided the assumptions of the wave model were accepted. Since Rayleigh's equation did not fit the facts, it was necessary to question the assumptions.

One assumption of the wave model was that an oscillating charge (an electron moving back and forth, for example) could create a wave having any amount of energy, depending on its amplitude. Planck hypothesized that the oscillators could not have just any amount of energy but, rather, could have only discrete amounts of energy. The amount of energy an oscillator could have was directly proportional to its frequency, and he called the proportionality constant h. Thus the energy of the oscillator had to be an integral multiple of $h\nu$. Planck called this quantity of energy a *quantum* (pl., *quanta*), from a Latin word meaning "how much?" since the size of a quantum was a crucial question. The proportionality constant, h, is now usually called Planck's constant, and the currently accepted value is 6.6256×10^{-34} J s.

Furthermore, when an oscillator produced an electromagnetic wave, the oscillator had to give up energy to the wave only by losing a whole energy quantum (or more than one). It could not lose a fraction of a quantum. The result was that the wave had to have an energy $E = h\nu$, that is, the energy of an electromagnetic wave was a function only of its frequency and increased as the frequency increased. In effect, Planck hypothesized that there were "particles" of energy and that a radiating body could give off one particle of energy or two particles of energy, but never one and a half particles of energy or, indeed, anything but an integral number of such entities.

How does Planck's quantum hypothesis solve the black-body dilemma? In the black-body furnace, the oscillators were in thermal equilibrium with one another at the furnace temperature, T, and there was only a certain amount of total energy available to be distributed among the oscillators. Planck's major field of expertise was thermodynamics, and he knew that the most favorable equilibrium distribution of energy in a system was to have the available energy spread among as many components of the system as possible rather than concentrating it in a few. Although there are more possible oscillations at high frequency than at low, a high-frequency oscillator would take a larger share of the energy than a low-frequency oscillator. With a given amount of energy, many low-frequency oscillators can have energy compared to only a few high-frequency oscillators. Thus, the energy distribution should greatly favor the low frequencies.

At low frequencies (as at all frequencies), the number of oscillators increases with increasing frequency of oscillation, and the amount of energy distributed among them increases with frequency because none of them is taking a disproportionate share of the total. Thus, as Rayleigh predicted and experiment showed, the amount of energy radiated increased with frequency, but because the energy distribution favors the lowest frequencies, the rate of increase was not as fast as predicted (see Figure 26). At high frequencies very few oscillators would have a quantum of energy, so the emission at high frequency would drop off rapidly, as experiment showed. At some intermediate frequency, the quantity of energy emitted would reach a maximum. If the temperature of the furnace were raised, the general amount of energy available for radiation would increase as the fourth power of the absolute temperature. More energy would be available for distribution among the oscillators and they all could have more, including an increased proportion of the higher-frequency oscillators. Therefore, as observed, the emission would increase at all frequencies, and the maximum in the energy emitted would move to higher frequency. At a temperature of 6000 K, the peak would be in the visible light region, but the larger quanta of ultraviolet would still be formed in minor quantities.

Using this model, Planck derived a relationship for the amount of energy emitted as a function of the frequency of the black-body radiation. The only unknown parameter in his equation was the proportionality constant, h. By choosing the appropriate value for this constant (modern value given above), he was able to reproduce the mathematical relationship he had found empirically to fit the black-body emission data.

If the energy of oscillators producing electromagnetic radiation was quantized, why had quantum effects not been seen previously? Using the relationship $c = \lambda\nu$, we find that orange light of wavelength 600 nm has a frequency $\nu = c/\lambda = (3.00 \times 10^8 \text{ m s}^{-1})/(600 \times 10^{-9} \text{ m}) = 5 \times 10^{14} \text{ s}^{-1}$. The energy content of a quantum of this orange light is $h\nu = (6.6256 \times 10^{-34} \text{ J s})(5 \times 10^{14} \text{ s}^{-1}) \approx 3.3 \times 10^{-19}$ J. This is an extremely tiny amount of energy. It is little wonder that no individual quanta of radiant energy or their effects were casually observed before the days of Planck.

Electromagnetic-Wave Units

The frequency of waves is often expressed in hertz (Hz) where 1 Hz = 1 s^{-1}. The unit is named in honor of the German physicist Heinrich Hertz (1857–94), who was the first to prove experimentally that Maxwell's electromagnetic waves existed and had the predicted properties. Electromagnetic waves in the radio-frequency range, which encompasses AM, shortwave, FM, VHF-TV, cell phones, UHF-TV, radar, and microwaves (see Figure 1), have frequencies from about 5×10^5 hertz (= 500 kilohertz [kHz]) to about 3×10^{11} hertz (= 300 gigahertz [GHz]) and are usually expressed in hertz (or 1000-fold multiples like these). At higher frequencies—infrared, visible, ultraviolet, and above—we usually choose to describe the waves in terms of their wavelengths in a vacuum, as we have been doing. Comparisons among commonly used units for these quantities are shown in Table 1.

The Initial Impact of the Quantum Hypothesis

When it was first announced, Planck's hypothesis created little stir. Planck himself did nothing with it at first but explain the distribution of black-body radiation, and physicists were not ready to accept so radical a change of view of energy just to achieve that one victory. Planck himself was dubious and at times tried to draw his quantum hypothesis as close as possible to classical notions by supposing that energy was only radiated in quanta and that it might be absorbed continuously.

And yet quanta helped explain a number of facts about absorption of light that classical physics could not. In Planck's time, it was well known that violet light was much more effective than red light in bringing about chemical reactions and that ultraviolet light was more effective still. Photography was an excellent example of this observation, for photographic film of the type used in the nineteenth century was very sensitive to the violet end of the spectrum and rather insensitive to the red end. In fact, ultraviolet light had been discovered a century before Planck because of its pronounced effect on silver nitrate. Would it

Table 1.   Comparison of commonly used units for wavelength, frequency, and energy of light.

Measure	Unit, symbol	Equivalence	Visible light range (red-violet)
Wavelength (λ)	nanometer, nm	$1 \text{ nm} = 10^{-9} \text{ m}$	700–400 nm
	Ångstrom,* Å	$1 \text{ Å} = 10^{-10} \text{ m}$	7000–4000 Å
Frequency (v)	second^{-1}, s^{-1}	$v = c/\lambda$	4.3×10^{14}–$7.5 \times 10^{14} \text{ s}^{-1}$
	hertz, Hz	$1 \text{ Hz} = 1 \text{ s}^{-1}$	4.3×10^{14}–$7.5 \times 10^{14} \text{ Hz}$
	wavenumber,** v	$1/\lambda = v/c$	14,000–25,000 cm^{-1}
Energy (E)	joule, J	$E = hv$	2.8×10^{-19}–$5.0 \times 10^{-19} \text{ J}$
	electron volt, eV	$1 \text{ eV} = 1.602 \times 10^{-19} \text{ J}$	1.8–3.1 eV

*The Ångstrom, once extensively used as the unit for wavelength, has now been largely replaced by the nanometer, but Ångstrom is still found in some sources.
**Wave numbers were commonly used by early spectroscopists but are now used mainly for frequencies in the infrared, about 10 to 13,000 cm^{-1}. Infrared spectrometers used for chemical analysis typically work in the middle of this range, from 400 to 4000 cm^{-1}.

not have been reasonable to suppose that the large quanta of ultraviolet light could produce chemical reactions with greater ease than the small quanta of red light? And would it not explain the facts only if energy was absorbed in whole quanta?

This argument was not, however, used to establish the quantum theory in connection with absorption. Instead, the German physicist Albert Einstein (1879–1955) made use of a very similar argument in connection with a much more recently discovered, and an even more dramatic, phenomenon.

The Photoelectric Effect

In the last two decades of the nineteenth century, it was discovered that some metals behave as though they give off electricity under the influence of light. At that time, physicists were beginning to understand that electricity was associated with the movement of subatomic particles called *electrons* and that the effect of the light was to eject electrons from metal surfaces. This is known as the *photoelectric effect.* (See Demonstration 12.18, The Photoelectric Effect.)

Quantitative study of the photoelectric effect produced results that the wave model could not explain. It seemed fair to assume that under ordinary conditions the electrons were bound to the structure of the metal and that a certain amount of energy was required to break this bond and set the electrons free. It also seemed that, if the light was made more and more intense, more and more energy would be transferred to the metal surface. Not only would the electrons then be set free, but also considerable kinetic energy would be available to them, so they would dart off at great velocities—the more intense the light, the greater the velocities. It seemed that the frequency of the light should make no difference, only the total energy carried by the light, whatever its frequency. The experiments told a different story.

The German physicist Philipp Lenard (1862–1947), after careful studies in 1902, found that for each surface that showed the photoelectric effect, there was a limiting *threshold frequency* above which, and only above which, the effect was observed. Let us suppose, for instance, that this threshold frequency for a particular surface is $5 \times 10^{14} \text{ s}^{-1}$, the frequency of orange light of wavelength 600 nm. If light of lower frequency, such as red light of $4.2 \times 10^{14} \text{ s}^{-1}$, falls on the surface, nothing happens. No electrons are ejected. It does not matter how bright the red light is nor how intense (how much total energy it carries).

However, as soon as the light frequency rises to $5 \times 10^{14} \text{ s}^{-1}$, electrons begin to be ejected, but with almost no kinetic energy. It is as though the energy they have received from the light is just sufficient to break the bond holding them to the surface but is not enough to supply them with any additional kinetic energy. Lenard found that increasing the intensity of the light at the threshold frequency produced more electrons but did not produce electrons with additional kinetic energy. The number of emitted electrons was proportional to the intensity (total energy) of the orange light, but all of them lacked kinetic energy.

If the frequency were increased still further and violet light of 10×10^{14} s^{-1} were used, electrons would be emitted with considerable kinetic energy. The number emitted would again be proportional to the intensity of the light but again all would have the same maximum kinetic energy. In other words, a feeble violet light would bring about the emission of a few high-energy electrons; an intense orange light would bring about the emission of many low-energy electrons; and an extremely intense red light would bring about the emission of no electrons at all. These results are summarized schematically in Figure 27.

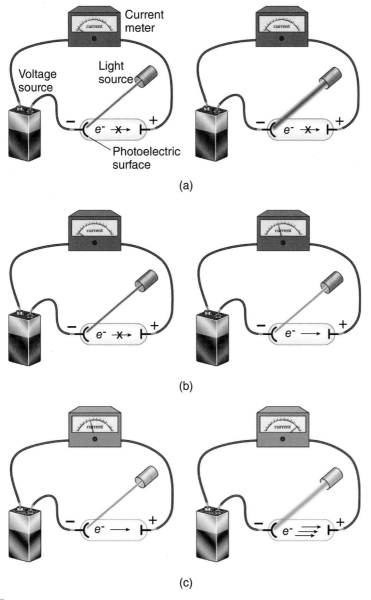

Figure 27. Pictorial summary of photoelectric effect observations. The thickness of the light beam represents its relative intensity (brightness). (a) With low-frequency light of different intensities, neither produces a current. (b) With low-frequency (red) and high-frequency (blue) light, only high-frequency light produces a current. (c) With high-frequency light of different intensities, higher intensity produces larger photoelectric current.

Quantum Explanation

The wave model of the nineteenth century could not account for these results. But in 1905, Einstein advanced an explanation that made use of Planck's quantum hypothesis, which was now five years old but still very much neglected. Einstein assumed that light was not only radiated in quanta, as Planck had maintained, but also that it was absorbed in quanta. When light fell upon a surface, the electrons bound to the surface absorbed the energy one quantum at a time. If the energy of that quantum was sufficient to overcome the forces holding it to the surface, it was set free—otherwise not.

Of course, an electron might conceivably gain enough energy to break loose after absorbing a second quantum even if the first quantum had been insufficient. This, however, is an unlikely phenomenon. The chances are enormous that before it can absorb a second quantum, it will have radiated the first one away. Consequently one quantum would have to do the job by itself; if not, merely multiplying the number of low energy quanta would not do the job. (With modern high-intensity laser sources, such multiple quantum absorptions are accessible, but not with ordinary light sources.)

The size of the quantum, however, increases as frequency increases. At the threshold frequency, the quantum is just large enough to overcome the electron bond to a particular surface. As the frequency and the energy content of the quantum increase further, more and more energy will be left over after breaking the electron bond and can provide the emitted electron with kinetic energy (Figure 28). For each substance, there will be a different and characteristic threshold energy, depending on how strongly the electrons are bound. For cesium, to which electrons are bound very weakly, the threshold frequency is in the infrared. Even the small quanta of infrared radiation supply sufficient energy to break that weak bond. For silver, to which electrons are bound more strongly, the threshold frequency is in the ultraviolet.

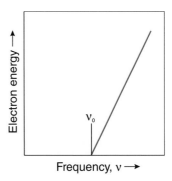

Figure 28.   Maximum energy of ejected photoelectrons as a function of light frequency. No photoelectrons are produced below the threshold frequency, v_0.

Furthermore, Einstein could explain quantitatively how the frequency dependence of the maximum kinetic energy of the emitted electrons (Figure 28) fit the quantum hypothesis. If the threshold frequency for emission is v_0, the energy of the quantum is hv_0. If light of a higher frequency, v, is used, its energy is hv. If all the extra energy goes into the kinetic energy of the emitted electron, then the maximum energy it can have is

$$E_{max} = hv - hv_0 = h(v - v_0)$$

This relationship tells us that the slope of the line in the figure is h. The value of h from the photoelectric effect was identical to the one Planck derived from the black-body analysis.

Einstein's explanation of the photoelectric effect was so elegant, and the identity of the value for h so compelling, that the quantum theory sprang suddenly into prominence. It had been hypothesized originally to explain the facts of radiation, and now, without modification, it was suddenly found to explain the photoelectric effect, a completely different phenomenon. This was most impressive, and physicists have universally accepted the quantum theory. It is now the general assumption that energy can be radiated or absorbed only in whole numbers of quanta and, indeed, that energy in all its forms is quantized, that is, energy at the molecular-atomic level must be considered as behaving as though it were made up of indivisible quanta.

Photons

Einstein carried the notion of energy quanta to its logical conclusion. A quantum seemed to be analogous to a particle of energy, so why not consider such particles to *be* particles? Light, then, would consist of particles, which were eventually called *photons* (again, as with other "photo" words, from the Greek for "light"). This notion came as a shock to physicists. The wave theory of light had been established just a hundred years before and for a full century had been winning victory after victory until Newton's particle theory had been ground into what had seemed complete oblivion.

Note, however, that it is in the conversion of the energy of matter into electromagnetic radiation (for example, by an oscillating electric charge) and the reverse, the conversion of electromagnetic radiation into the energy of matter (for example, providing energy to electrons in the photoelectric effect), that the photons reveal themselves. The propagation of electromagnetic radiation through space and its interactions with itself are still governed by the wave model, which was so beautifully developed by the nineteenth-century physicists, with the added property that the energy transmitted by a wave is a function of its frequency. The intensity of a light beam can be interpreted in terms of wave amplitude, but it is less confusing to think in terms of number of photons, so that the total energy of the beam is not confused with the fixed energy each photon can transmit when it interacts with matter. Thus, the present view of light is that it exhibits both wave and particle properties, depending on the phenomena being considered.

At this point, we have concluded our exploration of the properties of light and now need to turn to a consideration of the mechanisms by which matter changes quantum levels to produce light and in turn the mechanisms by which light interacts with matter to transfer its energy to the matter. Our focus will be mainly on those phenomena that create or absorb visible light, since these are the ones directly accessible to our unaided vision.

INTERACTIONS OF LIGHT AND MATTER

A very large amount of what we know about the world around us we learn by seeing. All that we see depends upon the way light interacts with our surroundings (being emitted, reflected, refracted, diffracted, and absorbed), upon how it is sensed by the chemical reactions in our eyes, and upon how it is interpreted by our brains. Light-sensing instruments, such as spectrophotometers, enhance our knowledge by providing evidence of the nature of the world that we cannot perceive directly, such as its atomic and molecular properties, as exemplified in the previous section by black-body radiation and the photoelectric effect. The remarkable array of products that we take for granted—cell phones, laptop computers, barcode readers, flat-screen televisions, and so on—all exploit our intimate knowledge of the molecular world. Besides information about nature, light also provides ways to manipulate matter and cause reactions that produce new materials.

Light interacts with matter via the motion of electric charges. Their movement in matter produces light, and light's oscillating electric field interacts with movable charges in matter. Hertz's initial production and detection of radio waves is an example. He created an oscillating electric current (moving electrons in a conducting metal wire) with a spark and then passed the current through a transmission antenna at a frequency of about one GHz (30-cm wavelength), which produced electromagnetic radiation of this frequency. The waves emitted from the antenna propagated through space and were detected by a receiving antenna (another conducting metal wire) of the appropriate length so that the electrons in the antenna were set oscillating at the same frequency and produced a spark at the receiver. Several decades later, over-the-air video signals were able to be transmitted at very high frequency (VHF, 30–300 MHz range, 10–1-m wavelength) and at ultra-high frequency (UHF, 300 MHz–3 GHz range, 1–0.1-m wavelength). Antennas for VHF signals were generally a few meters in length and situated outside a house, whereas those for UHF were usually rather small loops attached directly to the television set.

Refraction and Reflection

The interaction of light with electric charges in matter—gas, liquid, or solid—reduces the speed of light compared to its speed in a vacuum. The ratio of the speed of light in a vacuum, c, to the speed in a material, c_m, is the refractive index, n, of the medium. Since c_m and, hence, the refractive index are dependent on the wavelength of the light (which is, of course, why the refraction of white light by a glass prism produces a spectrum), the refractive index is usually given for the light from a sodium-vapor lamp at 589 nm. Refractive indices for a few representative materials that are transparent to visible light are given in Table 2.

Table 2. Refractive indices at 589 nm for some representative media.

Material	$n = c/c_m$
Vacuum	1 (exact)
Air	1.00029
Water	1.333
Ethanol	1.361
Benzene	1.501
Quartz	1.458
Diamond	2.419
Human eye, cornea	1.376
Lens of human eye	
Surface	1.386
Core	1.406

Even when the energy of the electromagnetic wave does not match the energy of one of the transitions in the matter, the oscillating electric field of the wave interacts with the electrons in the matter. The interaction can set the electrons in motion momentarily (sort of like causing them to twitch), before the motion causes the wave to be reemitted. Although this interchange of energy is almost instantaneous, it is a momentary pause in the progression of the wave, and the observable result is a slowing of the passage of light through the material. The stronger the interactions of the electric field with the electrons in the matter, the slower the light travels and the higher the refractive index.

Each of these interactions gives rise to a new electromagnetic wave that interferes constructively and destructively with those from other interactions, a process that, in a sense, recalls the Huygens wave model (Figure 13), in which the light wave is continuously re-creating itself, although, in the present model, through the intermediacy of the wave-electron interactions. Even in a transparent material, these interactions slightly attenuate the light wave and decrease its amplitude as it passes through. For a thin sheet of material, like a windowpane, the attenuation is not noticeable, but, as we noted previously, visible light cannot penetrate more than about a 30-m thickness of water. The interactions in this thickness of water reduce the amplitude of the light waves to zero.

Boundaries

When a light wave hits a boundary between two materials with different refractive indices (different velocities), we can use the Huygens representation, which is shown in Figure 29, to picture what is going on, keeping in mind that it is the interactions of the electric field of the wave with the electrons that create the new waves. (See Demonstrations 12.13, Disappearing Glass: Index of Refraction; and 12.14, Disappearing Gel: Index of Refraction.) For all transparent materials, there is both reflection and refraction of an incident light beam. Seeing a reflection in a store's plate-glass window or the reflection of a beautiful landscape on a calm body of water are familiar experiences. The ratio of reflection to refraction depends on the angle at which the light beam strikes the boundary (and on the difference in refractive indices of the two materials it is traveling through); the more oblique the angle (that is, the further from perpendicular to the boundary), the greater the amount of reflection.

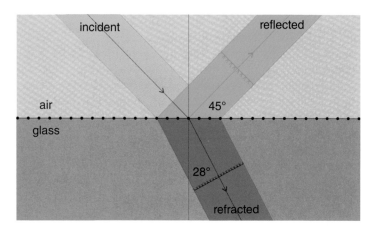

Figure 29.   Representation of an incident light wave (red) hitting a boundary between two materials with different refractive indices (different velocities of light) and being reflected (green) and refracted (blue).

The model just presented is for materials that are dielectrics—their electrons are associated with individual atoms or molecules and are not free to move about in the material. The interaction of light with the surface of a metal is different, because the conduction electrons in a metal (those responsible for carrying an electric current or transporting thermal energy) are relatively free to move about. An incident light beam can easily set these electrons in motion, which is equivalent to setting up an electric current at the surface of the metal. However, the laws of electrodynamics (which were developed by Maxwell) show that a net current cannot exist in a metal with "free" conduction electrons (unless, of course, there is an electrical potential difference [voltage] between two points in the metal), so an instantaneous current—motion of electrons—equal and opposite to that produced by the light beam is created. These moving electrons, in turn, create electromagnetic radiation that can propagate only away from the surface, as pictured in the Huygens model for light reflected from a boundary. Although there is some attenuation, almost all the intensity of light that strikes the surface is reradiated in the reflected beam, which is why smooth, shiny (highly reflective) metal surfaces are used to make mirrors.

If the material is colored, such as colored glass or a colored solution, then some incident wavelengths of light will be absorbed by the kinds of interactions discussed earlier and will not be available for the reflection and refraction discussed in this section. The unabsorbed wavelengths will be reflected and refracted just as in a transparent, colorless material, but we will observe color and characterize the material by the color we perceive. Almost all metals have "silvery" colors (when their surface is not oxidized or otherwise tarnished because of their reactivity, making them dull or, even, colored) when white light is reflected. The two notable exceptions are copper (reddish-brown) and gold. In these metals, there is an energy gap between the valence and conduction levels that is in the range of visible wavelengths, so these wavelengths are absorbed. Thus, all the wavelengths of white light are not reflected, and the metals are shiny but not silvery.

Atomic Spectra

The idea that matter was electrical in nature and consisted of heavy, positively charged parts balanced by light, negatively charged particles, called electrons, was well known at the beginning of the twentieth century. However, although there were conjectures, there was no experimental evidence to show how the positive and negative parts were arranged to form atoms and molecules. Two developments led to the present nuclear model of the atom: the interpretation of alpha-particle scattering and the unique atomic spectra of the elements.

In 1911 the New Zealand–born physicist Ernest Rutherford (1871–1937) published alpha-particle scattering results from his laboratory that showed that atoms seemed to be mostly empty space with almost all their mass concentrated in a very tiny, positively charged nucleus and with the remainder of the atom occupied by its electrons. This model posed a problem because no stable arrangement of *stationary* charges like these is possible, and *moving* electric charges would emit electromagnetic radiation and quickly lose their energy, thus causing the atom to collapse.

The Danish physicist Niels Bohr (1885–1962), who was working with Rutherford at this time, solved the problem by applying the new ideas of quantum physics and the nuclear model of the atom to the interpretation of atomic spectra. The now-familiar Bohr model of the atom pictured the electrons moving around the nucleus in circular (or elliptical) orbits, but not just *any* orbits. Bohr postulated that the angular momentum of the stable electron orbits had to be integral multiples of a universal constant and that emission from an atom that is in an excited state (a state in which the electrons are not all in their most stable orbits) could occur *only* when an electron moved from one orbit to another of lower energy. The energy of this emission, E, would be equal to the difference in energy between the two orbits and produce light with a frequency, $\nu = E/h$, based on Planck's quantum hypothesis. The model elegantly explained why, as we discussed above, excited atoms emit only a few colors or frequencies of light, not a full spectrum. (See Demonstration 12.5, Emission Spectra from Gas-Discharge Lamps.) Furthermore, Bohr could calculate the emission spectrum of the simplest atom, hydrogen, and get exact agreement with the experimental frequencies (wavelengths), if the universal constant for his orbits was $h/2\pi$, where 2π accounts for angular momentum.

Once again, as with Einstein's analysis of experimental results from the photoelectric effect, Bohr's atomic model required Planck's constant for a quantitative explanation of atomic spectra. Although quantitative application of Bohr's model was limited to one-electron atoms and ions (H, He^+, Li^{2+}, etc.), its explanation for the origin of discrete atomic emission spectra led scientists to accept the idea that the energies of electrons in atoms are quantized, that is, they can have only certain values and no others. One way to visualize these energies is in an energy-level diagram for the allowed electron energies, as shown in Figure 30 for the energies in a hydrogen atom.

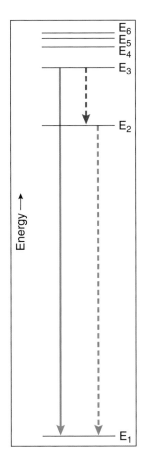

Figure 30. Electron energy levels in the hydrogen atom labeled with subscripts corresponding to Bohr's integers for the quantized orbits. Electromagnetic emissions are represented by the colored lines, with blue being the highest energy (shortest wavelength) and red the lowest energy (longest wavelength). The wavelengths corresponding to the blue and green lines are actually in the ultraviolet region of the spectrum. The wavelength corresponding to the red line is actually observed as a red emission at 656.3 nm.

Although the energy-level diagrams for multielectron atoms are more complex than for the hydrogen atom, the relationship shown here should be true for them as well. Note that subtracting the energy of the green line from the energy of the blue line gives the energy of the red line, $E_{blue} - E_{green} = E_{red}$ (or, using the Planck relationship, $\nu_{blue} - \nu_{green} = \nu_{red}$). Nineteenth-century scientists had searched for patterns in atomic emission spectra and had found many that fit frequency relationships like this. The Bohr model of quantized energy levels immediately provided an explanation for all of these patterns. The previous energy relationship can be written in terms of the energy levels as $(E_3 - E_1) - (E_2 - E_1) = (E_3 - E_2)$. If the electron-energy levels in atoms were quantized, then there would be many such relationships among emission energies (frequencies) and many patterns to be observed. Even if the Bohr model could not calculate the energy levels for multielectron atoms, knowing that they were there was enough for a satisfying explanation of the observations.

Quantized electron energies in atoms also readily explain atomic absorption spectra. When a broad range of light frequencies shines through a sample of atoms, a few frequencies of the electromagnetic waves have exactly the correct energies to match the differences in energies between the electron energy levels in the atoms. Consider one such energy-level difference and assume that there is an electron at the lower energy level. The electric field of a wave that matches the energy-level difference can interact strongly with the electron and provide exactly the energy required for the transition from the lower to the higher energy level. This is an all-or-nothing process. We have to account for the conservation of energy and the photonic nature of light and realize that the entire photon of this energy is used up in changing the energy of the electron. Thus, the electromagnetic wave corresponding to this photon disappears (is absorbed) from the spectrum of light and a gap appears in the spectrum of the transmitted light, as shown in Figure 4. Because only photons corresponding to energies that exactly match possible electronic transitions in the atoms are absorbed, these gaps are very narrow. Photons (waves) with energies only slightly above or below the match cannot interact with the atom and are not absorbed.

Electron Waves

The Bohr model of the atom was successful in explaining, at least qualitatively, the experimental observations on atomic spectra, but it rested on the unprecedented and untestable assumption that the angular momentum of electrons in orbits was quantized. Within about a decade, our present model, based on another startling, but testable, assumption was proposed. The change began with the work of the French physicist Louis de Broglie (1892–1987), who proposed on theoretical grounds in his doctoral thesis that atomic particles—in particular, electrons—should have wave, as well as particle, properties. From his mathematics, de Broglie could calculate the wavelength associated with an electron traveling with a given velocity, u: $\lambda = h/mu$, where h is Planck's constant and m is the electron mass.

Experimental verification of the de Broglie hypothesis came within two years, when two American physicists, Clinton Davisson (1881–1958) and Lester Germer (1896–1971), used the electron-wave model to explain the results of experiments they had done. A modern version of these results is shown in Figure 31. In a sense, the de Broglie electron-wave model creates a natural symmetry between particles having wave properties and electromagnetic waves having particle properties. And, once again, Planck's constant is the common factor in these phenomena.

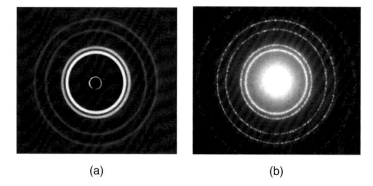

(a)   (b)

Figure 31.   (a) The pattern produced when x-rays—high-frequency, short-wavelength electromagnetic radiation—are diffracted by the closely spaced atoms in aluminum foil. (b) The identical pattern is produced by the diffraction of electrons that have been accelerated to a velocity at which they have a wavelength similar to that of x-rays.

Standing Waves

To apply the electron-wave model to atoms and molecules we need to examine the properties of *standing,* or stationary, waves. Most of the waves discussed so far are traveling waves propagating through space, and there is no restriction on their wavelengths. If a wave is constrained to remain in a particular space, only certain wavelengths, which depend on the constraints, are possible. We met this restriction in discussing the black body and the restrictions on the wavelengths that could "fit" in the oven. Concrete examples of such constraints are stringed instruments, where each string is constrained to vibrate with particular wavelengths that depend on the length of the string between its fixed endpoints, as shown schematically in Figure 32.

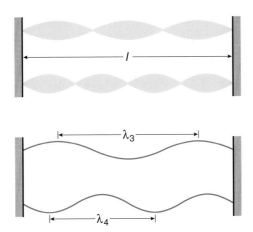

Figure 32. Two of many possible standing waves on a string of length l are shown as the blurred images we actually observe and as instantaneous "snapshots" at maximum amplitude. The constraint on the waves is that they must have a node at each of the fixed ends of the string so the wavelengths that fit on the string are determined by the equation $\lambda_n = 2l/n$, where n is a positive integer.

Standing Electron Waves

Soon after de Broglie postulated the electron-wave model, scientists began applying it to the electrons in atoms (they assumed the Rutherford nuclear-atom model, which Bohr had used so successfully). The most familiar application is probably that of the Austrian-born physicist Erwin Schrödinger (1887–1961), who applied the well-known mechanics of standing waves to the electron wave constrained in the atom by its attraction to the nucleus. In essence, Schrödinger's approach summed the potential and kinetic energies of the constrained electron to give its total energy. The solutions to the Schrödinger equation are a series of standing electron waves, each associated with a fixed energy, that is, a quantized series of energy levels, just as in the Bohr model. The shapes of the electron waves indicate a volume in which the electron is likely to be found, much like the blurred images of the string in the figure, but no instantaneous trajectory analogous to the "snapshots" of the waves in the figure is implied. To honor the contribution of Bohr's breakthrough model of quantized electron orbits, each of these waves is called an *orbital*.

When applied to the hydrogen atom (and hydrogen-like ions), the Schrödinger solutions for the energies are identical to those Bohr calculated and therefore identical to the experimental spectroscopic data. The strength of the Schrödinger approach (and others based on electron waves) is that the fundamental postulate, that electrons in atoms can be treated as waves, is based on the experimentally testable assumption that electrons have wave properties. Also, quantization arises naturally as a consequence of the standing-wave model of the electron.

The Schrödinger equation cannot be solved exactly for atoms with more than one electron, but excellent approximate solutions are possible using various iterative techniques that are readily implemented by modern computers. Similarly, computer programs using several other approaches based on the electron-wave model are available, and many are able to reproduce experimentally determined values to the same precision as the experiments.

Molecular Spectra

For most molecules the energy required for the transition from the ground to the first (lowest energy) excited state is in the range 100–600 kJ mol^{-1}. These are energies in the visible and ultraviolet range of electromagnetic radiation. (As a rough rule of thumb, the wavelengths between 200 nm [deep ultraviolet] and 600 nm [visible red] correspond to energies from 600 to 200 kJ mol^{-1}.) Thus, for example, the highest-energy electron(s) in molecules whose first excited state is about 200 kJ mol^{-1} above the ground state can interact with the red wavelengths in a beam of visible light to absorb their energy and produce excited state molecules. If we are looking at the beam of light transmitted through a sample of these molecules, the red wavelengths will be missing and the sample will be seen as blue.

Just as for atoms, there are also many computational techniques available that can be applied to electron waves in molecules. The results from this modeling also compare favorably with experimental observations and in an increasing number of cases are used to point experiments in interesting, productive, and sometimes surprising directions. Here, we will focus on some simple systems and mostly qualitative models that help us understand the interactions of electromagnetic waves with molecules.

Electronic Energies

The usual approach to modeling molecules is to imagine the nuclei held a fixed distance apart, calculate the energy (and, if possible, the arrangement and shape) of the electron waves that surround them, then move the nuclei to a new separation and repeat the calculation—and repeat. The results are a series of energies calculated as a function of nuclear geometry as the computer searches for the lowest total energy, that is, the most stable arrangement of the nuclei and electrons. The results are easily visualized graphically for the simplest molecules—diatomics such as H_2, O_2, CO, HCl, etc.—that can be represented by potential energy curves as a function of internuclear separation, the distance between the nuclei, as shown in Figure 33.

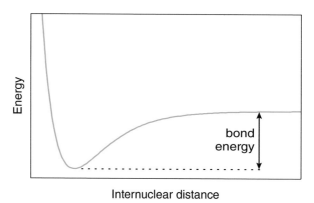

Figure 33.   Electronic energy curve for the ground state of a diatomic molecule.

At the far right, the separation is so large that the two atoms are essentially independent of one another. As they approach, the attraction of the electrons in both atoms to both nuclei simultaneously leads to electron waves that lower the energy, and the system is more stable than the separated atoms. At some separation, a minimum in the energy is reached, the lowest point of the curve. The figure represents the energy of the molecule with all the electrons in their lowest possible energy states, the *ground state* of the molecule. The double-headed black arrow in the figure represents the stability of the molecule relative to its separated atoms. As the nuclei are brought even closer, at the far left, the repulsion between the positive nuclei becomes the dominant interaction, and the energy rises sharply and becomes repulsive. These curves are often called *potential energy wells.*

Essentially the same kind of calculation can be carried out for the molecule in its excited states, with the results shown in Figure 34, where the upper curve is the energy of the molecule with one of its electrons in an energy state above its ground state. (In order to conserve energy, at least one of the atoms represented at the far right must be in an excited state, since the molecule is in an excited state.) There can be many excited states of a molecule, but the usual ones of interest are those for which an electron in the highest level of the ground state has been raised in energy to the next highest level. This is the energy difference, which is represented by the double-headed red arrow in the figure. Or, thinking about it somewhat differently, the next highest level is the lowest energy level that does not describe an electron wave in the ground state.

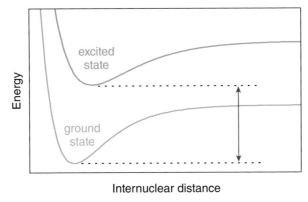

Figure 34.   Electronic energy curves for the ground and an excited state of a diatomic molecule.

Electron orbitals are descriptions of electron waves and do not exist as physical entities in the absence of the electron. However, we can *calculate* the energy of any orbital (or a molecule) as though it did describe an electron wave, and the common nomenclature treats orbitals somewhat like boxes that are occupied or not occupied by electrons. Thus, another description of the transition between the ground and first excited state is that an electron has absorbed enough energy to go from the highest occupied molecular orbital (HOMO) to the lowest unoccupied molecular orbital (LUMO). Note that once this transition has occurred, the LUMO is no longer unoccupied and the description of the actual molecule must include this energy level. What happens to the energy absorbed by a molecule will be discussed below.

Vibrational and Rotational Energies

Atoms behave mechanically as essentially one-dimensional points that can move about (translational motion) but undergo no other motions. Molecules, however, are either two-dimensional (linear molecules) or three-dimensional (all others). A two- or three-dimensional object (molecule) can rotate in space, as well as translate. A molecule can also vibrate, because the bonds between atomic cores are not rigid and the cores are always in motion with respect to one another. The energy required to set a tiny molecule in motion to rotate or vibrate more rapidly should be less than that required to excite one of its electrons. Therefore, if electromagnetic waves can provide the energy required to excite rotation or vibration, we would expect that low energy light, that in the infrared or microwave region, would do the trick. Figure 35 shows the absorption of infrared radiation by a sample of gaseous hydrogen chloride, HCl, a diatomic molecule. This is a vibration-rotation spectrum in which the location of this absorption envelope (the imaginary enclosure that contains all the peaks shown) in the infrared

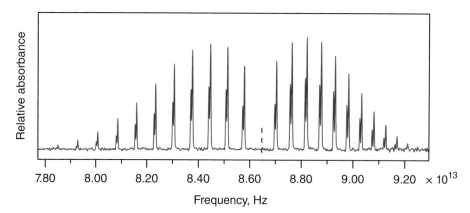

Figure 35.   Infrared spectrum of HCl(g). The center of the spectrum (marked by the dashed line) at 8.65×10^{13} Hz is about 35 kJ mol^{-1}, which is the energy spacing between the lowest two vibrational energy levels. Spacing between the peaks varies from about 0.5 to 0.8×10^{13} Hz, which is about 2 to 4 kJ mol^{-1}, the energy spacing between rotational levels. These data can also be used to determine the bond length of HCl.

region corresponds to the energy difference between the first and second vibrational energy levels of the molecule. The peaks within the envelope are due to absorptions of energy between different rotational energy levels in the first and second vibrational energy levels.

The sharp peaks are an indication that energies of rotation and vibration are quantized, that is, only certain energies are absorbed. The rotational and vibrational energy levels are usually illustrated by superimposing them on the electronic energy state diagrams for the molecule, as shown in Figure 36. Because the charge distribution in a vibrating molecule is changing, the electrons can interact with appropriate frequencies of electromagnetic radiation to cause transitions from one level to another. As we reasoned above, the differences in energy between vibrational levels are much smaller than the differences in energies between electronic states illustrated in Figure 34. Vibrational energy transitions only require light in

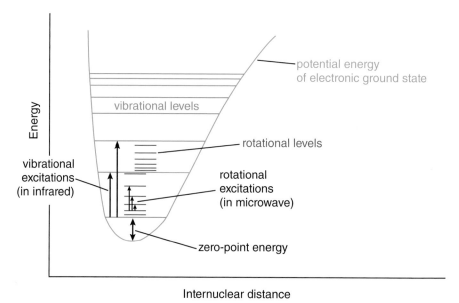

Figure 36.   Vibrational and rotational energy levels for a molecule superimposed on its electronic energy curve. To make the diagram less complicated, only rotational levels associated with the two lowest vibrational levels are shown.

the infrared part of the spectrum, energies between about 10 and 100 kJ mol^{-1}. Note that the lowest vibrational energy level, the lowest energy for this electronic state, is not at the very bottom of the energy curve in Figure 36. The energy difference between the bottom of the well and the lowest vibrational level is called the *zero-point energy,* which is present in all molecules and remains even at 0 K.

Rotational energy levels are even more closely spaced than vibrational energy levels, as shown in Figure 34, where the rotational energy levels associated with the first two vibrational levels are shown. If a molecule is polar, the rotation of the molecule is equivalent to moving an electric charge. Rotational motion can interact with appropriate frequencies of electromagnetic radiation to cause transitions from one rotational level to another. It does not take much energy to get a molecule to rotate, so these energies are in the microwave region of the spectrum, the range about 1–10 kJ mol^{-1}.

In the infrared absorption spectrum of gaseous hydrogen chloride (Figure 35), the peaks shown are like the gaps in the absorption spectra of atoms, that is, frequencies of the infrared light beam that are missing after passing through the sample. In this case, the light at these frequencies is not entirely missing because the number of molecules in the sample is not large enough to absorb all the photons of the interacting frequencies. The absorptions here are from rotational energy levels of the lowest vibrational state to the rotational energy levels of the first excited vibrational state. Thus, the spacing between successive peaks is a measure of the energy difference between rotational energy levels.

Peaks farther from the center of the spectrum (marked by the dashed line) are transitions from higher rotational energy levels of the lowest vibrational level. Since there is an unequal distribution of molecules among the rotational levels, with fewer of the molecules in the higher levels, and hence fewer to absorb energy, the absorptions become weaker farther from the center. Because two isotopes are present, $H^{35}Cl$ and $H^{37}Cl$ molecules (in about their natural three-to-one ratio), each peak is actually two overlapping peaks. These isotopic molecules have almost the same vibrational energy levels, but they differ in their rotational levels because it is a little more difficult to get the molecule with the heavier end rotating.

In many cases, gas-phase electronic spectra can provide information about the spacing of vibrational energy levels in the molecules that absorb energy from the analyzing light beam as well as about the energy spacing between electronic states. An example is illustrated in Figure 37, which shows the absorption spectrum of gaseous benzene in the near ultraviolet.

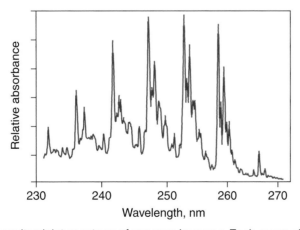

Figure 37. Near-ultraviolet spectrum of gaseous benzene. Each group of peaks is an absorption from the lowest vibrational level of the ground electronic state to higher and higher vibrational levels of the first excited electronic state. The separate peaks within a group are due to the rotational levels associated with these vibrational levels. The lowest energy group, at about 267 nm, indicates that the excited state is about 450 kJ mol^{-1} above the ground state. The energy spacing between the groups, about 12 kJ mol^{-1}, is the energy spacing between vibrational levels in the excited state.

Although gas-phase spectra are a rich source of information about the structure of and energy levels in molecules, the substances we deal with most are liquids and solids. For comparison with the gas-phase spectrum of benzene, the near-ultraviolet spectrum of liquid benzene is shown in Figure 38. Although a good deal of structure remains in the spectrum, the groups of separate sharp peaks in the gas-phase spectrum are blurred together into single broad peaks, and careful inspection shows that the positions of the groups have shifted a bit to longer wavelengths (lower energies).

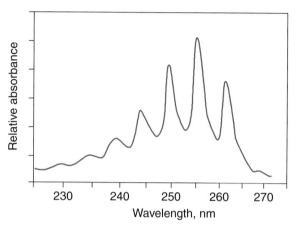

Figure 38.   Near-ultraviolet spectrum of liquid benzene.

Both of these effects can be attributed to the influence of electrons in neighboring molecules. In the gas phase, molecules are quite far apart and interact only weakly with one another, but in liquids and solids, the electron waves in molecules are in contact with one another. The contacts are always shifting as molecules move about, so the influence of its neighbors on any molecule is a sort of average of many interactions, each of which alters its electronic wave energies (and vibrational and rotational energy levels) slightly. Therefore, at any instant, as an electromagnetic wave interacts with it, this molecule will absorb a photon of slightly different energy than it does when isolated. The consequence is that a broader range of photons will have the appropriate energies to be absorbed by molecules in the sample, and the absorption peaks will broaden and overlap to produce a less-structured absorption spectrum.

Some effects of the medium are going to be in the same direction for all the molecules in it. Any shift of the electronic energies due to these effects will be observed in the spectrum as a shift of all the peaks toward higher or lower energy levels compared to their positions in another medium, as was observed for the peaks in the benzene spectrum in the gas and liquid phases. The positions of spectral peaks for a substance will often be different when it is dissolved in different solvents. (See Demonstration 12.45, The Effects of Solvents on Spiropyran Photochromism and Equilibria.)

Absorptivity: The Beer-Lambert Law

To obtain the most information from spectra, it is best to study samples that do not absorb all the light in the wavelength range of interest, so that the wavelengths at which the peaks occur can be accurately determined, as illustrated in the previous examples. The amount of light transmitted (not absorbed) by a sample at a particular wavelength is its *transmittance, T.* If the intensity of light reaching the spectrophotometer detector in the absence of sample is I_0 and the intensity with the sample in the light beam is I, then $T = I/I_0$. (Transmittance is often expressed as a percentage, $\%T = 100T$.) A more useful measure is the *absorbance, A,* of the sample, where $A = \log(1/T)$. Note that, when the sample absorbs no light, $T = 1$ and $A = 0$, and when T is very small, most of the light is absorbed and A is large.

Two principal factors determine how much of any particular wavelength will be absorbed: the number of molecules in the path of the light beam and the probability that the electromagnetic wave can be absorbed and cause the electronic energy level to change. The number of molecules in the light beam is often expressed as the concentration of the sample (c, usually in molarity, mol L^{-1}, for samples in solution) multiplied by the path length of the light beam through the sample (l, usually in centimeters, since the path lengths of commonly used sample cells are in the range 0.1–10 cm).

The probability that a photon of a particular energy will be absorbed to cause an energy-level change depends on how the electron-wave descriptions of the two levels are related to one another. Calculations based on these descriptions can be applied for simpler molecules, but an empirical approach is common for complex molecules. In this approach, the probability is expressed in an absorptivity parameter called the *molar absorption coefficient, ε,* (other names for this factor are *molar extinction coefficient* and *molar absorptivity*), which is determined from the Beer-Lambert law, $A = \varepsilon l c$, a relationship at a specified wavelength among absorbance, the molar absorption coefficient, path length, and concentration. Since A has no units, the units of the molar absorption coefficient are L mol^{-1} cm^{-1}.

The Beer-Lambert law indicates that the absorbance of a sample at a specified wavelength should be a linear function of the sample concentration, if the path length is held constant. Many compounds obey this relationship and their molar absorption coefficients can be obtained from the slope of the line generated by plotting A versus c. Given a value of ε (or the slope of the A-versus c line) for a compound, an absorbance measurement on a solution of unknown concentration can be used to determine its concentration. Numerical values of ε range from near unity for very weak absorbers to tens of thousands for strong absorbers, such as dye molecules.

Many factors, including the pH of the solution, can influence the value of ε for a compound. Familiar examples are acid-base indicators where the acid (protonated) and base forms of the indicator molecules have quite different absorption spectra and therefore different molar absorption coefficients at almost all wavelengths. The Beer-Lambert law is also not always followed over a wide range of concentrations, especially at high concentrations where the molecules of interest are forced to be close together and can interact with one another to disturb one another's electronic energy levels. Such disturbance can change the probability of the absorption of light and hence change molar absorption coefficients, which must be constant for the Beer-Lambert law to hold. Nevertheless, even with their limitations, spectrophotometric analyses are powerful tools for determining both the identity and amount of substances in a wide variety (chemical, biological, geological, etc.) of samples.

Reemission of Energy

We have discussed the emission of electromagnetic radiation by gaseous atoms that have been excited to energy states above their ground state by being heated or subjected to an electrical discharge (electrons fired from a negative to a positive electrode through the gas). The glow from an incandescent lamp filament that has been heated by an electrical current passing through it can be analyzed like the emission from a black body. But what happens to the energy in molecules that have absorbed photons from a beam of light and are now in excited states? The various pathways that molecules can follow back to their ground states are the topic of this section. In the next section, we will look at photochemistry, the chemical reactions that these light-energized molecules can undergo to produce new molecules.

Electron Spin

To understand some of these pathways, we need to consider spin, another property of electrons, and its consequences for the absorption and emission of electromagnetic radiation by molecules. Electron spin arises as an intrinsic property of electrons in a relativistic approach to quantum physics and is a purely quantum phenomenon with no classical analog. The *electron spin angular momentum* is quantized and can take only two values, $\pm\frac{1}{2}(h/2\pi)$, which are usually designated as $+\frac{1}{2}$ and $-\frac{1}{2}$ and are often visualized as an up or down arrow. In an atom or molecule, two electrons with opposite spin can occupy the same region of space (that is, be described by the same electron wave) and are said to be spin paired. Using the orbital nomenclature for the electron waves, we say that an orbital may be occupied by a maximum of two electrons of opposite spin. When all of the electrons in a molecule are spin paired, their angular momenta cancel one another to give an overall electron spin angular momentum of zero, $S = 0$, which is called a *singlet* state. Most ground-state molecules are singlets, with all their orbitals doubly occupied by spin-paired electrons.

If a singlet ground-state molecule absorbs energy and one of its electrons is raised in energy, say from its HOMO to its LUMO, the molecule now has two orbitals that are singly occupied by electrons. If the electrons in the two orbitals have opposite spins, the spin angular momentum is still zero, $S = 0$. Electrons in two different, singly occupied orbitals may have the same spin (according to the Pauli principle), and in this case their spins combine to give the molecule a total spin of one ($S = \frac{1}{2} + \frac{1}{2} = 1$). This spin angular momentum is a vector quantity that is quantized—the components of the vector can have three values, so this state is called a *triplet* state. This can be thought of as the vector taking on three orientations (often labeled $+1$, 0, and -1) with respect to an external magnetic field. The total energy of a triplet-state molecule is almost always lower (more stable) than that of the corresponding singlet. (This is an example of a regularity knows as Hund's rule.) A wave mechanical explanation is that the spatial arrangement of the separate orbitals occupied by the same-spin electrons in the triplet state strengthens the net coulombic attraction between nuclei and electrons and lowers the overall energy.

Excited Molecules

Some of the changes that excited molecules can undergo are illustrated in Figure 39, which is called a *Jablonski diagram*. When a ground-state molecule, usually denoted as S_0 (the lowest possible singlet state), absorbs a photon to raise the energy of an electron to the first excited electronic state, the electron spin is unchanged. The excitation, shown as pathway 1 in the figure, forms an excited singlet state, S_1. In general, there are wave mechanical restrictions that make electronic transitions between singlet and triplet states highly improbable.

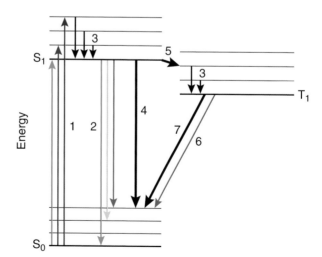

Figure 39.   A Jablonski diagram illustrating some of the light absorption and emission processes (colored arrows) and radiationless processes (black arrows) possible in a molecular system. The three sets of horizontal lines represent the first few vibrational energy levels in the three electronic states, S_0, S_1, and T_1.

This is because an electron spin needs to flip, and doing so is difficult in a transition, so direct excitation of S_0 by absorption of electromagnetic radiation to form the corresponding excited triplet state, T_1, is "forbidden." If, however, some energy levels in the two states coincide, it is often possible for molecules in the S_1 state to cross (change from one system of energy levels to another) to the T_1 state, that is, change the electron spin in one of the singly occupied orbitals. This radiationless *intersystem crossing* (the arrow labeled 5 in the figure) can be rapid and efficient for some molecules.

Fluorescence

After the absorption of light by S_0 to form S_1 (arrows labeled 1 in Figure 39), light (energy) may be rapidly reemitted from S_1 to return to S_0. This rapid reemission (shown by the arrows labeled 2) from the excited state is called *fluorescence* and occurs within one microsecond (10^{-6} s), often within a few nanoseconds (10^{-9} s), following absorption. Since the reemission is so fast, fluorescence emission is observed only when the excitation light is on and then disappears instantly (as far as our eyes can tell) when the light is turned off. (See Demonstration 12.41, The Fluorescence of Molecular Iodine Vapor.) Many fluorescent substances require excitation energies in the ultraviolet region, as is the case with minerals and fabrics that emit visible light (glow) when irradiated with "black light," which is near-ultraviolet radiation. (See Demonstration 12.35, Photoluminescence.) Detergents often contain brighteners that stick to the laundry and fluoresce, especially in sunlight, to make white fabrics appear "whiter than white," because they are not only reflecting the incident light but also emitting extra light.

The absorption or excitation is generally from the lowest vibrational level of S_0 to any of the vibrational levels of S_1. Thus, initially, many vibrational levels (and their accompanying rotational levels) of S_1 are populated. In a very short time, before emission occurs, the excited molecules collide with less energetic neighboring molecules and transfer their excess vibrational energy to them to end up in the lowest vibrational level of S_1. This process, called *internal conversion,* is shown by the arrows labeled 3 in Figure 39.

Observe in Figure 39 that the lowest energy excitation wavelength is from the lowest vibrational level of S_0 to the lowest vibrational level of S_1 (green arrow up). The figure also shows that fluorescence emission from S_1 is at lower energies (longer wavelengths) than the light absorbed by S_0. The highest energy emission wavelength is from the lowest vibrational level of S_1 to the lowest vibrational level of S_0 (green arrow down). That is, the lowest energy excitation and highest energy emission are the same and occur at the same wavelength. The absorption spectrum of a substance superimposed on its fluorescence emission spectrum overlaps at this wavelength, as shown in Figure 40 for the dye fluorescein (named for its strong yellow-green fluorescence, which is quite visible even in a lighted room).

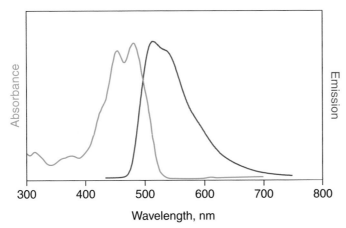

Figure 40.   Fluorescein absorption and emission spectra superimposed. Note the overlap around 500 nm and the rough mirror-image relationship of the two curves. The absorption and emission vertical scales are different and are adjusted to make comparisons easier.

Figure 39 shows that the structure in the electronic absorption spectrum reflects the vibrational levels in the excited state, S_1. Structure in the emission spectrum reflects the vibrational levels in the ground state, S_0. If the molecular geometry in the ground and excited states are similar, the spacing of the vibrational levels will be similar and the structures of the absorption and emission spectra will be similar. This circumstance results in absorption and emission spectra that are roughly mirror images of one another, as shown for fluorescein in Figure 40, where two broad peaks in each spectrum are separated by 25 nm.

Radiationless Decay

The arrow labeled 4 on the Jablonski diagram (Figure 39) represents the return to the ground state without the emission of light. This radiationless decay of the excited state arises from crossing between the energy levels of the excited and ground states followed by internal conversion of the excess energy in the ground state to bring the molecule to its stable lowest energy. Internal conversions transform energy that was absorbed from the exciting light to increased rotation, vibration, and translation of the molecules in the system, thus increasing the temperature of the system. Under the usual conditions of relatively weak light sources, the amount of energy transformed into molecular motions is small compared to the heat capacity of the medium and dissipation into the surroundings, so we are usually not aware of this energy transformation. For very intense light sources, the capacity of the system and surroundings to dissipate thermal energy can be exceeded and the system's temperature will increase appreciably.

Radiationless decay reduces the number of molecules in the excited state, S_1, so they are no longer able to emit radiation. This reduces the amount of fluorescence (and any other emission) compared to what would be observed if every photon absorbed by S_0 resulted in a photon of emitted radiation (not necessarily of the same energy). The ratio of the number of photons emitted to the number absorbed is called the *quantum yield* for the fluorescence. For strong emitters, fluorescence quantum yields can approach one, their maximum value. For most compounds, fluorescence quantum yields are generally much lower because the competing processes, such as intersystem crossing and radiationless decay, remove molecules from the excited state, S_1. (See Demonstration 12.36, The Halide Quenching of Quinine Fluorescence.)

Phosphorescence

Much of what we have said about the fate of the S_1 excited state is also applicable to the excited T_1, triplet state, which is formed by intersystem crossing from S_1. A major difference, though, is that emission from T_1 to S_0, like the reverse process (absorption from S_0 to T_1), is forbidden because of the change in electron spin multiplicity and the need for the spin to flip during the transition. Emission can occur, but, because it is improbable, it is quite slow. This slow emission, from a few microseconds to seconds following excitation, is called *phosphorescence*. Phosphorescence can be distinguished from fluorescence because the emission persists for at least a short time after the excitation source is turned off. (See Demonstration 12.37, Differentiation of Fluorescence and Phosphorescence.)

The Jablonski diagram shows that phosphorescence emission (arrow 6), like fluorescence, is at lower energy than the light required for excitation to S_1. In general, the phosphorescence spectrum is at even lower energy (longer wavelengths) than fluorescence (if any) from the same compound and does not overlap the absorption spectrum, as shown in Figure 41. Quantum yields for phosphorescence are usually small, because the emission lifetime of T_1 is relatively long, and radiationless processes (including chemical reactions) compete successfully with emission. Phosphorescence from compounds in solution is often hard to detect, because quantum yields are low. Phosphorescence quantum yields and, hence, intensity of emission can sometimes be increased by embedding the compounds in solid matrices, such as frozen solvents or polymers, where they are relatively isolated and no longer likely to encounter other species that interact to remove T_1 before it can emit.

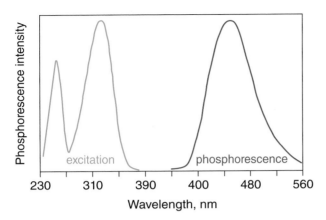

Figure 41.   The excitation and phosphorescence emission spectra of salicylic acid adsorbed on filter paper. The excitation spectrum is obtained by monitoring the phosphorescence emission as the exciting wavelengths are scanned. An excitation spectrum is similar to, but not identical to, the absorption spectrum for the same compound.

Almost everyone has seen glow-in-the-dark objects like the plastic stars used to decorate a nursery ceiling or signs showing emergency exits in public buildings. These objects absorb short-wavelength radiation from the sun or room lights and can emit longer wavelength light for several hours after the excitation is turned off. One of the most common phosphorescent materials (phosphors) in these objects is crystalline zinc sulfide, ZnS, to which tiny amounts of other substances (dopants), often copper compounds, have been added. ZnS is a semiconductor, which means that its valence electrons are normally tightly bound to the atomic cores in the crystal and the material is an electrical insulator. But these electrons can be excited to higher energy levels in which they are free to move about in the crystal, which then becomes an electrical conductor. Figure 42a represents the unexcited state of ZnS containing a dopant with electrons at an energy level intermediate between the valence band (electrons bound) and the conduction band (the energy levels where the electrons are free).

The excitation of a valence-band electron to the conduction band leaves a hole in the valence band (Figure 42b), that is, a site where there is a positive charge not balanced by a negative charge. An electron from the dopant quickly fills the valence-band hole (Figure 42c), leaving a hole in the dopant. The electron in the conduction band cannot return to the filled valence band, and, in this system, a small activation-energy barrier has to be overcome for the electron and hole to come together to make the radiative transition to the dopant level (Figure 42d). This activation energy barrier slows the recombination of electrons and

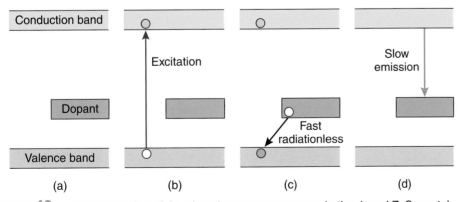

Figure 42.   Representation of the phosphorescence process in the doped ZnS crystal described in the text.

holes, so it takes a long time for all the electrons that have been excited to the conduction band to make the radiative transition and recombine with holes in the dopant levels. Thus, the phosphorescence emission may last for hours. Thermal motion of all the particles in the system normally supplies the activation energy required to get over the barrier. At higher temperatures, the system has more thermal energy, the transition is faster, and there is more emission per unit time (the glow is brighter but lasts for a shorter time). (See Demonstrations 12.38, Phosphorescence Excitation: Energy and Color Relationship; 12.39, Quenching Phosphorescence with Light; and 12.40, Quenching Phosphorescence with Thermal Energy.)

Lasers

The details of laser construction and the materials required are quite different for different kinds of lasers, including the familiar ones in bar-code readers, laser pointers, CD and DVD players, and in Blu-ray technology, as well as in medicine (laser surgery, for example) and a host of other applications. However, the basic principle of almost all lasers is the same. The interactions of electromagnetic waves with matter that have been discussed so far have been absorptions of photons to raise the energy of a system from a lower level to a higher level. However, if the system is already at the higher level, its interaction with an electromagnetic wave can stimulate the system to emit a photon that is exactly the same energy (frequency) as the stimulating wave and exactly in phase with it. Thus, there are now two waves (photons) traveling together where there was one before the interaction.

Because the distribution of energies in a system populates just the lowest energy levels, the absorption of light usually overwhelmingly predominates and any stimulated emission is negligible. However, consider a three-state system like the Jablonski diagram above or the simplified diagram in Figure 43. If a large number of photons of the appropriate energy are absorbed by this system, a majority of the ground-state molecules in level 1 will be excited to level 3 and then rapidly go to level 2, where they are trapped because emission back to level 1 is slow (forbidden). This is a *population inversion,* which has a larger population of molecules in the excited state than in the ground state (in the figure, $N_2 > N_1$, where N is the number of molecules). Population inversion is the essence of lasers, because, if a few of the level-2 molecules emit, they can stimulate the emission of others, which can stimulate the emission of still others, and so on, in a cascade. Thus, the emission of just a few photons is multiplied many times over to yield *l*ight *a*mplification by *s*timulated *e*mission of *r*adiation—*laser* action.

To enhance the amplification, the lasing system is usually confined between two mirrors (an optical cavity), so that light is reflected back and forth several times. One of the mirrors is not a perfect reflector and allows the transmission of some light. The transmitted light is a polarized, coherent (all waves in phase), monochromatic (single wavelength) beam of light with all waves traveling in essentially the same direction (along the line that is the axis of the lasing system). (See Demonstration 12.24, Laser Light Is Polarized.) Such a collimated, monochromatic beam of light can be easily focused to a very small area, which is why lasers are so useful in applications, like surgery, that require precise control of the location where the energy is needed.

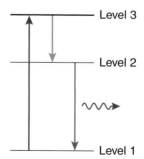

Figure 43.   Diagram of three energy states that may lead to laser production.

In the model just described, the burst of laser emission returns all molecules to the ground state and the population inversion disappears. In order to get another burst of laser emission, the excitation source must be used again to produce another inversion. The result is pulsed laser emission. To get continuous laser emission, the inversion must somehow be maintained even as the excited state is being depleted by the stimulated emission. One way to do this is illustrated by the He-Ne laser discussed below, where the input of excitation energy is an electric discharge instead of electromagnetic radiation. Electrical energy is also used to excite emission in other devices.

Light-Emitting Diodes and Injection Laser Diodes

The continuous lasers commonly used in consumer devices are *injection laser diodes,* which are closely related to light-emitting diodes (LEDs), both of which are electrical semiconductor devices. These solid-state devices pair two kinds of semiconductors, *n-type* and *p-type,* which get their names from the negative or positive charges on the mobile carriers. The electrons injected into the n-type and the holes in the p-type meet at the interface between the two semiconductors, and the energy released in their recombination is emitted as electromagnetic radiation. To illustrate what n-type and p-type semiconductors are and how an LED works, we will start with a pure semiconductor, silicon, which has four valence electrons per atom and at normal temperatures has almost no electrons in its conduction band, which is about 1.12 volts above the valence band.

To prepare an n-type semiconductor (Figure 44a), the silicon can be doped with, for example, a small number of phosphorus atoms, perhaps one part in a million. The phosphorus atoms take the place of silicon atoms in the crystal and bring with them five valence electrons, one of which must go into the conduction band, because the valence-band energy levels are filled with four each from all the atoms in the crystal. Thus, the number of conduction electrons is greatly increased (even though the number is still quite small compared to the number of conduction electrons in a metallic conductor). These electrons leave behind phosphorus atoms with only four valence electrons to balance their +5 atomic core charge, so positive dopant centers are created in the crystal.

Similarly, a p-type semiconductor (Figure 44b) can be prepared by doping silicon with a small number of boron atoms. The boron atoms take the place of silicon atoms in the crystal but bring with them only three valence electrons. This creates a hole (missing electron) in the valence band of the crystal, so an electron from an adjacent atom can move in to fill the hole, thus leaving another hole. Thus holes can move through the crystal, although not as rapidly as electrons in a conduction band. These holes leave behind boron atoms with four valence electrons to balance their +3 atomic core charge, so negative dopant centers are created in the crystal.

n-type 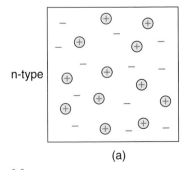   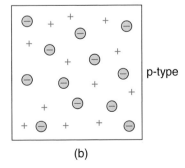 p-type

(a)   (b)

Figure 44.   Schematic illustrations of (a) n-type and (b) p-type semiconductors. The plus and minus signs inside the circles represent the positive and negative dopant centers, and those outside the circles are the mobile charge carriers—holes and electrons.

When an n-type and a p-type semiconductor are brought together, the mobile charge carriers move into the opposite type semiconductor and combine with one another, as shown in Figure 45a. This leaves the dopant charge centers without compensating mobile charges and creates regions of positive and negative charge, which produce an electric field (Figure 45b) that opposes the further movement of the mobile charges. Holes (positive charges) from the p-type semiconductor would have to move against the field to cross into the n-type semiconductor and vice versa for the electrons from the other direction. Thus, an electrical potential is set up at the interface between the two semiconductors (the depletion region where there are no longer mobile charge carriers). In the doped silicon system here, the electrical potential is about 0.6 volt. In an electrical circuit, a diode allows electron flow in only one direction, and because this semiconductor combination allows electron flow only from the n-type side to the p-type side, it is a diode.

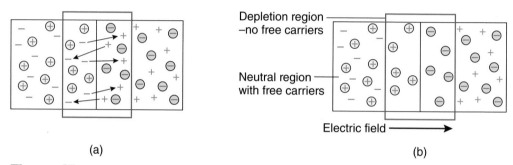

(a)   (b)

Figure 45.   (a) Interaction of n-type and p-type semiconductors in which mobile charges (electrons and holes) combine creating a depletion region. (b) The depletion region without mobile charge carriers and with the resultant electric field created by the fixed dopant charge centers.

To create light emission from this diode, an electrical potential must be impressed across the device, as shown in Figure 46. Electrons injected at the n-type side are driven by this external potential field (which is opposite of the one internally created at the p-n junction) toward the junction. The external field also drives the holes from the p-type side toward the junction where they meet and combine with the electrons. The combination is equivalent to the transition from the conduction band to the valence band and emits photons having an energy of about 1.12 eV (a radiation of about 1100 nm in the near-infrared region). Different emission colors, including visible, are produced by semiconductor materials with different band gaps and electrical inputs that provide electrons and holes with sufficient energy to emit photons of the desired color—from ~1.8 eV, red, to ~3 eV, blue, for the visible. (See Demonstration 12.7, Light-Emitting Diodes: Voltage and Temperature Effects.)

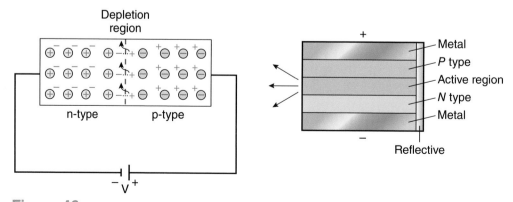

Figure 46.   Illustration of the operation of an LED and schematic of its structure. The curved arrows in the left-hand diagram represent the emission of photons as the electrons and holes combine.

Injection laser diodes (Figure 47) are constructed very much like LEDs. In laser devices, the semiconductors used provide a short time delay before recombination of electrons and holes at the junction; this is equivalent to a population inversion, because the electrons and holes are poised to combine and emission of a few photons triggers the laser cascade. Reflections at the ends of the semiconductor sandwich form the optical cavity in which light amplification occurs. Since the current source continuously creates the electrons and holes, the output is continuous.

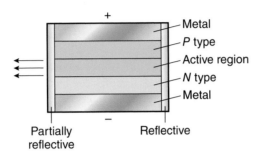

Figure 47.   A schematic diagram of an injection laser diode.

Although lasers often output high energies (warnings about damage to eyes are posted in laser laboratories), the energy input required to produce the laser emission is even larger. Some of the excitation energy is always lost, usually as heat, as shown by the radiationless process shown in Figure 39, so high-energy lasers need special cooling systems. Tiny laser diodes also emit energy as heat and are usually in contact with materials that can conduct the energy away, so they do not overheat.

Energy Transfer

In addition to various internal radiationless transfers of energy between states, an excited molecule can also transfer energy to another molecule in the system, if the acceptor molecule has an energy state that closely matches that of the donor molecule. The energy transfer often leaves the donor molecule in its ground state and the acceptor molecule in an excited state that may undergo all of the possible processes that could occur if it had been directly excited, for example, by absorbing a photon of the appropriate energy. Thus, the excited acceptor molecule might emit electromagnetic radiation at wavelengths characteristic of its excited states (Figure 48).

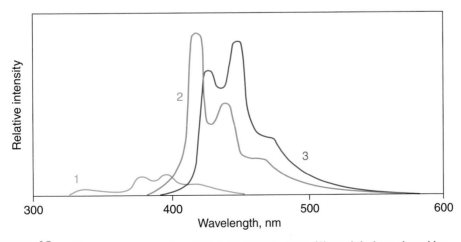

Figure 48.   Fluorescence spectra of (1) pure naphthalene, (2) naphthalene doped by anthracene, and (3) pure anthracene.

In the process illustrated in Figure 48, the emission from the donor naphthalene disappears when anthracene is added, and the emission of the acceptor anthracene appears. However, in many cases, the acceptor molecules do not emit, and the evidence that there is energy transfer comes from the *quenching* (loss) of emission from the donor. The amount of quenching depends on the concentration of the acceptor (quencher) molecules and can often be studied by varying their concentration and following the change in emission intensity from the donor. (See Demonstration 12.36, The Halide Quenching of Quinine Fluorescence.)

An especially useful form of quenching is *fluorescence resonance energy transfer* (FRET; also sometimes called Förster resonance energy transfer because it was extensively investigated by the German chemist Theodor Förster [1910–74]). In this form of energy transfer, the fluorescence emission spectrum of the donor overlaps the excitation (absorption) spectrum of the acceptor. If the donor and acceptor are closer together than about 10 nm, the resonance (exact correspondence of frequency) between the electric dipoles that change upon donor emission and acceptor excitation results in a transfer of energy without emission or absorption of electromagnetic radiation. If the acceptor is not near the donor, the excitation of the donor produces donor fluorescence. If the acceptor is near enough to the donor, the excitation of the donor produces less donor fluorescence, and FRET causes the acceptor to be excited and emit acceptor fluorescence. FRET is extensively used to study biological molecules that have donor and acceptor molecules attached. For example, to discover the conditions under which two biomolecules come together to form their active structure, one may be labeled with a donor and the other with an acceptor and FRET is observed when they come together.

Phosphorescence Quenching by Oxygen

Because the lifetimes of excited states that fluoresce are so short, there is little time for quenching, and nonresonant fluorescent quenching is less common than quenching of the much longer-lived triplet-state phosphorescence. Oxygen molecules are of particular importance for phosphorescence quenching. The ground state of oxygen molecules is a triplet, that is, two of the electrons are unpaired and have the same spin. Ground-state triplet oxygen is less reactive than might be predicted from the great stability of its reaction products (CO_2 and H_2O, for example), because spin restrictions make its reactions with singlet molecules, including organic molecules, quite slow. Interaction of an excited triplet molecule with oxygen can result in a spin exchange that promotes the oxygen to an excited singlet state and leaves the donor molecule in its singlet ground state. Thus, phosphorescence is often difficult to observe in systems that contain oxygen, which usually must be carefully degassed or studied in solid solutions or at low temperatures in frozen solvents where diffusion of molecules to meet one another is very slow.

Singlet oxygen is very reactive, primarily because there are no spin restrictions on its reactions with singlet molecules. Some forms of photodynamic therapy (therapy based on the action of light) take advantage of this reactivity. For example, to treat basal cell carcinoma (a form of skin cancer), aminolevulinic acid, a compound necessary for the synthesis of myoglobin and hemoglobin, is applied to the cancerous area and the fastest dividing (cancerous) cells quickly assimilate it and begin to synthesize the porphyrin molecules required for the oxygen transporters, myoglobin and hemoglobin. Then light in the red region of the spectrum is shined on the area to excite the porphyrin molecules, which cross to their excited triplet state. Spin exchange with oxygen in the cells creates singlet oxygen molecules, which destroy proteins and nucleic acids and lead to cell death. (See Demonstration 12.47, The Photobleaching of Carotene.)

He-Ne Gas Laser

Another example of use of energy transfer is the He-Ne gas laser, one of the first continuous lasers and one that is common in research and instructional laboratories (Figure 49).

The lasing medium is a low-pressure mixture of He and Ne gases with He in large excess. An electric discharge through the gases is used to excite the He atoms and produce a state that is almost exactly the same energy as a long-lived, excited (metastable) Ne state. Energy exchange produces the metastable Ne state, and its continued production results in population inversion and stimulated emission at 633 nm. Before the invention of the much smaller, less expensive, and more rugged solid-state lasers, this gas laser was used in commercial applications such as barcode readers.

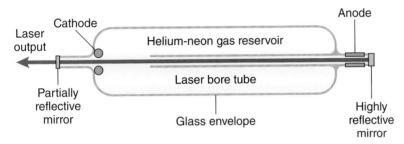

Figure 49.   Schematic diagram of a He-Ne gas laser.

Photochemistry

Almost all chemical reactions must overcome an energy barrier for the reactant molecule(s) to be transformed to a new product molecule (or molecules). For many reactions the necessary energy is gained from collisions and energy exchange with surrounding molecules. However, other reactions obtain the necessary energy through the absorption of light, and the molecules react from their excited states. Such reactions are called *photochemical reactions.* There are a great variety of these photochemical reactions. (See Demonstrations 12.42, The Reversible Photochemical Bleaching of Thionine; 12.43, Photochromic Methylene Blue Solution; 12.44, The Photochemical Reaction of Chlorine and Hydrogen; 12.47, The Photobleaching of Carotene; 12.48, Making a Cyanotype; 12.49, An Iron(III)-Oxalate Actinometer; 12.50, The Photoreduction of Silver Halide; 12.51, Photochemistry in Nitroprusside-Thiourea Solutions; and 12.53, The Photodissociation of Bromine and the Bromination of Hydrocarbons.) Two of the most important are the reactions in your eyes that allow you to read these words and the reactions in green plants that use sunlight absorbed by chlorophyll to begin the photosynthetic (synthesis by light) pathway for the production of glucose and oxygen from carbon dioxide and water, the source of essentially all the food we eat and oxygen we breathe.

Photoisomerization

The electron distribution of molecules in their excited states is often different enough from the ground state to allow structural/bonding changes that lead to isomers of the starting compounds when the excited species return to their ground states. (See Demonstrations 12.45, The Effects of Solvents on Spiropyran Photochromism and Equilibria; and 12.52, Photochromism in Ultraviolet-Sensitive Beads.) A well-studied example is the trans-cis isomerization of stilbene that occurs when it is photoexcited by near-ultraviolet light (Figure 50). In such diagrams, it is often useful to show the electrons localized in the excited state as a kind of extreme example of the difference between the ground and excited state distributions. In this case, we see that the electrons from the ground state π (pi) bond between the central carbons are separated in the excited state. One electron remains in the π bonding orbital, and the other is promoted to the π^* ("pi star") antibonding orbital. These electrons are no longer paired in a bond between the carbons. Energetically, this leaves essentially a single bond between the carbons and allows relatively free rotation of the ends of the molecule with respect to one another.

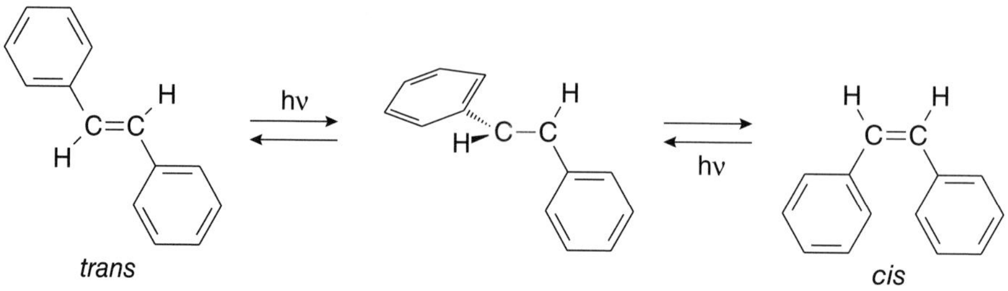

trans   *cis*

Figure 50.   Stilbene photoisomerization. The symbol *hv* is commonly used in reaction equations to denote that light of the appropriate energy is required to cause the reaction in the direction of the arrow. Although only one intermediate structure is shown, the reaction passes continuously through rotation about the C–C bond in the intermediate.

Calculations suggest that the most stable form of the excited state shown in the figure is with the rings at the ends of the molecule at approximately a right angle to one another. Thus, there is about an equal chance of the molecule twisting back to the trans form before returning to the ground state or twisting on to the cis form before deactivation. As shown in the figure, *cis*-stilbene can also absorb light to lead to the same excited intermediate state. After a long enough period of irradiation, a *steady state* is reached with the overall reactions proceeding at the same rate in both the forward and reverse directions. At that point, the reaction mixture is an approximately equal mixture of *trans*- and *cis*-stilbene. Substituted stilbenes (compounds with different substituent groups bonded to one or both rings) often produce steady-state amounts of the trans and cis forms that are not equal and as a result have been studied to provide information about the energies and electronic structures of the excited state(s).

If the internal twisting of the stilbene molecule that is required to proceed to the intermediate stage is physically restricted, emission from the excited state competes favorably with the nonradiative pathway, and fluorescence, instead of isomerization, is observed. When stilbene is frozen at low temperature in solid solutions, for example, the internal rotation is restricted and emission is observed. Clefts and pockets in some folded protein molecules can trap appropriately substituted stilbene molecules, and emission (or lack thereof) from the photoexcited stilbene can be used as a probe of the strength of its interaction with the protein and hence of the structure of the binding site.

Photodissociation

Even simple molecules can be broken into fragments by absorbing ultraviolet radiation of high enough energy. There are several ways to characterize the different wavelength ranges of ultraviolet radiation. The one that is probably most familiar, because of its connection to tanning and burning of human skin, divides the radiation into three ranges: UV-C, 100–280 nm (12.4–4.43 eV); UV-B, 280–315 nm (4.43–3.94 eV); and UV-A, 315–400 nm (3.94–3.10 eV).

UV-C radiation can destroy DNA and other biomolecules and is lethal to most organisms. UV-B radiation is responsible for sunburn and can damage DNA, which causes mutations and skin cancer. UV-A radiation causes sun tanning and is required for the natural production of vitamin D in our skin.

Stratospheric Oxygen Photochemistry

We owe our existence to two photodissociation processes in the earth's atmosphere that absorb almost all of the sun's most damaging ultraviolet radiation, UV-C and UV-B, before it can reach the surface. The reactions of interest are:

$$O_2 \xrightarrow{\text{UV-C}} 2O \qquad\qquad \text{slow}$$

$$O + O_2 \longrightarrow O_3$$

$$O_3 \xrightarrow{\text{UV-C, UV-B}} O + O_2 \qquad\qquad \text{fast cycle}$$

$$O + O_3 \longrightarrow 2O_2 \qquad\qquad \text{slow}$$

The outcome of these reactions is illustrated by the plot of the ozone, O_3, concentration as a function of altitude above the earth's surface (Figure 51). The first reaction, photodissociation of oxygen molecules, O_2, to give oxygen atoms, O, is slow because the amount of UV-C in the sun's radiation is relatively small. The sun resembles a black-body and emits very little at this high energy (see Figure 26). The reaction is critical, however, to absorb much of the UV-C and to produce the oxygen atoms required to initiate the fast cycle of the second and third reactions. The result, represented schematically in Figure 51, is that almost all of the UV-C radiation is absorbed above about 40 km (24 mi) and does not reach the surface.

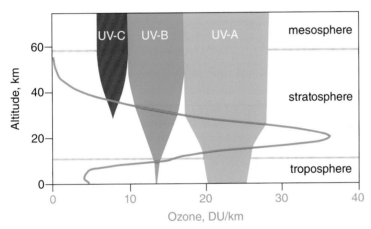

Figure 51.   Variation of atmospheric ozone concentration with altitude and depth of atmospheric penetration of UV-A, UV-B, and UV-C solar radiation. UV-C is absorbed by molecular oxygen, producing an increase in ozone concentration. Ozone, in turn, absorbs UV-B and, to a lesser extent, UV-A, which convert it back to molecular oxygen.

The second and third reactions are a cycle, forming ozone from oxygen molecules and atoms and subsequently photodissociating ozone to re-form these reactants. This fast cycle of reactions, plus the fourth reaction, the combination of oxygen atoms and ozone molecules to give molecular oxygen, produce steady-state concentrations of oxygen atoms and ozone molecules that vary with altitude. The steady state at a particular altitude depends upon the concentration of oxygen molecules (which increases closer to the surface) and the amount of UV-B radiation (which decreases, because of absorption, closer to the surface). The result is a maximum in the steady state concentration of ozone in the stratosphere at an altitude of about 20 km (12 mi; Figure 51). Ozone concentrations, even at the maximum, are only a few parts per million of all the gas molecules present. However, this tiny amount is essential to absorb most of the UV-B radiation, so that only a small amount reaches the surface of the earth, where it can damage DNA. The release of ozone-depleting molecules into the atmosphere is a great concern, because the result is that more damaging UV-B radiation reaches the surface.

Consider the fast cycle of the second and third reactions involving O, O_2, and O_3. Since this is a cycle, there is no net change in any of the species (at the steady state), and the only change in each round of the cycle seems to be the absorption of some UV-B radiation. But energy cannot be destroyed, so the energy lost by the radiation has to appear somewhere, and that will be as thermal energy, faster motion, of the atoms and molecules. That is, absorbed radiation energy warms the gases and is the major source of the energy that causes the temperature of the stratosphere to increase with altitude. Because warmer gas is less dense than cooler gas, this temperature profile creates a stable condition with layers of decreasing density lying on top of one another and little vertical mixing of the gases. The stratosphere is stratified (hence the name) largely because of the interaction of electromagnetic radiation with the gases there.

Ground-Level Ozone and Photochemical Smog

Ozone in the stratosphere is life saving, but ozone at ground level can be life threatening. Not only is ozone toxic, but it is also one of the important actors in the complex series of reactions that produce photochemical smog, a mixture of ozone, nitrogen oxides, volatile organic compounds, and noxious organic compounds derived from the other components. The origin of the nitrogen oxides is mainly from high-temperature reactions of air in some industrial processes and principally from the internal combustion engines of cars and trucks. A simple series of photoinitiated reactions involving the nitrogen oxides is

$$NO_2 \text{ (reddish-brown gas)} \xrightarrow{h\nu} NO + O$$

$$O + O_2 \longrightarrow O_3$$

The volatile organic compounds from many sources—unburned fuel, evaporation from fuel tanks, dry-cleaning, etc.—can react with the ozone formed in the second reaction (the same reaction that forms stratospheric ozone) to produce reactive oxygenated organic radicals that go on to form the noxious mixture, including increased amounts of ozone, that is harmful, especially to humans who have respiratory problems, but also to healthy animals and plants.

A natural source of volatile organics is trees, especially the terpenes we appreciate for their pleasant "piney" odors. For example, the name Native Americans gave the Great Smoky Mountains of Tennessee and North Carolina meant the "land of the blue mist," and early colonists from Europe adapted that meaning in the name we now use. The natural haze is a photochemical product of sunlight, the volatiles from the trees, and oxygen in the air. The advent of increased tourist automobile traffic in the mountains, as well as upwind urban areas, has added nitrogen oxides to the mix and produced photochemical-smog conditions comparable to those in large cities, with dire consequences for the trees, some species of which have become extinct in the area.

Photobromination

This reaction is a simpler result of a photodissociation that is easy to observe in the laboratory. (See Demonstration 12.53, The Photodissociation of Bromine and the Bromination of Hydrocarbons.) Although the bromine molecule, Br_2, is relatively reactive, it is not reactive enough to replace hydrogen with bromine atoms in most alkane molecules. Visible light has enough energy to photodissociate Br_2, which is reddish-orange because it absorbs at the blue end of the spectrum. The products are two bromine atoms, each with an unpaired valence electron, which makes them quite reactive free radicals that can initiate reaction chains like these (where R–H represents an alkane and only the unpaired electrons are shown):

$$Br-Br \text{ (reddish-orange)} \xrightarrow{h\nu} Br^{\bullet} \qquad \text{initiation}$$

$$R-H + Br^{\bullet} \longrightarrow R^{\bullet} + H-Br \qquad \text{chain propagation}$$

$$R^{\bullet} + Br-Br \longrightarrow R-Br + Br^{\bullet} \qquad \text{chain propagation}$$

$$R^{\bullet} + Br^{\bullet} \longrightarrow R-Br \,, \; R^{\bullet} + R^{\bullet} \longrightarrow R-R, \; Br^{\bullet} + Br^{\bullet} \longrightarrow Br_2 \qquad \text{termination}$$

In the dark, a solution of Br_2 in alkane remains reddish-orange for a long time, because the thermally activated reaction is very slow. However, it rapidly decolorizes in the light as the photobromination proceeds, and all of the Br_2 reacts to form the colorless products. In this free-radical chain reaction, the bromine-atom reactant produced in the photochemical initiation reaction (initiation) is regenerated by the two central reactions (chain propagations) as a consequence of formation of the brominated alkane product. These two reactions can go through many cycles, using up many bromine molecules and forming many brominated alkane molecules, before the cycle/chain is terminated by reactions of the bromine atom and/or alkane radicals with one another (termination). Since photodissociation often leads to free radicals, chain reactions and regeneration cycles like these are a characteristic of many such reactions.

Excited-State Reactants

Because their electronic structures are different, an excited-state molecule can sometimes react with other molecules in ways that are unavailable to the ground state. For example, benzophenone in its ground state does not react with alcohols to abstract (remove) a hydrogen atom. A well-studied reaction often used in undergraduate chemistry laboratories is the hydrogen abstraction by the photoexcited triplet state of benzophenone. Usually sunlight is used to excite benzophenone to its lowest singlet excited state, which then forms the triplet state by rapid intersystem crossing (ISC):

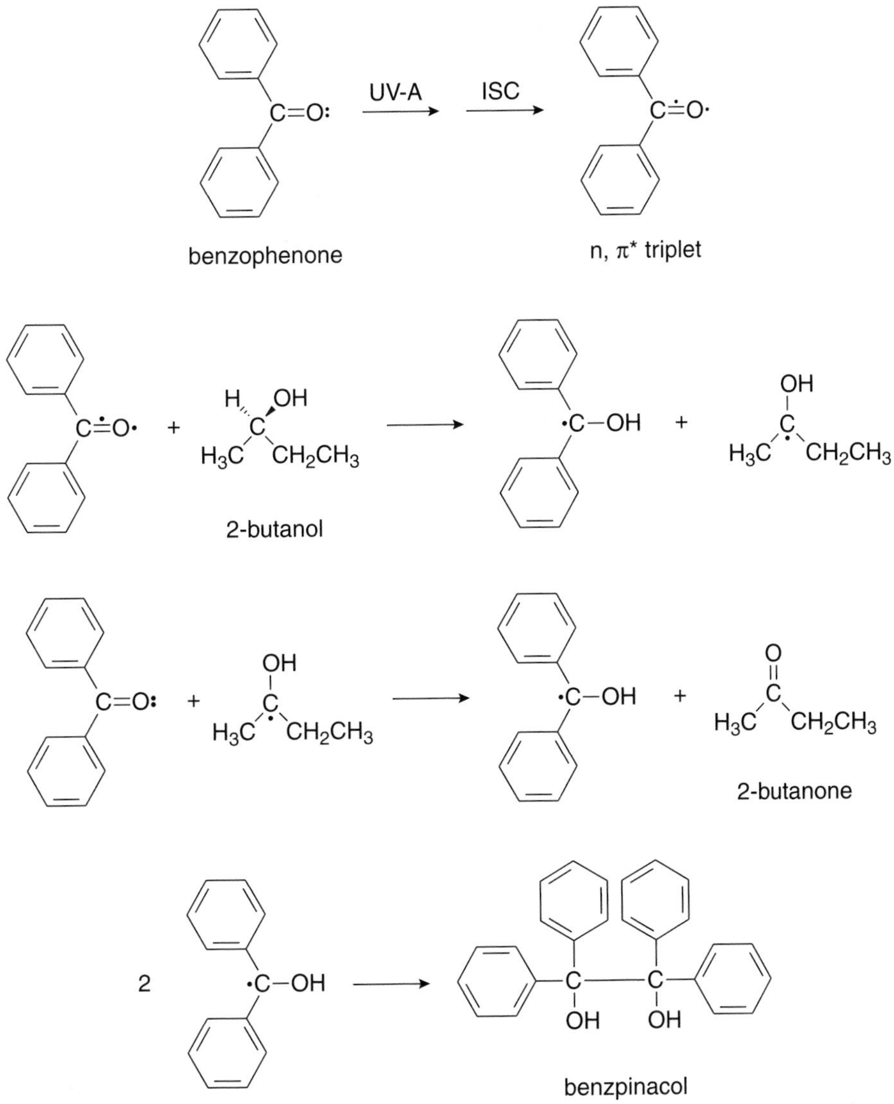

benzophenone   n, π* triplet

2-butanol

2-butanone

benzpinacol

The electronic state formed by absorption of light and intersystem crossing is an n, π* ("en pi star") triplet state in which a ground-state, nonbonding electron on oxygen is in a pi antibonding orbital in the excited state. The remaining nonbonding electron on oxygen and the extra electron in the pi system of the carbonyl are represented by localized dots in the reaction product. This representation is meant to show the radicallike structure of the n, π* triplet, which abstracts a hydrogen from the solvent alcohol in the second reaction (as the bromine atom abstracts a hydrogen atom from an alkane, as described above) to form two radicals of similar structure. The alcohol-derived radical from this reaction can go on to donate a hydrogen atom to another benzophenone molecule, forming a stable ketone product, 2-butanone, and another benzophenone-derived radical. Two of the benzophenone-derived radicals combine to form the relatively stable benzpinacol, the product of the final reaction.

The sum of these reactions is the oxidation of a secondary alcohol molecule to a ketone and the reduction of two benzophenone molecules to an alcohol product. If these reactions are 100 percent efficient, one photon of light absorbed leads to the loss of two benzophenone molecules, that is, a quantum yield of two for the disappearance of benzophenone. When oxygen is excluded from the reaction mixture, the experimental quantum yield is very close to two. As with the stilbene photoisomerization discussed above, the photochemistry of many substituted benzophenones has been explored to obtain more information about the excited states. For example, if a naphthyl (naphthalene) ring replaces one of the phenyl rings in benzophenone, no photoreduction is observed. This result shows that the lowest energy excited state is not a radicallike n, π* state, but rather one in which the electronic excitation is in the pi electron system of the naphthyl ring.

Photovoltaic Cells

In some cases, a photoexcited electron can be transferred to another molecule or another part of a crystal, in the process creating a charge separation and the potential to do useful work if the energy released when the charges recombine can be captured. Solar-energy cells, for example, are designed to capture energy from the sun's radiation to produce an electric current that can provide power to run small electronic devices like your calculator, as well as to provide solar-energy panels with enough power to light, heat, and cool buildings and the power to make some extended missions in space and to other planets possible. (See Demonstration 12.46, A Copper Oxide Photocell.) In nature, the most important photovoltaic reaction is photosynthesis, the photoredox reaction (electron-transfer reduction-oxidation reaction caused by the absorption of light) by which green plants produce the oxygen we breathe and the food we eat.

In a sense, a photovoltaic cell is like an LED run in reverse: light in results in electric current out. The great majority of the solar-energy cells in use now are based on the semiconductor properties of silicon, which were described above in the discussion of LEDs. Figure 52 illustrates the components and operation of a simple silicon-based photovoltaic cell. The n-p semiconductor is sandwiched between two metal electrodes that are the contacts for the external electric circuit, represented here by the electric motor and its conducting-wire connections to the cell. The top electrode, the one exposed to the light, is a grid with holes to allow the light to enter the semiconductor. The bottom electrode is continuous and often is a shiny aluminum layer that both conducts an electric current and reflects light back into the semiconductor to increase the number of electron-hole pairs that are formed. To minimize loss of light by reflection from the surface of the silicon, the top surface is covered with a layer of material that reduces light reflection.

In Figure 52, the mobile electrons and holes, respectively, in the n- and p-type semiconductors and the fixed dopant charge, including those in the p-n junction region (see Figure 45), are omitted for clarity, but the essential electric field created by these fixed charges in the junction is indicated. A hole-electron pair (+ − in the figure) is created by a photon of appropriate energy absorbed by a silicon atom to promote a valence electron to the conduction

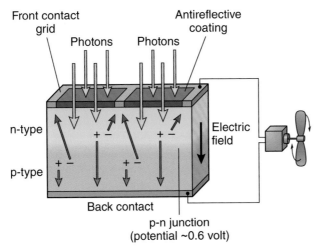

Figure 52. Schematic illustration of the construction and operation of a silicon-based solar-energy cell.

band. These mobile charge carriers are quickly moved in opposite directions by the electric field in the junction. The result is a build up of negative charge in the n-type semiconductor and positive charge in the p-type semiconductor (both of which were electrically neutral before light entered the cell). The charge separation drives electrons from the n-type semiconductor through the external circuit, where the current can be harnessed to do useful work, to the p-type semiconductor where they recombine with holes. As long as light shines on the cell, electron-hole pairs will continue to form and current will continue to flow. The electric potential across the p-n junction provides the charge separation and driving force for the flow, so the voltage from this cell is about 0.6 volt.

A fundamental limitation of silicon-based solar-energy cells is that a good deal of the energy in the sun's electromagnetic radiation is wasted. The most effective wavelengths for exciting the silicon are in the near-infrared and the long-wavelength (red) end of the visible spectrum, but the emission from the sun peaks in the green, thus much of the shorter-wavelength light is either not absorbed or some of its energy is lost as heat as the very energetic conduction-band electrons it creates lose their energy by collision within the crystal.

Another kind of photovoltaic cell is based on organic molecules. In these cells there are donor (D) and acceptor (A) molecules that have appropriate energy levels to undergo these reactions:

$$D + h\nu \longrightarrow D^*$$
$$D^* + A \longrightarrow D^+ + A^-$$

The result is a separation of charge and under appropriate conditions (often requiring an electrically conducting organic polymer), the electrons from A^- flow from one electrode attached to the organic system through an external circuit to a second electrode, where they recombine with D^+ molecules. At present, organic photovoltaic solar-energy cells are less efficient than silicon (or other semiconductor-based) cells in their conversion of sunlight to electrical energy. However, they are generally much less expensive to prepare and can be fabricated as flexible sheets that make them useful in applications where molding to conform to odd or changing shapes is required, for example, to power "electronic paper" displays.

Photosynthesis

Reactions in the chloroplast-containing cells of green plants (and other photosynthetic organisms) bear some resemblance to those in organic photovoltaic cells but are more complex and involve a number of electron donor-acceptor pairs. The light absorbing molecules are chlorophylls (Figure 53), named for their source, the green leaves of plants (from the Greek yellow-green, the same root as the element chlorine, a yellow-green gas, plus the Greek *phyll,* meaning "leaf," as in phyllo dough, the paper-thin leaves [layers] in many Mediterranean pastries). (See Demonstration 12.35, Photoluminescence.) The chlorophyll molecules are stacked together in a special thylakoid (from the Greek meaning "sack-like," which describes the appearance of these structures) membrane enclosure in the interior of the chloroplast. There are two kinds of these stacks, *photosystem I* (PSI) and *photosystem II* (PSII). In either photosystem, when one of the stacked chlorophyll molecules is excited by absorption of a photon from sunlight, the excitation moves rapidly by Förster resonance transfer to a special pair of chlorophyll molecules where ionization occurs.

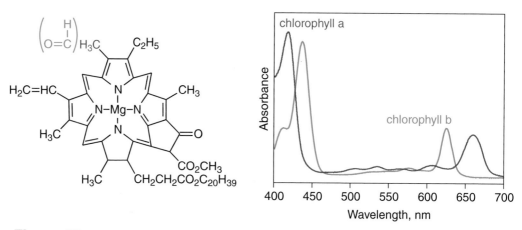

Figure 53.   Structures and spectra of chlorophyll *a* and chlorophyll *b* in solution. The spectra differ somewhat when the molecules are stacked in the thylakoid membranes.

In PSI, the excited electron is transferred to an iron-containing protein, where Fe^{3+} is reduced to Fe^{2+}, and then on to a molecule of nicotinamide adenine dinucleotide phosphate ($NADP^+$). Absorption of two photons in PSI leads to reduction of $NADP^+$ to NADPH:

$$2\ PSI(chl) + 2h\nu + NADP^+ + H^+(aq,\ outside) \longrightarrow 2\ PSI(chl^+) + NADPH$$

PSI(chl) represents the chlorophyll molecules in PSI that donate electrons to reduce the $NADP^+$. H^+(aq, outside) is a hydronium ion from the solution outside the thylakoid membrane enclosure. NADPH is the reducing agent used to reduce carbon dioxide, CO_2, to carbohydrate in the dark reactions (not photolytic) that follow these photoreactions.

In PSII, the excited electron is transferred to a quinone molecule (Q). Absorption of a second photon by chlorophyll in the thylakoid membrane leads to transfer of a second electron to the quinone. Subsequent reaction of the quinone dianion with hydronium ions from the cell solution produces a hydroquinone, H_2Q. These processes can be summarized as

$$2\ PSII(chl) + 2h\nu + Q + 2H^+(aq,\ outside) \longrightarrow 2\ PSII(chl^+) + H_2Q$$

In a chain of oxidation-reduction reactions, the electrons from H_2Q are transferred to $PSI(chl^+)$ to return it to its neutral form ready to reduce more $NADP^+$. The net reaction is

$$H_2Q + 2\ PSI(chl^+) \longrightarrow 2\ PSI(chl) + Q + 2H^+(aq,\ inside)$$

Note that this reaction has released its hydronium ions to the interior of the thylakoid membrane, so the quinone has acted as both an electron and a hydronium-ion transporter.

In a protein complex associated with PSII, there are four manganese atoms (ions) with two water molecules bound to them. $PSII(chl^+)$ is a powerful oxidizing agent, and manganese has a number of oxidation states that are used (in an as yet poorly understood mechanism) to transfer four electrons, one at a time, from two water molecules to $PSII(chl^+)$. Each electron returns the PSII to its neutral form, from which it is again photoexcited and transfers an electron to Q. After the four electrons have been transferred to PSII, the oxygen atoms from the water form an oxygen molecule (O_2). This process is the origin of essentially all the oxygen in our atmosphere. The overall reaction (writing four $PSII(chl^+)$ but remembering that it is the same PSII used four times) is

$$2 H_2O + 4 PSII(chl^+) \longrightarrow 4 PSII(chl) + O_2 + 4H^+(aq, inside)$$

We get the net photochemical reaction of photosynthesis by adding these four reactions (with the first three doubled to account for the four PSII in the fourth reaction):

$$2 H_2O + 6 H^+(aq, outside) + 2 NADP^+ + 8 h\nu \longrightarrow O_2 + 8 H^+(aq, inside) + 2 NADPH$$

The energy from the eight photons has been used to remove four electrons from water (an energetically difficult reaction) to produce oxygen and the reducing reagent, NADPH, which provides the electrons to reduce carbon dioxide to carbohydrates.

The energy from the photons has also been used to move hydronium ions from the outside to the inside of the thylakoid membrane, producing a concentration difference across the membrane (essentially a stored energy source). The hydronium ions move down this concentration gradient (from inside to outside) through a protein complex that captures the energy by synthesizing adenosine triphosphate (ATP) from adenosine diphosphate (ADP) and phosphate ion. The ATP is then used as an energy source for the dark reactions required for the conversion of carbon dioxide to carbohydrates. Thus, the photoredox reactions in photosynthesis have produced the electrons and the energy for the dark reactions that produce the food we eat and, as a consequence, also the oxygen we breathe.

Finally, note that the chlorophyll molecules do not absorb light in the middle of the visible spectrum where sunlight has its maximum energy output. It might seem odd that photosynthesis evolved using molecules that do not efficiently use the most abundant photons in the solar spectrum. But this may be a protective mechanism to avoid photodegradation by the production of too many excited molecules (and electrons) and to permit better control of the above reactions, which are tightly coupled to produce the necessary products without wasting much energy as heat.

Chemiluminescence

In the previous photochemistry section, we discussed several of the mechanisms by which energy in the form of electromagnetic radiation, light, causes chemical reactions. *Chemiluminescence,* the other side of the coin, is light emitted as a consequence of the release of energy by chemical reactions. All objects warmer than their surroundings emit some energy as infrared radiation (the basis of night-vision goggles), but this is not usually considered to be chemiluminescence. All chemiluminescent reactions that produce visible or ultraviolet radiation are oxidation-reduction reactions. (See Demonstrations 12.8, Electrogenerated Chemiluminescence; and 12.9, Chemiluminescence.)

Flames

These common chemiluminescent reactions are familiar to everyone. Flames involve the oxidation of a fuel, usually by oxygen from the air or from a pure oxygen source, and emit a good deal of thermal energy as well as light. (See Demonstration 12.3, Incandescence from the Combustion of Iron and of Zirconium.) Very fast flame propagation and explosions also emit mechanical energy in the form of sound waves. (See Demonstration 12.10, Chemiluminescence from the Explosive Reaction of Nitrous Oxide and Carbon Disulfide.)

The light emitted by flames from hydrocarbon fuels comes from two sources, emission from glowing carbon particles (the source of soot) and electronic-state transitions from excited molecular fragments (usually free radicals).

It is easy to observe both forms in a Bunsen burner flame from a natural-gas (mostly methane) laboratory source. When the air intake at the base of the burner is closed, air has to move into the gas flow from the flame's exterior; as a result, combustion is not complete, the flame is mostly yellow, and a cool porcelain dish held near the top of the flame quickly darkens as soot particles are deposited. The yellow emission from these particles, which are aggregates of unburned carbon, heated to about 1200–1400 K by the energy released in the oxidation of the fuel is from excited states of these particles. (See Demonstration 12.1, The Emission Spectrum from a Candle Flame.)

Among the substances in the soot are fullerene molecules, including C_{60}, buckminsterfullerene (Figure 54; named for Richard Buckminster Fuller, the architect whose geodesic domes this spherical molecule resembles), which, like diamond and graphite, is an allotropic form of pure carbon. Although these molecules have been present in soot ever since fires have burned, they were only discovered and characterized in 1985.

Figure 54.   The structure of C_{60}, buckminsterfullerene, nicknamed "buckyball."

When the air intake of the burner is open and properly adjusted, air is premixed with the gas before it reaches the flame, the flame color is blue (dark blue inner cone and lighter blue outer envelope), and no soot is deposited from the flame. There are many chemical species in the flame, including the final products, CO_2 and H_2O, but the blue emission is largely from excited CH and C_2 radicals (excited OH radicals are also abundant, but emit in the ultraviolet and do not contribute to the visible flame color). These free radicals are also present and emitting in the unmixed yellow flame, but there are fewer of them and the luminescence from the soot particles is brighter and overwhelms them.

Electrical Discharges

Another familiar example of emission that can be characterized as chemiluminescence is the light you see when you scuff across a carpet on a dry day and see (and feel) a spark jump between your finger and a metal object. Lightning bolts between clouds or from the ground to clouds during a thunderstorm are more spectacular examples of the same phenomenon. When a sufficiently large difference in charge builds up between two objects, for example, between a cloud and the ground below it, the electric field produced can ionize a few of the molecules in the air between them. This creates a conducting pathway for electrons to begin to flow from the more negative to the more positive object, and these energetic electrons can ionize more molecules in a cascade that produces a rush of electrons and a great deal of molecular ionization and excitation. Emission from the excited molecules and ions produces the light we see, much of it from excited nitrogen molecules (see Figure 56 below). Stretching the point just a bit, the excitation by a spark is the result of an oxidation-reduction reaction as the electrons jump the gap between the charged objects and reduce the positive species.

Cold Light

Often the term chemiluminescence is applied more narrowly only to those reactions that emit light but very little thermal energy, so-called cold light. Examples that are familiar to many are the lightsticks, necklaces, and other adornments often available at popular music concerts, the glow of fireflies on a summer evening, and, familiar to some, the emission of light when a Wint-O-Green Life Saver candy is crushed. The pathways to emission in the first two examples bear some similarities to one another, while the third is quite different. All, however, share the characteristic that the energy from some process produces a molecule in an excited state rather than leaving the system as heat.

Lightsticks

The reactants in lightsticks are an oxalic acid diester, hydrogen peroxide (H_2O_2), and a fluorescent dye (D), which has loosely held electrons in its pi orbitals and is the actual light emitter in the system. A diester commonly used in these systems is *bis*(2,4,6-trichlorophenyl) oxalate (TCPO), which reacts with H_2O_2 in a multistep reaction to produce 2,4,6-trichlorophenol and 1,2-dioxetanedione:

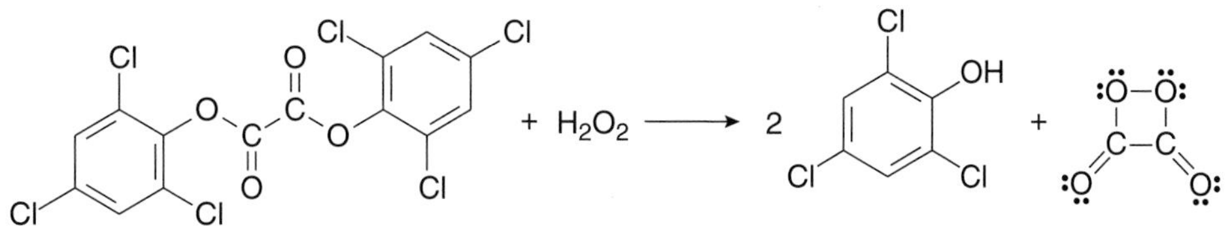

The dioxetanedione forms a charge-transfer complex with D, and the complex subsequently loses carbon dioxide:

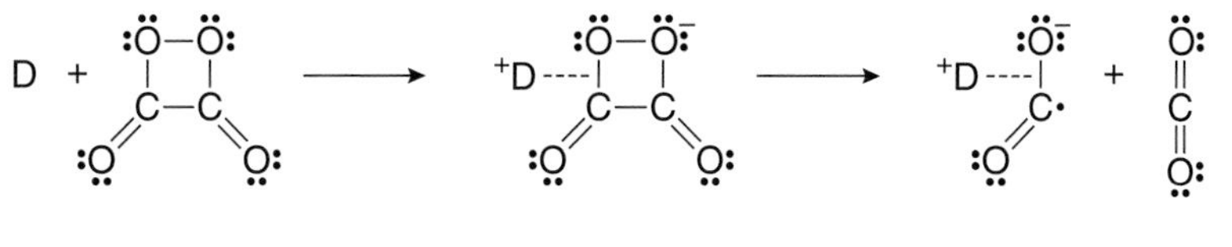

An electron is then transferred back to the dye, and the overall reaction is energetic enough to put the dye in an excited singlet state, from which it emits light:

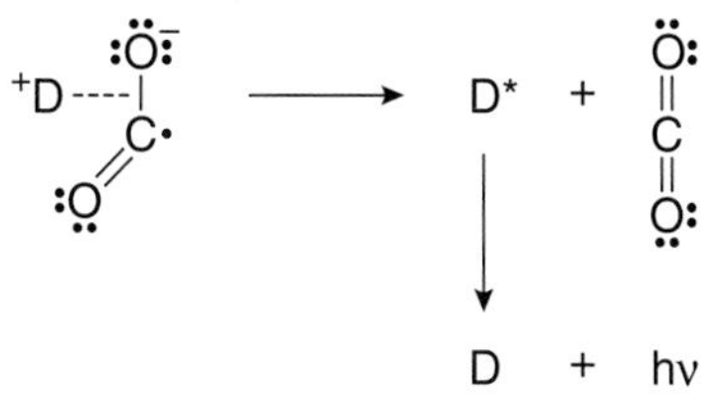

Relief of strain in the four-membered dioxetanedione ring and the great stability of carbon dioxide (which is stabilized by electron delocalization in the pi electron systems) provide the energy required to put the dye in its excited state.

Oxalyl diesters that incorporate a donor fluorescent dye (Figure 55) are being studied as possible tools for analyzing hydrogen peroxide concentrations in living cells. The charge-transfer-complex donor and acceptor are produced together, which could lead to a higher probability of forming the complex and an increased sensitivity for the analysis.

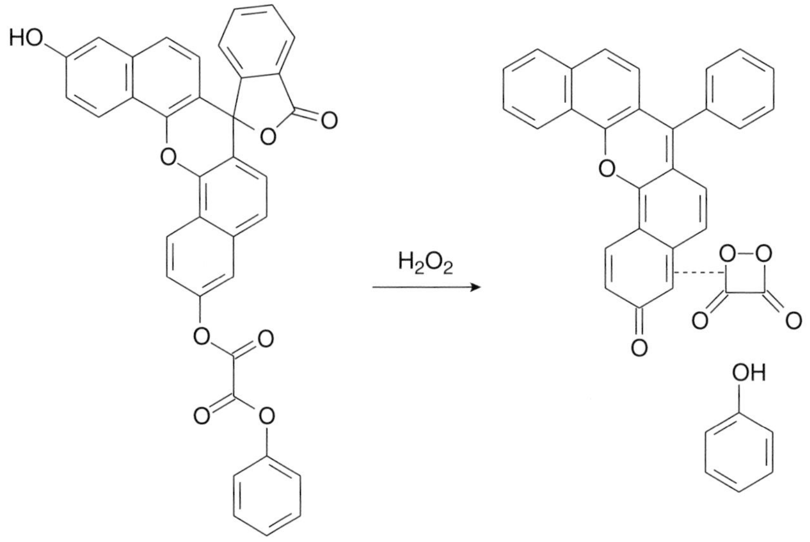

Figure 55.   An oxalyl diester that combines a fluorescent dye in a dioxetanedione-producing structure for possible analysis of H_2O_2 in cells.

Bioluminescence

Light emission from living organisms, such as fireflies, anglerfish, and dinoflagellates, is called *bioluminescence.* Fireflies have been extensively studied, both to understand the origin of the emission and to use the reaction analytically to measure levels of adenosine triphosphate (ATP) and the activation of particular genes in biological systems. The species required for the reaction are the enzyme luciferase, its substrate luciferin, ATP, and oxygen (O_2). The overall oxidation reaction, which occurs in several steps at the active site of the enzyme, is

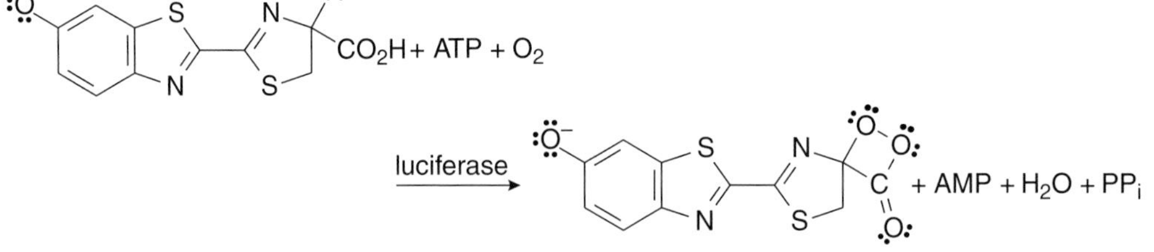

luciferin (anion form)   luciferin dioxetanone

The hydrolysis of ATP (actually, the reaction of the nucleophilic oxygen from the luciferin carboxyl group –OH) to yield AMP and pyrophosphate (PP_i) provides the energy for this reaction. Note the similarity of the four-membered-ring ketone structure in the luciferin dioxetanone to the diketone in the lightstick reaction in the previous section.

Evidence from studies of this system indicates that the light-emitting reaction is chemically initiated electron-exchange luminescence (CIEEL), which involves an internal electron transfer followed by the loss of carbon dioxide to produce the excited singlet state of oxyluciferin, the emitting molecule:

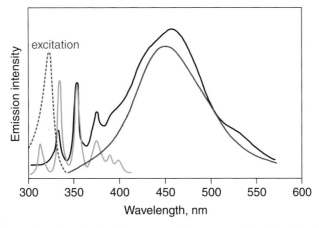

oxyluciferin (excited singlet)

This reaction is analogous to that of the lightstick, except that, here, charge (electron) transfer occurs internally within the molecule to put the oxyluciferin in its excited emissive state, so the fireflies can locate their mates.

Triboluminescence

Light emission for which the excitation energy is provided by rubbing or crushing some material, usually a crystalline substance, is called *triboluminescence,* from the Greek for "rubbing." The example most familiar to many (and available on multiple Internet Web sites) is the flash of light emitted when a wintergreen-containing hard candy, such as a Wint-O-Green Life Saver, is crushed, perhaps by biting it (in a darkened room, because the emission is momentary and weak). When sugar crystals are crushed, an electric-charge separation occurs at the site of fragmentation. (This is similar to the separation of charge when a length of sticky tape is torn from a smooth surface, producing an electrically charged tape and oppositely charged surface.) If enough charge separation occurs, an electric spark jumps between the fragments as electrons return to neutralize positive charges (like the lightning bolt in a thunderstorm). The electric discharge excites nitrogen molecules that are in the air or adsorbed on the crystal surface, and they emit light in the near-ultraviolet and faintly in the violet. The blue curve in Figure 56 is identical to the emission spectrum from nitrogen gas excited in an electric-discharge tube.

Figure 56 is a composite designed to explain the emission from a crushed wintergreen hard candy. The wintergreen flavor is produced by methyl salicylate, a fluorescent molecule that is excited by radiation in the near-ultraviolet and emits in the blue region of the visible

Figure 56. An overlay of emission from crushed sugar crystals (blue) due to excited nitrogen molecules, excitation (dashed red) and emission (solid red) spectra of methyl salicylate—the wintergreen flavoring compound—and emission from a crushed wintergreen hard candy (black).

spectrum, as shown by the red curves in the figure. The black curve is the spectrum of the light emitted by the crushed candy; the identification of the broad emission from methyl salicylate is obvious. Also obvious are the emissions from excited nitrogen molecules, but take note of the relative intensities of the emission peaks compared to those emitted by a crushed sugar sample without the fluorescent flavoring, the blue curve. The peaks at about 357 and 380 nm are essentially the same in the two cases, but the shorter-wavelength emissions, 315 and 338 nm, are substantially reduced in the presence of the fluorescent flavoring. These emissions are at wavelengths that excite the methyl salicylate and are partially absorbed by the flavoring, causing it to fluoresce, as observed.

VISION

The retinas of the eyes are part of the brain (Figure 57). During fetal development, a branch of the neural cells that were to become the brain grew to the front of the head and developed into the retinas, which cover about two-thirds of the interior surface of each eyeball (Figure 58).

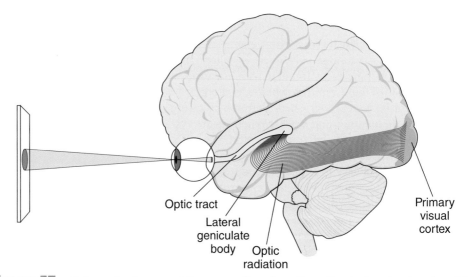

Figure 57.   Schematic diagram of the connections from the retina to the visual cortex of the brain.

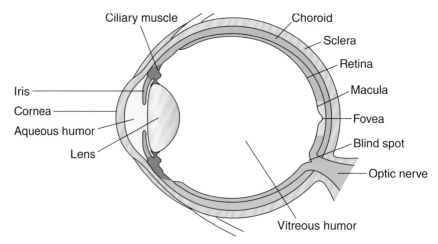

Figure 58.   Structure of the eye shown in horizontal cross section of the right eye, viewed from above.

The retina is a very active part of the brain and is where a good deal of processing of the light that enters the eyes takes place before signals are even sent along the optic nerves to the lateral geniculate nuclei (the bent structures named for their shape from the Latin for "knee"—as also in genuflect; Figure 57), which are located in the thalamus (about the middle of the brain), for further processing, ultimately by the primary visual cortex at the very back of the brain.

The Photochemistry of Vision

Vision begins with the absorption of photons of light by molecules in the retina. The absorbed light starts a cascade of events that begins with the photoisomerization of *cis*-retinal to *trans*-retinal (called *retinal* because it is found in the retina of the eye), as illustrated by the models in Figure 59. In the eye, *cis*-retinal is covalently bonded to the amino

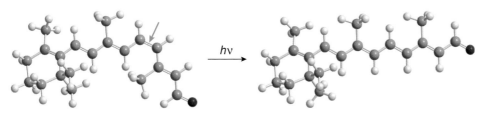

Figure 59. Photoisomerization of *cis*-retinal to *trans*-retinal. The colored arrow points to the double bond where the cis-to-trans reaction caused by absorption of visible light occurs.

side-group of the amino acid lysine in proteins called *opsins* (from the Greek for "sight"):

$$\text{retinal—CH}{=}\text{O} + \text{H}_2\text{N—opsin} \longrightarrow \text{retinal—CH}{=}\text{N—opsin} + \text{H}_2\text{O}$$

Opsins are members of a large family of membrane-associated proteins—G-coupled protein receptors—that are responsible for detecting stimuli outside the cell and transmitting the information to the interior. The senses of taste, hearing, touch, and smell and the action of many hormones involve other members of this family. The retinal-opsin combination in the rod cells of your retina is called *rhodopsin* (from the Greek for "rose"), which has a peak absorbance in the green part of the visual spectrum that gives it a reddish-blue (purplish) color. Retinal absorbs relatively low energy, visible light because the six π bonds in the molecule are conjugated (alternating double and single bonds). The resulting delocalization of the π electrons lowers their energy and gives a HOMO-LUMO energy transition in the visible range. This is a specific example of the general case that the more conjugated systems with greater electron delocalization absorb lower-energy light and often are visibly colored. (See Demonstration 12.47, The Photobleaching of Carotene.)

Note the striking change in overall shape of retinal on photoisomerization—from the bent cis form to the long, straight trans form. In the rhodopsin molecule, this change in shape forces the protein to change its overall shape (Figure 60). The change in shape of rhodopsin causes changes in other closely associated proteins, resulting in a cascade of reactions that cause the rod cell to change its electrical polarization (due to differing ion concentrations inside and outside the cell) and hence to trigger the neural sequence that signals that photons have been detected at this site on the retina. The photoisomerization of the retinal molecule also leads to its dissociation from the opsin (since it no longer fits in the normally folded protein). Dissociation is followed by removal of the *trans*-retinal and opsin from the rod cell, a series of energy-requiring reactions to regenerate the *cis*-retinal that again binds to an opsin, and return of the new rhodopsin to a new rod cell.

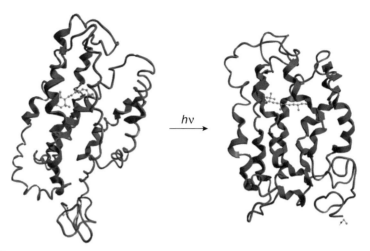

Figure 60.   Photoisomerization of *cis*-retinal to *trans*-retinal in rhodopsin causes a change in the shape of the protein.

Wavelength Sensitivity in Vision

The rod cells generally allow us to see under relatively low light conditions and are not sensitive to differences in the color of the light causing the photoisomerization (which is why we see everything in gray scale in a dimly lit room or at dusk). Color vision, which was briefly discussed earlier, uses essentially the same processes, but it is carried out in the retinal cone cells. There are three types of cone cells, which are sensitive, respectively, to long (L, red-yellow), medium (M, green), and short (S, blue) wavelengths of visible light (Figure 61). The retinal-opsin molecules in these cells are often called *iodopsins* (from the Greek for "purple"; the same root as for the element iodine). The electronic energy levels of retinal are influenced

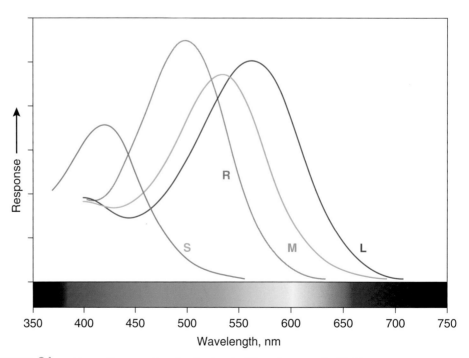

Figure 61.   Absorption spectra of retinal rods (R) and cones (L, M, S).

by the polarity of their surroundings, and polarity in the immediate vicinity in iodopsins is determined by nearby amino acid side chains. Slight differences in the amino acid sequences of the opsin proteins in the three kinds of cone cells are responsible for differences in the absorption characteristics of the bound retinal molecules in the iodopsins. The wavelength of maximum absorption for the rod cells and the three kinds of cone cells are given in the figure. These wavelengths vary somewhat from one individual to another, due to minor mutations in the opsin proteins, so you will find slight differences in wavelengths in different reference sources. More profound mutations are responsible for several forms of color blindness.

Now we are in a position to answer a question posed near the beginning of this introduction: is it chance that has limited the wavelengths of electromagnetic radiation we can see, or is there some fundamental reason for the limitation? To be detected by an organism, light has to have enough energy to produce a "permanent" but reversible change in some molecule. Radiation in the infrared or longer wavelengths can change the vibrational energies of molecules, but it does not have enough energy to bring about such changes as isomerizations. Radiation in the ultraviolet has enough energy to bring about isomerizations, as we saw above for stilbene, but, as we have seen, it also has enough energy to cause chemical bonds to break and destroy any molecule. Life is essentially impossible in the presence of high intensities of ultraviolet and shorter wavelengths of light, because complex molecules cannot survive. Although some organisms can detect light a little less or a little more energetic than the visible region, the fundamental nature of the interactions of light with matter limit most to the 700 to 400 nm range.

What Do Honeybees See?

Many insects, including honeybees, have true color vision with three visual pigments (Figure 62a), but they differ from humans and most mammals because one of the photopigments is sensitive to near-ultraviolet light, having a peak sensitivity at 360 nm, and they lack a photopigment sensitive to red light. Although their compound eyes allow a wide field of vision, the resolution of the images is relatively poor. Honeybees, however, need to find the center of the flowers from which they collect the nectar they use to make honey. And, conversely, many flowering plants require that bees land at the center of their flowers to get some pollen brushed off onto their bodies and carried to other flowers to pollinate them. Many flowering plants have evolved to take advantage of bees' ultraviolet vision to help them target the center of the flower (Figure 62b). In the example shown, pigments at the base of the petals strongly absorb ultraviolet wavelengths in sunlight and appear dark to a bee. The dark central area, with no reflected ultraviolet light, surrounded by the ultraviolet reflecting outer part of the petals forms a bull's-eye-like target for bees to home in on. Since

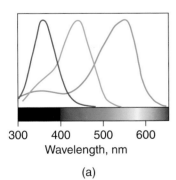

(a)

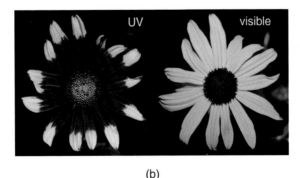

(b)

Figure 62.   (a) Absorption spectra of the honeybee photopigments normalized to the same absorbance. (b) Photographs of a black-eyed Susan flower taken with illumination by ultraviolet light and visible light.

we cannot detect reflection (or lack of reflection) of ultraviolet, the petals appear uniformly colored to us.

The Front of the Eye

For light to be used and interpreted effectively, a clear image of the part of the world you are looking at must be correctly focused on the retina, and these are the tasks of the structures at the front of your eye. In order that the image on the retina be clear (not blurred), the cornea, lens, aqueous humor (fluid between the cornea and lens), and vitreous humor (gel that fills the interior chamber) must be transparent and free of surface imperfections and inclusions that might scatter the light. The cells and their contents that make up the cornea and lens are stacked together in such a way as to make them transparent. Corneal wounds, such as scratches, cataracts in the lens, and diseases like river blindness create corneal and lens opacity and interfere with vision. (Small bits of matter, dead cells, for example, are usually present in the vitreous humor and you can usually see these as "floaters" drifting across your field of vision, if you stare at a clear blue sky or a uniform light-colored wall.)

The Cornea and Lens

Focusing an image of the world on your retina is the task of the cornea and the lens and depends upon the refraction of light as it passes from one medium to another. There are four of these interfaces—air–cornea, cornea–aqueous humor, aqueous humor–lens, and lens–vitreous humor—as light travels from the front to the back of the eye. About two-thirds of the focusing is at the air–cornea interface, because of the large difference in refractive indices between the air and the cornea (Table 2, page 32) and the high degree of curvature of the cornea. (If you swim and open your eyes under water you cannot see very well, because the refractive index difference between the water and cornea is small, so the necessary refraction to focus images on the retina is lacking. If you snorkel or scuba dive, you wear a mask that traps a layer of air in front of the cornea and allows you to see well under water.) The refractive index differences at the other interfaces are much smaller, so their contributions to the focusing are smaller.

The lens is flexible, and in its "relaxed" state it is relatively thick, that is, more spherical. By action of the ciliary muscle that surrounds it, the lens can "fine tune" the eye's focus by changing shape, becoming thinner with flatter surfaces (for distance vision) or fatter with rounder surfaces (for near vision, like reading). These changes are called *accommodation,* since they allow us to accommodate our vision to both near and far scenes. The ciliary muscle is essentially a ring of muscle bundles surrounding the lens and attached to it by flexible ligaments. When the muscles stretch (lengthen), the ring gets larger, and the pull on the lens through the attached ligaments makes the lens flatten out (in an adult's eye, to about the size and shape of an M&M candy). When the ciliary muscles contract (shorten), the ring gets smaller and the tension of the ligaments is relaxed, so the lens returns to its thicker, more rounded shape. As we age, the lens becomes less flexible and accommodation becomes more difficult. Many of us have to wear glasses, some with bifocal, trifocal, or continuously variable lenses, to compensate for the loss of lens accommodation.

Images in the Eye

In discussions of its optical characteristics, the eye is often compared to a camera. They both have a way to detect the image of an object: the retina in the eye and the film in the camera (or electronic device in a digital camera). The eye and the camera also have curved, refractive surfaces to focus light rays from the object to form an image on the detector. And your eye and cameras with adjustable settings both have an iris that opens and closes to provide a larger or smaller aperture for light to enter (the pupil, in the eye). The iris is important for controlling the amount of light entering and the depth of field in the camera and in your eye. The role of the iris in focusing light at a distance and close up is diagrammed in Figure 63.

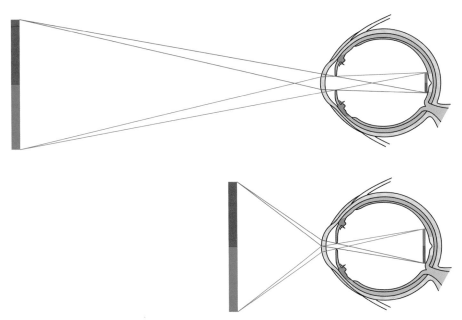

Figure 63.   Simplified schematic diagrams viewed from the top of the right eye focusing the image of a distant and close-up object on the retina. All refraction is shown taking place at the air-cornea interface. The roles of the shape of the lens and other interfaces are neglected although accommodation by the lens is required to focus the image at the retina. The images are out of scale compared to their actual size on the retina, where they would, if possible, be focused on the fovea, several-fold smaller than represented here.

Light beams from every part of the objects we look at enter every part of the cornea, and the beams from any particular part must all be focused at the same point on the retina to form a clear image. This process is illustrated schematically in Figure 63, where two beams are shown entering the cornea from each end of the object. (Many more beams could be shown from each end, but they would make the figure more difficult to interpret.) Any other beam between the two shown from each end would also be focused on the retina at the same point. When you look at close objects, such as text on a page, the iris contracts to make your pupil smaller, and the ciliary muscle contracts to allow the lens to bulge more. These coordinated changes compensate (like adjusting the lens and aperture on a camera) to keep the image sharply focused at the fovea (see below), which is necessary for you to see fine detail, like text.

Images on your retina are upside down, as they must be with this focusing system. This is one of the first tasks your brain faces in interpreting retinal images, but it has effortlessly and unconsciously turned the world right side up for you since infancy. (In experiments where participants wear special glasses that produce a right-side-up image on the retina, it takes only a few hours for most people to adjust and for their brain to interpret the image correctly.)

The image of an object that is farther away is smaller than the image of the same object closer to the eye. Image size is one of the clues your brain uses to help judge how far away an object, say the car driving in front of you, is from you. Another clue your brain uses is the angle between the eyes' optic axes, the imaginary lines drawn from the center of each cornea to the very back of the eye. Your eyes turn slightly toward one another as you look at close, compared to distant, objects. Additional clues are the slightly different images and positions of the images on your right and left retinas. You can easily observe the different images by focusing on an object a few meters in front of you and then covering first one and then the other eye. With your left eye covered, you will observe that you cannot see part of the scene on your left (without moving your head), and vice versa when the right eye is covered. All these clues, as well as experience, are what provide you with depth perception in a three-dimensional world. Artists (e.g., painters) who work in two dimensions on canvas or other flat surfaces use a number of strategies to "trick" your eyes and brain into seeing depth that is actually not present in the work.

The Back of the Eye: The Retina

Much of the interior of the eye is covered by the retina, a complex structure of various kinds of neural cells that is about as thick (and as fragile) as a piece of tissue paper. A schematic diagram of a very tiny portion of the retina is shown in Figure 64a. The layered structure illustrated in the diagram is also shown clearly in Figure 64b, which is a photomicrograph of a cross section of part of an actual retina. The cells were stained with a dye that is preferentially taken up by the cell nuclei, so the nuclei of the receptor (rod and cone) cells and the signal-transmitting (bipolar and ganglion) cells show prominently. Note that, in both the diagram and the photomicrograph, light strikes the retina at the bottom, which is the side that faces the interior of the eye. Directions within the eye are given with respect to its center, so the cells at the bottom in the figures are in inner (internal) layers, and those near the top are in outer (external, but not outside the retina) layers. The segments of the receptor cells are likewise labeled as inner and outer, nearer or further, respectively, from the center of the eye. (Omitted from the diagram are the fine blood vessels that are present in the retina to furnish the inner cells with oxygen and the nutrients they need to survive.)

Perhaps the most striking (and counterintuitive) part of the retinal structure is that it seems to be "inside out." The photosensitive parts of the receptor cells are at the very outside of retina, so light must pass through all the intervening blood vessels and neural cells, including the inner segments of the receptors themselves, before it can be detected. Fortunately, the intervening cells are relatively clear and colorless, and so do not much affect the incoming light. One reason for this structure is that the photosensitive parts of the photoreceptors (at least the rods) need to be in direct contact with the retinal pigment epithelium (RPE). It is within the RPE cells that *trans*-retinal(ol)—which is exported with opsin from the rods after a photon has been detected there (see the section The Photochemistry of Vision above)—is converted to the 11-*cis* form, recombines with opsin to form rhodopsin, and is then transferred back to the rods to be incorporated into the photoreceptors.

The RPE cells also contain a great deal of melanin (the dark spots shown in the schematic diagram), which makes them very dark (like the flat black paint on the inside of a camera), so that light that is not absorbed by the photoreceptors is not reflected back into the eye to confuse the images. (However, to increase the sensitivity of their night vision, some nocturnal animals have a reflective layer in the back of their eyes that reflects the dim light back to the photoreceptors for a second chance to be detected. This reflection from the back of the eye is why the eyes of a cat or a deer caught at night in a beam of light seem to glow.)

Neural Connections and Communication

The diagram in Figure 64a may look complicated, but it barely hints at the complexity of the interconnections among the neural cells in the retina. There are several kinds of neural cells in your body and each typically has a cell body (containing the nucleus); an extension called an *axon*, through which signals from the cell body are sent; and several highly branched extensions called *dendrites* (from the Greek for "tree"), through which incoming signals are sent toward the cell body. The signals that travel through axons and dendrites are electrical impulses created by changes in ionic concentrations triggered by opening and closing membrane channels through which cations (usually Na^+ and K^+) rush into or out of the cell. These signals are transmitted at about 100 meters per second. Axons can be very long (up to a meter from your spinal cord to your big toe, for example) and are often branched at their ends, although not as much as dendrites.

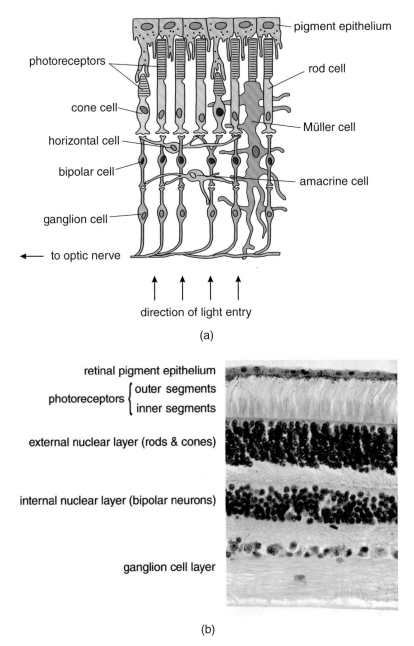

photoreceptors

pigment epithelium

rod cell

cone cell

Müller cell

horizontal cell

bipolar cell

amacrine cell

ganglion cell

to optic nerve

direction of light entry

(a)

retinal pigment epithelium

photoreceptors { outer segments
                 inner segments

external nuclear layer (rods & cones)

internal nuclear layer (bipolar neurons)

ganglion cell layer

(b)

Figure 64.   (a) Schematic diagram of a very tiny portion of the retina. Müller cells are essential scaffolds to maintain the positioning of the neural cells and possibly part of the biochemical processing required to regenerate the 11-*cis*-retinal required by cone cells. (b) Photomicrograph of a stained cross section of a small piece of retina.

Neural cells communicate with one another at *synapses* (Figure 65), where the end of one branch from an axon and one branch from a dendrite are connected via interactions (hydrogen bonding, for example, not covalent bonds) between the extracellular parts of protein molecules that are embedded in the membranes of each neural cell. These connections hold the membranes of the two cell branches within about 20 nm of one another. When an electrical signal reaches the end of an axonal branch, it triggers the release of neurotransmitter molecules (from where they are stored in vesicles just inside the axon cell membrane)

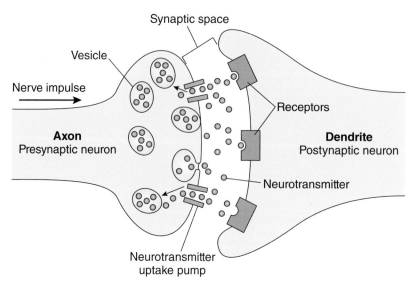

Figure 65.   Schematic diagram of a synapse. After a nerve impulse has triggered release of neurotransmitter molecules into the synaptic space, enzymes in the space begin to degrade them and molecular pumps in the axon membrane begin to take them back, in order to turn off the signal to the dendrite, so it stops firing.

into the synaptic space between the axon and dendrite. The neurotransmitters diffuse across the gap between the two cells in about 0.5 milliseconds and bind to receptor proteins in the membrane of the dendrite. Binding causes changes in the receptors' geometry that triggers other changes inside the dendrite (compare this action to the change in shape of rhodopsin, which signals absorption of a photon in a rod or cone) that ultimately produce the electrical signal carried by the dendrite.

Neurotransmitters are small molecules such as acetylcholine, epinephrine, and glutamate (the neurotransmitter from rods to bipolar cells in the retina). The same neurotransmitter can produce different actions in different dendrites, depending on the particular receptor each dendrite has. Chemical signaling at the synapses, which involves a great variety of neurotransmitters and receptors, is responsible for the enormous complexity and exquisite differentiation of signals among neural cells throughout your brain and body.

Each photoreceptor cell in the retina is connected via a synapse to at least one bipolar cell, which might be interconnected with others via horizontal cells. Ultimately these networks are all connected to ganglion cells, from which axons extend across the inner layer of the retina and are gathered together as the optic nerve that extends through the retina (producing the blind spot in each eye where there are no photoreceptors) and back to the lateral geniculate nuclei in the thalamus. To get some idea of the interconnectivity within the retina, consider that there are about 120 million rod cells and about six million cone cells in each of your eyes, but only about one million ganglion cells (each optic nerve is composed of about one million axons). Thus, on *average,* the signals from more than 100 receptors are gathered and processed before the result is sent to the brain as a single signal. The signals sent to the visual centers of the brain are not a one-to-one mapping of retinal "pixels" (individual photoreceptors) but instead are complex combinations of the signals from the pixels.

The Distribution of Photoreceptors in the Retina

Figure 66 shows that the rods and cones are not uniformly distributed across the retina. At the center of the back of the eye there is a small depression about three millimeters in diameter called the *macula,* which has mostly cones and almost no rods. At the center of the macula is a tiny area, about 0.3 mm in diameter, called the *fovea,* into which about 30,000 cone cells are densely packed (Figure 67). To get a sense of the size of these structures, the image of a full moon on your retina has a diameter of about 0.2 mm, that is, the area of light from the moon would cover about $[(0.2 \text{ mm})^2/(0.3 \text{ mm})^2]100\% = 44\%$ of the area of the fovea. Although there is a large concentration of cones in the macula, about 90% of the cones are scattered throughout the large area of the retina outside the macula. Since your peripheral vision is in color, you already have direct evidence for cones everywhere in the retina.

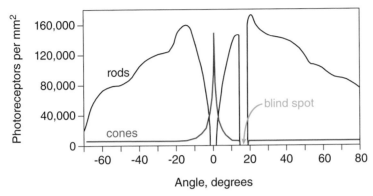

Figure 66.   Distribution of rods and cones in the retina as a function of position on the retina of the left eye. Zero on the horizontal axis corresponds to the position on the retina where a light ray from the center of the pupil strikes the center of the fovea. Positive angles (toward the nose) and negative angles (toward the ear) represent the position where light rays at these angles from the central ray strike the retina.

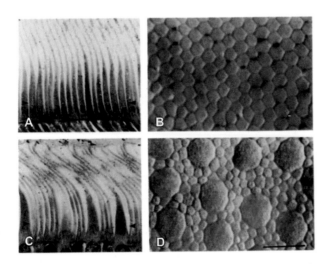

Figure 67.   (a) Transmission electron micrograph (TEM) of a cross section of the fovea and (b) scanning electron micrograph (SEM) of the surface of the fovea; (c) and (d) are comparable TEM and SEM images of the retina away from the macula. The magnification is approximately the same for all images.

The cone cells in the fovea are about 2.0 μm in diameter, which is thinner than the cones elsewhere in the retina and comparable to the size of rod cells. For comparison, the larger cells in Figure 67d are cones in the peripheral retina, outside the macula, and the smaller cells are rods. The conical structure of the cone cells shown in Figure 67c, which is also evident in Figure 67a, tells you why they were so named. The dense packing of these thin cone cells makes the fovea the area of the retina with the highest resolution, so it is the area where you have the ability to see the finest detail. To aid the resolution even further, no other neural cells or blood vessels overlie the cone cells in the fovea. Dendrites from bipolar cells must extend to communicate with each cone cell, but the bipolar cell bodies and the ganglion cells with which they communicate lie outside the foveal area. Thus, interference with and distortion of the incoming light rays are minimized. Also, there is a one-to-one correspondence of foveal bipolar and ganglion cells with the cones, so part of the information leaving the eye is essentially a direct mapping of the image on the fovea.

The Range of Photoreceptor Sensitivity

One of the astounding properties of your eyes is the range of light intensities over which they operate to provide a view of your surroundings. On a cloudy night, somewhere away from the glow of artificial illumination, you can detect the shapes of objects, although you have lost color vision in this dim light. At the other end of the range, you can see in color a landscape of freshly fallen snow in (painfully) bright sunlight. (Ultraviolet as well as visible light is reflected by snow and can result in snow blindness, if the exposure is long enough to cause sunburn of the cornea. In extreme cases permanent damage and vision loss can occur.) These endpoints represent a range of light intensities spanning about 10 orders of magnitude (10 log units) (Figure 68).

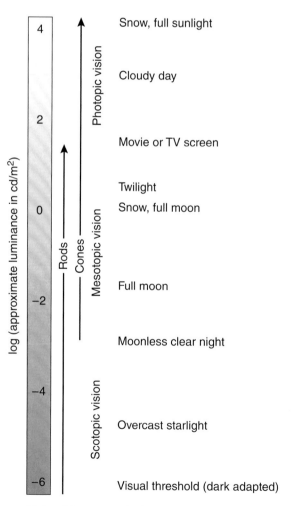

Figure 68.   The sensitivity of the eye to the intensity of light can be measured semi-quantitatively in standard luminance units, candela meter^{-2}, as indicated by this logarithmic scale. Since most of us have no feeling for what these units mean, the various values are related here to more familiar experiences.

A small part of the adjustment for changing intensity is accomplished by the automatic variation in the size of your pupil—from about 2 to 8 mm in adults, a factor of about 16 in the area allowed for light to enter your eye. The major contributor to the wide range of illumination over which your eye works is, however, the differential sensitivity to light of the rods and cones. We said previously that the rods are very sensitive to light. Although absorption of one photon of visible light triggers a response in a single rod, a physiologically observable neural signal requires absorption of five to ten photons (still a very small number) within about 0.1 sec. If single photons were observable, our brains would likely be overwhelmed by the random input from stray and scattered photons, so the requirement for more photons is perhaps a protective mechanism. (See Demonstrations 12.32, The Persistence of Vision; and 12.34, The Pulfrich Phenomenon: Perception of Motion.) At high intensities the rods become bleached—that is, much of the visual pigment, rhodopsin, has reacted and dissociated to opsin and *trans*-retinal. Regeneration of rhodopsin is a relatively slow process (see below), so the rod photoreceptors become inactive at about the level of light from a movie or TV screen (Figure 68).

Cones are about six times less sensitive than rods, so brighter (more intense) light is required to generate signals from them. They begin to produce observable signals at the approximate intensity of a moonless clear night (Figure 68). Vision in the range of intensities where *only* the cones are active is called *photopic vision* (another "photo" word for this range of bright light). At intermediate intensities, generally after sundown with no clouds to obscure the moon and stars, both rods and cones are active and we retain our color vision. Vision in this range is called *mesotopic vision* (from the Greek *mesos* for "middle"). At very low intensities, a cloudy night, only the rods are active, we have no color vision, and our visual acuity is low (that is, we cannot make out fine detail), because the receptive fields of the rods (see below) are relatively coarse grained. Vision in the range of very dim intensities where only rods are active is called *scotopic vision* (from the Greek word for "darkness").

Dark Adaptation

You have probably noticed that when you enter a dim space (like a darkened movie theater) from a brightly lit space, you can't see very well, but soon you will have adapted to the lower light level (and can make your way to an empty seat). Adaptation to a total absence of light takes longer and is a function of the brightness of the light before it is turned off, as shown in Figure 69.

The first phase of the adaptation to total darkness is a relatively quick increase in sensitivity (so signals of lower intensity become detectable) to a threshold where color vision is

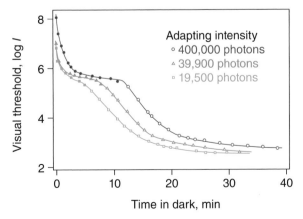

Figure 69. Adaptation to total darkness. Filled symbols indicate that the subject saw the color of the test light (violet), and empty symbols indicate that the detected light was not perceived as colored.

lost. The eye is still not very sensitive at this point, and it takes a reasonably intense signal to elicit a response from the subject. The second phase is slower, but ultimately ends up being sensitive to a very weak signal. The difference in sensitivity between the first and second thresholds in the figure is about three orders of magnitude (three log units). This difference is consistent with the information in Figure 68 that shows a difference of about three orders of magnitude between the lowest intensities detectable by rods and cones. This is one piece of evidence that the first phase of dark adaptation represents a relatively rapid adaptation by cones, while the adaptation of rods is much slower and is responsible for the second phase. This interpretation is strengthened by the observation that color vision is retained in the first phase, before the rods have adapted and the eye still requires intensities of light that are high enough to be detectable by the cones. Once the rods have adapted sufficiently to be sensitive to intensities below the first threshold, the test signal is detected at levels where the cones are inactive and color vision is lost.

What is happening when your eye adapts to darkness? As Figure 69 shows, adaptation time is related to the brightness of the light from which you enter the darkness. Brighter light (more photons) causes more of the photoreceptor molecules to absorb a photon so, the brighter the light, the more bleached the rod and cone photoreceptors become. (See Demonstration 12.31, Saturation of the Retina: Afterimage.) When you enter the dark, the reactions that regenerate the photoreceptors have a chance to catch up, but it takes longer to restore them if they were more bleached to begin with. Thus, during dark adaptation, your retina is being brought back to full sensitivity by regenerating the photoreceptor molecules and getting them in place in the rods and cones. The results indicate that recovery of the cones is a good deal faster (about 20 times faster) than for the rods. Since the cones are designed to work at relatively high light levels, it makes sense that the regeneration of their photoreceptors should be fast. There is evidence that the mechanism for the regeneration is different than that for rods and might not involve the retinal pigment epithelium as the primary mechanism.

PERCEPTION

The retina is organized into receptive fields, each of which can be visualized as a small circular array of rods (or cones) surrounded by a ring of other rods (or cones) (Figure 70), all of which are connected via a neural network to a single ganglion cell. These *center-surround receptive fields* overlap one another, so a particular photoreceptor will be part of several receptive fields and connect to the separate ganglion cells through different neural networks. The two primary types of receptive fields are called *center-on* and *center-off* to describe the output of the ganglion cell when the central part of the field is illuminated.

The left-hand part of Figure 70 illustrates four experimental illumination conditions for a center-on field and the associated output of electrical pulses from its ganglion cell. The first experiment is illumination of the central part of the receptive field, which produces a rapid series of signals from the ganglion cell that trigger a response in the brain. In the second experiment, the surround ring is illuminated and produces no signals from the ganglion cell. The third condition is no illumination (completely dark in the figure) and, as you might expect, there is no response by the ganglion cell when no light hits the photoceptors. The fourth experiment is illumination of the entire field, and the output from the ganglion cell is only a weak signal. When both parts of the field are stimulated at the same time, they inhibit one another and the output from both is suppressed.

The right-hand part of the figure illustrates the results for exactly the same four experimental conditions in a center-off receptive field. You see that the results for experiments one and two are just the opposite of those for the center-on field. The center-off field ganglion cell does not fire when there is light on the central field (hence the names center-*off* and center-*on*) but does when the surround ring is illuminated. Once again, condition four, the

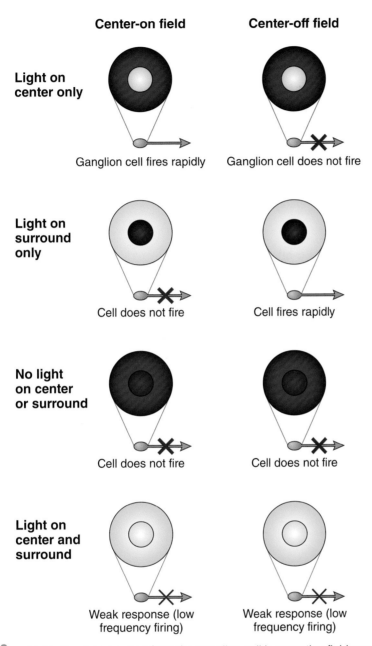

Figure 70.   Light stimuli and output from the ganglion cell in receptive-field experiments. Input to the ganglion cell combines the signals from the receptive-field center and surround and produces an output signal to its axon.

two parts of the field inhibit one another and illumination of both fields at once produces none or only a feeble signal. The mechanisms that cause ganglion cells to fire or not depend upon the synaptic connections in the neural networks between the photoreceptor cells and the ganglion cells. If you are interested in this level of detail, many cell physiology textbooks provide descriptions of these mechanisms.

The center-surround receptive fields vary greatly in size from those in the fovea, where the center of each field is probably a single cone cell, to those further out toward the periphery of the retina, where the center encompasses thousands of photoreceptor cells. Thus, as we have said previously, our ability to distinguish fine detail, read text for example, is greatest at the center of the retina (in the fovea) and decreases toward the periphery. (See Demonstration 12.33, The Imprecision of Peripheral Vision.)

The Detection of Shapes

Why is the retina organized in these mutually inhibitory center-surround receptive fields? In part, the organization reduces redundant signals to the brain. If part of a scene you are looking at contains an area that is all the same color, say white, then where this part of the scene is focused on your retina, the information from each of the rods (or cones) is the same, so they add nothing to one another. The receptive fields that are in this part of the retina are uniformly illuminated and, whether they are on-center or off-center, will be in the condition represented by the fourth experiment in the figure, so they will not be sending redundant data to the brain. The same is true, of course, if the receptive fields are in a part of the retina that is detecting a dark area in the scene and are not illuminated.

At the edges of the white area focused on your retina, the receptive fields will not be uniformly illuminated, and the differences in illumination on the centers and the surrounds will trigger signals to the brain. This condition is illustrated in the simplified schematic diagram of a center-on receptive field in Figure 71. In this simplified case, the output from the inhibitory surround does not cancel the center output when the field straddles a boundary between two areas of different brightness. There is an output signal to the ganglion cell that is interpreted as an edge, a change in intensity of illumination on the retina. (See Demonstration 12.28, The Hermann-Grid Illusion.)

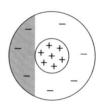

Figure 71.   An idealized, partially illuminated center-on receptive field. Illumination of the center produces a relative +7 signal, and illumination of the surround inhibits the output by a relative value of −7. The output from this partially illuminated receptive field is not totally inhibited so there is a signal from the ganglion cell that would not occur if the field were uniformly illuminated (or totally dark).

Thus, your brain will sense the shape of the white area and interpret it (probably correctly) as the blank sheet of white paper on the table or the white house next door. In a way, the compression of a computer image to jpeg format mimics how the eye-brain combination uses information efficiently. To create the jpeg, the color (digital code) of each pixel is compared to surrounding pixels. If the differences are large enough, this information is retained, but if the differences are small or nonexistent, the information for the whole area is combined and only a single code has to be saved for this area. If you compare the size of jpeg files (at the same resolution) for a simple image with not too much detail to one for a complex, finely detailed image of the same dimensions, you note that the latter file is larger, since there are more edges and smaller uniform areas.

Sensitivity of Edge Detection

Experiments to determine the intensity difference required for the retina to detect an edge reveal another important property of receptive fields. In these experiments subjects look at a screen to adapt their retina to a uniform level of light while one of their ganglion cells is monitored for activity. After adaptation, a spot of light is flashed on the screen at a location such that its image will fall on the center of the receptive area of the ganglion cell. The intensity of the flashing light is increased slowly until the subject indicates that she sees the light. At the same time that the subject visually detects the light, the ganglion cell also begins to send its signals to the brain, so we know that the cell signal is required to produce the physiological sensation.

One result of these experiments is that, to sense its presence, a spot of light about two percent higher in intensity (brighter) than the surrounding light is required. A second result is that this percentage difference in detectable intensity does not depend upon the intensity of the background. That is, no matter what the intensity of the background (as long as it is within the normal range of intensities to which the retina is sensitive), a signal about two percent more intense than this surrounding light will be detected. Conversely, you can reason that an area of a retinal image that is about two percent less intense than the surrounding area will appear less bright than the surroundings. (See Demonstration 12.27, The Perception of Brightness Is Relative.)

Thus, an edge is detectable whenever adjoining areas of light on the retina differ in intensity by about two percent, and the physiological perception of difference becomes even more pronounced as the difference in intensity becomes greater. This phenomenon helps to explain how you perceive black areas in a television image. A television screen when the set is turned off is visible by the light it reflects, and it appears grey. This is the least intense light you can ever detect from the screen, since, when the set is on, the only thing it can do is emit light from the screen, never take it away. If an area of the screen is not emitting but the surrounding areas are, the intensity of light from the nonemitting area will be so much less that it will *appear* to you as essentially an absence of light, that is, black. A light meter measuring the intensity of light from the nonemitting area will read the same as before the set was turned on (assuming that the illumination in the room is still the same), but your retina-brain combination, which is very sensitive to *differences* in intensity from one part of a scene to another, perceives the area as black, relative to its surroundings.

Camouflage

Many animals have evolved camouflage to stay hidden from predators (or prey), and humans also use camouflage for both military and recreational (hunting) purposes. The idea of camouflage is to reduce the number of edges between the background and the camouflaged animal or object. Common examples are organisms whose coloration or texture blend with their usual habitats, such as the fish shown in Figure 72a. Human use of camouflage in battle has changed over time—from the use of natural materials, leaves, and branches, as when Birnam Wood comes to Dunsinane in Shakespeare's *Macbeth* (act 5, scenes 5 and 6), to modern digital camouflage, which is designed mainly for clothing used in woodlands and rain forests (Figure 72b).

(a)   (b)

Figure 72.   (a) The fish blends well into its background. (b) Modern military camouflage fabric patterns are generated digitally.

Some animals can change color and may do so to blend in with their surroundings (Figure 73a). These animals usually have layers of different-colored cells (*chromatophores*) just under the outer layers of their bodies. By changing the relative sizes of the different chromatophores, the animal can change its color. Chameleons are such well-known examples of these animals that we sometimes call a person with a changeable temperament a chameleon. This is an apt analogy, because chameleons generally change color not for camouflage but, rather, in response to environmental factors, such as temperature and light, or "mood," that is, at mating or as warning (Figure 73b).

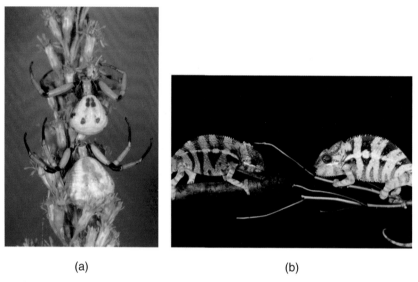

(a)   (b)

Figure 73. (a) Goldenrod spiders can turn white or yellow to match their background and make it harder for prey to see them. The upper spider has been on the flower long enough to blend with its color. (b) The panther chameleon on the left has ventured too close to the territory of the one on the right, which has turned yellow as a warning not to come nearer.

Another protection for animals is mimicry, that is, when they look like something else. Walking-stick insects look like small twigs, and the viceroy butterfly has markings that look very much like a monarch butterfly, which birds avoid because the poisons in its diet of milkweed make it unpalatable. In mimicry, unlike camouflage, the prey is not hiding but is easily seen and is mistaken for a nonfood item or an undesirable meal. There are also predator mimics, which are prey that take on the appearance of their predators and so are disguised from those predators.

Retinal Blind Spots

In dim light, when your vision is scotopic, the cones in the fovea are inactive, thus producing a blind spot where, in brighter light, you have the most acute vision. This is sometimes called the physiological blind spot, because it is caused by the physiology of the photoreceptors, not by their absence. People who work in conditions of dim light, for example, astronomers and night-navigating airplane and ship pilots, do not look directly at an object, such as a star or a distant beacon, but rather look at it at a slight angle, so the image on the retina falls a bit outside the fovea. Note in Figure 66 that the density of rods is at a maximum at an angle of about 20 degrees from the fovea. There is a high-density ring of rods surrounding the fovea, and it is on this area that you have the most acute scotopic vision. Although not acute enough to make out very fine detail like print, vision in this region makes it easy for you to move about in a very dimly lit room without tripping over the furniture. High sensitivity to movement at the periphery of your vision in dim light has an evolutionary advantage.

Whether you are the hunter or the hunted, being able to see the prey or predator out of the corner of your eye is important.

The other blind spot that is always present in each eye is the area where the axons from the ganglion cells gather to form the optic nerve and pass through the back of the eye on the way to the brain. Here, there are no photoreceptors (Figure 66) and no possibility of vision. These areas are often called physical, or anatomical, blind spots, because they are caused by the physical structure of the eye. Why is it that you do not have blank spots corresponding to these anatomical blind spots in your images of the world? A major reason lies in your binocular vision. The images of the world focused on your right and left retinas are displaced from one another. The part of the image that falls on the right eye blind spot does not fall on the left eye blind spot and vice versa. Combining the images provides the brain enough information to ignore the lack of signals from the blind spots.

Demonstration 12.29, Finding the Blind Spot, shows how easy it is to locate the blind spots in your eyes. An analogous activity is included here to make one final point about the intricate retina-brain connection. Close or cover your left eye and, with your right eye, focus on the **X** at the top of Figure 74. Keeping your focus on the **X**, move your head further from (or closer to) the page until the right-hand black circle disappears. The image of the circle is now on the blind spot in your right eye. With your eye still focused on the **X**, see how much farther you have to move your head in order for the image of the circle to reappear. This gives you some indication of how the size of the image of the circle on your retina and the size of the blind spot compare. Test your left eye in the same way, except it is the left-hand circle that you will observe to disappear. For a further test of your blind spots, repeat the procedure of the previous paragraph, except use the lower diagram in Figure 74 and observe when the blank spaces in the line disappear.

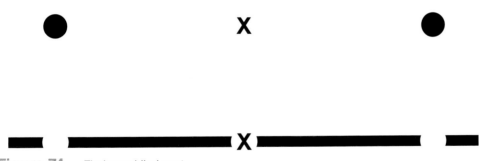

Figure 74.   Find your blind spot.

What did you observe when the circle disappeared? There was no "hole" in the image you saw, but rather the circle disappeared and the image at that spot appeared simply to be a uniform field of the surrounding color. But this information was not being transmitted to the processing centers in the brain, because there are no photoreceptors in this area. Your brain, in the absence of information about this area of the image, was "filling in" with information from the immediately surrounding area. This filling-in response can be more sophisticated, as you see from your second experiment where, when the blank space "disappears," the line appears continuous with no break. Your brain has filled in the area from which it gets no signal with a reasonable interpolation of the line as unbroken.

"It's all in your head." This is another way to express the message from the blind-spot activities above. Unfortunately, this expression is often used in a negative way to suggest that what is in your head is somehow not "real" (or true). But the connections of the compelling examples from demonstrations, chemical or otherwise, to the concepts underlying the "reality" are constructed by your brain. Demonstrations of perception, like the blind-spot activities, strip away all but the most relevant input to our senses, so that we are compelled to observe the input-brain connection(s) and interpretation(s). Chemical demonstrations are messier, because there are almost always many things going on at once. Focusing on and getting to the essence of these demonstrations is more complex, but you can use the lessons learned about perception to help. Chemistry is about transformations—some of which, highlighted in this book, produce light and others of which are produced by light. But chemistry is not only about characterizing reactants and products. Chemistry is the understanding—at the level of invisible entities, atoms, and molecules—what, how, and why these particular transformations occur. This understanding "is all in your head."

Thus, things are not always what they seem when you view the world, and "fooling" the brain (which includes the retina) into constructing or perceiving something different from the image that is actually present is what optical illusions are all about. The "mistakes" that the brain makes in processing these illusions are a result of its attempts to make sense of the scenes you are looking at. Optical illusions are one of the tools that physiologists and psychologists use to try to determine how the visual system interprets the often ambiguous information it receives from the world around you. A few of these illusions and their interpretations in terms of our visual processing of the actual images are included in the demonstrations in this volume to give a sense of how they are used to illuminate our sense of sight.

THE DEMONSTRATIONS

The foregoing background in color, light, vision, and perception introduced the concepts that are the basis for the demonstrations that follow. Among the many reasons for using demonstrations (gaining attention, sparking interest, exhibiting demonstrator enthusiasm for the subject, entertainment, and so on), a most fundamental one is to illustrate or, better, introduce concepts via concrete, compelling examples. Providing an audience the optimal opportunity to learn from a demonstration requires careful consideration of the context in which it is presented. What does the audience already know? What are they anticipating learning? How much (or how *little*) introduction must you provide to tell them what to be alert for without telling them what is going to happen?

The detailed procedures for preparing and presenting each demonstration are designed to provide excellent opportunities for your audience to observe the consequences of the underlying concepts as unambiguously as possible. The discussions that accompany each demonstration almost all begin with a paragraph that ties the observations from the demonstration to the basic concepts found in this introduction. These are often elaborated and extended in the further discussion, but essentially all hark back to the background included above. Thus, we intend this introduction as a resource you may refer to as you consider how to frame your own introduction to and follow-up discussion of a particular demonstration. As you delve more deeply into the rich concepts of color, light, vision, and perception, we hope you will continue to find it useful and a fruitful source of ideas.

The Production of Light

12.1

The Emission Spectrum from a Candle Flame

The brightest part of a candle flame appears yellow. Spectroscopic examination of the flame helps to determine what species are responsible for this emission.

Three options—a diffraction grating (Procedure A), a compact disc (Procedure B), and a video system with either of these diffraction techniques (Procedure C)—are presented for observing the visible emission spectrum of a candle flame. The spectrum appears to be continuous. A white porcelain dish held in the luminous part of the flame collects a deposit of black soot.

MATERIALS FOR PROCEDURE A

handheld, single-axis transmission diffraction gratings, one for each audience member (See Procedure A for description.)

candle with exposed flame, e.g., a taper

candle stand

matches

white porcelain dish at least 75 mm in diameter

MATERIALS FOR PROCEDURE B

compact discs, one for each audience member

candle with exposed flame, e.g., a taper

candle stand

matches

white porcelain dish at least 75 mm in diameter

MATERIALS FOR PROCEDURE C

video projection system*

handheld, single-axis transmission diffraction grating mounted as a slide, or a compact disc

candle with exposed flame, e.g., a taper

candle stand

matches

white porcelain dish at least 75 mm in diameter

---

* Various types of suitable video projection systems are described on page xxxiii. Some video projection systems are not capable of displaying a continuous spectrum but instead show a spectrum of several distinct colors. Test the system before using it to be sure that a continuous spectrum is displayed.

PROCEDURE A

Preparation

Each member of the audience should have a single-axis transmission diffraction grating. This grating material is available as a thin, flexible plastic sheet and needs to be mounted in a rigid holder for convenient use. Suitable gratings are available already mounted as 2-in × 2-in slides. Alternatively, bulk grating material can be used to prepare suitable gratings. A simple way to do this is to use a hole punch to make a round hole in an index card and tape a 1-cm × 2-cm piece of grating material over the hole (see Figure 1). (The audience members may also be instructed in how to make these themselves as part of the presentation.)

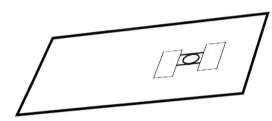

Figure 1.   Piece of grating material taped over hole punched into an index card.

Presentation

This demonstration needs to be done in a room that can be darkened as completely as possible, including closing blinds, shutting doors, and so on.

Distribute the handheld transmission diffraction gratings to the audience.

Place a candle where it is easily visible to all members of the audience. Light the candle. Instruct the audience members to look through the diffraction gratings directly at the candle flame. The full spectrum of colors from the flame will be seen to either side of the flame. If the spectrum appears above and/or below the flame image, rotate the grating by 90 degrees. Darken the room. The spectrum appears continuous, showing all colors from red, through orange, yellow, green, and blue, to violet.

With the lights on, display the white porcelain dish to the audience, showing that the outer surface is clean. Hold the bottom of the dish in the white region near the top of the flame for a few seconds. Show the audience the black spot of soot that has deposited on the bottom of the dish.

Extinguish the candle.

PROCEDURE B

Preparation and Presentation

This demonstration needs to be done in a room that can be darkened as completely as possible, including closing blinds, shutting doors, and so on.

Distribute compact discs to the audience.

Instruct the audience members to hold the compact disc in a position so that they see the reflection of the candle flame on the shiny side of the disc. Darken the room. When the white reflection of the flame is visible, the spectrum of the flame can be viewed to either side of the white flame. They may need to tip the disc a bit to see the spectrum. The spectrum appears continuous, showing all colors from red, through orange, yellow, green, and blue, to violet.

With the lights on, display the white porcelain dish to the audience, showing that the outer surface is clean. Hold the bottom of the dish in the white region near the top of the flame for a few seconds. Show the audience the black spot of soot that has deposited on the bottom of the dish.

Extinguish the candle.

PROCEDURE C

Preparation and Presentation

This demonstration needs to be done in a room that can be darkened as completely as possible, including closing blinds, shutting doors, and so on.

When using a transmission diffraction grating, tape the grating in front of the camera lens. Focus the camera on the flame and then turn the camera slightly to the side until the spectrum is displayed on the screen. Darken the room. The spectrum appears continuous, showing all colors from red, through orange, yellow, green, and blue, to violet.

When using a compact disc, display the spectrum of the candle flame via video projection by mounting the disc behind and somewhat to the side of the flame (from the perspective of the audience). Mask the flame from the audience, so only its reflection from the disc can be seen. Focus the camera on the reflection from the disc and tip the disc sideways a bit to display the spectrum on the screen. Darken the room. The spectrum appears continuous, showing all colors from red through orange, yellow, green, and blue to violet.

With the lights on, display the white porcelain dish to the audience, showing that the outer surface is clean. Hold the bottom of the dish in the white region near the top of the flame for a few seconds. Show the audience the black spot of soot that has deposited on the bottom of the dish.

Extinguish the candle.

HAZARDS

Be careful with the open flame and be sure no flammable liquids or other materials are nearby, and extinguish the flame as soon as the audience has had time to explore the spectrum.

DISPOSAL

Clean the porcelain dish. Materials may be saved and used in future presentations.

DISCUSSION

Until only a little more than a century ago, all artificial light was produced by fire, which is the result of a chemical reaction in which a fuel combines with oxygen in the air. Numerous types of fuels have been used, including wood and vegetable oils. Lamps that use liquid fuels are particularly convenient for producing light because a liquid can be used with a wick. Via capillary action, the wick delivers a steady, controlled flow of fuel to a flame, thereby producing a light of constant brightness. A disadvantage of liquid fuel is that it can be spilled, and a spilled fuel can lead to an uncontrolled fire.

The candle is a technological marvel that exploits the advantages of a liquid fuel but dispenses with its disadvantages. A candle has two components, wax and a wick. Wax is a solid fuel that melts at a low temperature, and because it is normally a solid, there is no danger of spilling fuel. The wax used in candles has several different sources, and its composition depends on its source. Most candles are made from either beeswax or paraffin wax.

Beeswax is a mixture of hydrocarbons, alcohols, fatty acids, and esters [1], but 95% of it is in the form of chains of methylene, CH_2, groups [2]. Paraffin wax is a petroleum product consisting of a mixture of predominantly straight-chain hydrocarbons, also consisting of CH_2 groups, with more than 20 carbon atoms.

How a candle works was famously explained by Michael Faraday in a series of six lectures to an audience of young people at the Royal Institution of Great Britain in London during the Christmas holidays of 1860–61 and published as the book *The Chemical History of a Candle* in 1861 [3]. When the tip of the wick is ignited, the heat of the flame melts a small amount of the wax, and the liquid wax is carried by capillary action up the wick into the flame. Beeswax melts at about 60°C, and paraffin wax melts in the range 52 to 57°C. The molten wax enters the flame, is vaporized by the heat, and burns. The heat from the burning wax melts more wax, which sustains the flame.

 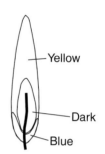

Cross-section diagram of candle flame

The flame of a candle is not uniform. At least two regions can be easily observed: a pale bluish region near the bottom of the flame and a bright yellowish region from the center to the top. Less easily visible is a dark region at the center near the wick. When molten wax moves up the wick, some of it is vaporized near the bottom of the flame. This region is rich in oxygen from the air.

The wax vapor and oxygen react to give off the same blue glow as any gas burning in ample oxygen. The glow is produced mainly by excited C_2 and CH molecules in the gas (see the section *Flames* in the introduction, page 61) and is identical to the blue glow when natural gas burns in a kitchen stove or propane burns in a gas grill. The blue region in the candle flame extends up around the outside of the entire flame, where there is ample oxygen, but, the blue is obscured by the bright light from the upper regions of the flame.

Some of the liquid wax moves up the wick, above the bottom of the flame, and vaporizes near the top of the wick. Here, there is little oxygen, because most has been used up below. As the wax vapor moves up and away from the wick, it is heated to a high temperature by the flame, but because there is no oxygen with which it can combine, the molecules in the vapor decompose in the heat, forming very tiny particles of solid carbon, as well as hydrogen gas. If a cool object, such as a porcelain dish, is held in the center of the flame, some of these carbon particles can be collected, producing a black spot of carbon soot on the object held in the flame. In addition to bits of graphite, this collected soot also contains small amounts of buckminsterfullerenes, ball-shaped molecules of C_{60}, C_{70}, and their fragments. (See the section *Flames* in the introduction, page 61.) The tiny particles of carbon and other products of the thermal decomposition of wax vapor drift upward and outward from the wick, where they encounter more oxygen, and there they undergo combustion.

The tiny particles of carbon that form in the yellow region of a candle flame are very hot, about 1000°C [*4*], and anything that hot emits visible light. The light emitted by the carbon particles gives the flame its bright yellowish glow, but the hot particles of carbon in the candle flame actually emit light of various colors. This emission can be observed by separating the colors using a diffraction grating, as shown in this demonstration. The spectrum of light emitted by a candle flame appears to be a continuous spectrum, with every wavelength from deep red to pale violet represented, although with different intensities. The light is brightest in the orange and yellow regions of the spectrum, and very dim, at best, in the blue and violet regions.

All matter constantly emits electromagnetic radiation. Objects at room temperature emit radiation mainly in the infrared region of the spectrum. As the temperature of an object increases, it emits more energy, and the wavelengths of the emitted radiation become shorter. As the object gets hotter, more and more of the emitted radiation is visible, and the object begins to glow. First, at around 700°C, it glows red, the color corresponding to the lowest energy of visible light. As the temperature increases, visible light of higher energy is emitted, from orange to yellow, to green, to blue, and to violet. When an object is hot enough to be emitting all visible wavelengths, its glow appears white, and we can say that it is "white hot." The emission of visible light by a hot object is called *incandescence.*

The electromagnetic radiation emitted by objects as a result of their temperature is called *black-body radiation* because it does not include reflected radiation; a black object is black because it does not reflect light. (See the section *Black-body Radiation* in the introduction, page 23.) The distribution of the wavelengths in black-body radiation is determined by temperature. Thermometers that display the temperature of an object at a distance detect black-body radiation and convert it to a temperature reading. Physicians commonly use this kind of thermometer to measure an infant's body temperature by placing a probe into the child's ear. The thermometer detects the radiation from the ear drum and displays the corresponding temperature.

If the emission from the flame of a candle were purely black-body radiation, the emission could be used to determine its temperature. Such a determination indicates the temperature to be about 1550°C [*5*]. Emission from a solid incandescent object at 1550°C looks yellow to an observer. However, measurements of the flame temperature using other methods show that the flame is at most 1400°C in the blue region, and in other regions, considerably less than that. Hence, the spectrum of radiation given off by a candle flame is not purely black-body radiation; it is, instead, the combination of numerous line spectra emitted by many different tiny particles of carbon.

The carbon particles in the flame are small enough to behave as molecules having distinct excited states, which are attained through the absorption of thermal energy in the flame. As the particles return to their ground states, they emit a narrow band of radiation. Because each particle has numerous excited states and there are numerous different particles with different excited states, the light emitted contains many closely spaced wavelengths, which give the appearance of a continuous spectrum.

The use of a video projection system to display the spectrum in this demonstration presents an opportunity to point out the difference between physical and physiological color perception. (See the section *Color Perception* in the introduction, page 4.) Consider the color yellow observed directly in the candle flame and in the projected spectrum. The color perceived from the flame is the physical color resulting from the action of wavelengths in the yellow region of the spectrum on the pigments in the retina of the observers' eyes. The yellow color perceived in the video-projected spectrum, on the other hand, is produced by a combination of red, green, and blue emissions in the projection system. This combination acting on the pigments in the observers' retinas provides the physiological perception of yellow, even though there are no yellow wavelengths in the projected light.

REFERENCES

1. A. P. Tulloch and L. L. Hoffman, "Canadian Beeswax: Analytical Values and Composition of Hydrocarbons, Free Acids, and Long Chain Esters," *J. Am. Oil Chemists' Soc.,* *49,* 696–699 (1972).

2. T. Kameda, "Molecular Structure of Crude Beeswax Studied by Solid-State ^{13}C NMR," *J. Insect Sci.,* *4,* 29–33 (2004).

3. M. Faraday, "A Course of Six Lectures on the Chemical History of a Candle: To Which Is Added a Lecture on Platinum," published as *The Chemical History of a Candle,* W. Crookes, ed., Griffin, Bohn, and Co.: London (1861). *The Chemical History of a Candle* is no longer copyrighted in the United States and many versions are available electronically.

4. A. G. Gaydon and H. G. Wolfhard, *Flames: Their Structure, Radiation and Temperature,* 4th ed., p. 155, Chapman and Hall: London (1979).

5. G. W. Stewart, "The Temperatures and Spectral Energy Curves of Luminous Flames," *Phys. Rev.* (Series I), *15,* 306–315 (1902).

The Temperature Dependence of the Emission Spectrum from an Incandescent Lamp

The color of the emission from any hot object depends on its temperature. Very hot solids, such as the filament in an incandescent lamp, closely approximate the properties of a black body and can be used to examine the temperature dependence.

Three options, a diffraction grating (Procedure A), a compact disc (Procedure B), and a video system with either of these diffraction techniques (Procedure C), are presented for observing the visible emission spectrum from an incandescent lamp. The spectrum appears to be continuous. As the lamp is dimmed, the intensity of the entire spectrum fades and the blue end disappears completely.

MATERIALS FOR PROCEDURE A

handheld, single-axis transmission diffraction gratings, one for each audience member (See Procedure A for description.)

tubular incandescent bulb with a linear filament (such as that for an aquarium lamp)

fixture to hold bulb vertically

dimmer or rheostat for lamp

MATERIALS FOR PROCEDURE B

compact discs, one for each audience member

tubular incandescent bulb with a linear filament (such as that for an aquarium lamp)

fixture to hold bulb vertically

dimmer or rheostat for lamp

MATERIALS FOR PROCEDURE C

video projection system*

handheld, single-axis transmission diffraction grating mounted as a slide, or a compact disc

tubular incandescent lamp with a linear filament (such as that for an aquarium lamp)

fixture to hold lamp vertically

dimmer or rheostat for lamp

---

* Various types of suitable video projection systems are described on page xxxiii. Some video projection systems are not capable of displaying a continuous spectrum but instead show a spectrum of several distinct colors. Test the system before using it to be sure that a continuous spectrum is displayed.

PROCEDURE A

Preparation

Construct a base with the lamp fixture and power cord incorporating the dimmer switch or rheostat as shown in this photograph.

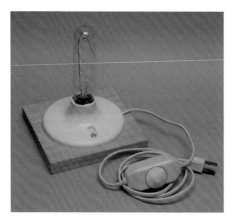

Each member of the audience should have a single-axis transmission diffraction grating. This grating material is available as a thin, flexible plastic sheet and needs to be mounted in a rigid holder for convenient use. Suitable gratings are available already mounted as 2-in × 2-in slides. Alternatively, bulk grating material can be used to prepare suitable gratings. A simple way to do this is to use a hole punch to make a round hole in an index card and tape a 1-cm × 2-cm piece of grating material over the hole (see Figure 1). (The audience members may also be instructed in how to make these themselves as part of the presentation.)

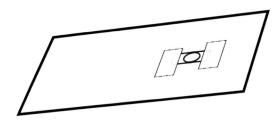

Figure 1.   Piece of grating material taped over hole punched into an index card.

Presentation

Distribute handheld transmission diffraction gratings to the audience.

Place the lamp where it is visible to all members of the audience. Turn on the lamp to full brightness. View the lamp through the diffraction gratings. When the lamp is visible straight through the slide, the spectrum is visible to either side of the filament. (If the spectrum appears above or below the lamp, rotate the grating by 90 degrees.) The spectrum appears continuous, showing all colors from red, through orange, yellow, green, and blue to violet. Dimming the room lights will make the observed spectrum more vivid.

While the audience members are viewing the spectrum, gradually dim the lamp. The blue end of the spectrum will dim more than the red end, and blue will disappear completely when the lamp is quite dim. When the lamp is dim, look directly at the filament. As the lamp is dimmed further, filament gradually yellows, then fades to orange, and finally becomes a dull red before going out.

PROCEDURE B

Preparation

Construct a base with the lamp fixture and power cord incorporating the dimmer switch or rheostat as shown in the photograph in Procedure A.

Presentation

Distribute compact discs to the audience.

Place the lamp where it is visible to all members of the audience. Turn on the lamp to full brightness. View the reflection of the lamp on the shiny side of a compact disc. When the white reflection of the lamp is visible, the spectrum of the filament is visible to either side of the lamp. The spectrum appears continuous, showing all colors from red, through orange, yellow, green, and blue, to violet. Dimming the room lights will make the observed spectrum more vivid.

While the audience members are viewing the spectrum, gradually dim the lamp. The blue end of the spectrum will dim more than the red end, and blue will disappear completely when the lamp is quite dim. When the lamp is dim, look directly at the filament. As the lamp is dimmed further, the filament gradually yellows, then fades to orange, and finally becomes a dull red before going out.

PROCEDURE C

Preparation

Construct a base with the lamp fixture and power cord incorporating the dimmer switch or rheostat as shown in the photograph in Procedure A.

Presentation

Display the lamp and its spectrum via video projection, using either a diffraction grating or a compact disc.

When using a transmission diffraction grating, tape the grating in front of the camera lens. Focus the camera on the filament and then turn the camera slightly to the side until the spectrum is displayed on the screen. Dim the room lights. The spectrum appears continuous, showing all colors from red, through orange, yellow, green, and blue, to violet.

When using a compact disc, display the spectrum of the filament via video projection by mounting the disc behind and somewhat to the side of the lamp (from the perspective of the audience). Mask the lamp from the audience, so only its reflection from the disc can be seen. Focus the camera on the reflection from the disc and tip the disc sideways a bit to display the spectrum on the screen. Dim the room lights. The spectrum appears continuous, showing all colors from red, through orange, yellow, green, and blue, to violet.

While the audience members are viewing the spectrum, gradually dim the lamp. The blue end of the spectrum will dim more than the red end, and blue will disappear completely when the lamp is quite dim. The filament, too, changes color as the lamp is dimmed. It starts white, but as the lamp is dimmed, its filament gradually yellows, then fades to orange, and finally becomes a dull red before going out.

HAZARDS

The lamp can get quite hot, so be careful to avoid touching it until it is off and has had time to cool.

DISPOSAL

Materials may be saved and used in future presentations.

DISCUSSION

Throughout most of human history, artificial light has been produced through heat. Burning wood in a fireplace gives off a glow whose color depends on how vigorously the wood is burning, which determines its temperature. When the fire is raging hot, the light it produces is yellowish, almost white. As the burning slows and the wood gradually cools, its color fades to dark yellow and then to orange, and then to a red glow. With the invention of oil lamps and candles, fires of nearly constant temperature, and therefore color, could be conveniently produced.

Any object, when heated hot enough, will emit visible light. The glow produced by a hot object is called *incandescence*. Iron heated in a blacksmith's forge becomes hot enough to glow red; the iron is said to be "red hot." When heated more, the iron begins to glow orange. Heated further, it glows yellow. If heated even more, its melting point is passed, and the molten iron glows white-hot.

With the development of electrical technology, it became possible to use electrical energy to heat objects. When a current passes through an electrically conductive material, such as a metal, the material gets warm, and some of the electrical energy is converted to heat. With some metals, enough heat can be produced to make the metal glow. This is what happens in the coils of an electric range and in the filament of an incandescent light bulb. Electrical energy flows through the coil or the thin wire filament, and each becomes so hot that it glows: the coils of an electric range become red-hot, but the filament of a standard light bulb can become white-hot.

All matter constantly emits electromagnetic radiation. Objects at room temperature emit radiation mainly in the infrared region of the spectrum that is invisible to us. As the temperature of an object increases, it emits more energy, and the wavelengths of the radiation emits become shorter. As the object gets hotter, more and more of the emitted radiation is visible, and the object begins to glow. This happens at around 700°C. First, it glows red, the color corresponding to the lowest energy of visible light. As the temperature increases, visible light of higher energy is emitted, from orange to yellow, to green, to blue, and to violet. When an object is hot enough to be emitting all visible wavelengths, its glow appears white, and we can say that it is "white hot." The emission of visible light by a hot object is called *incandescence*. (See the section *Black-body Radiation* in the introduction, page 23.)

The electromagnetic radiation emitted by objects is called *black-body radiation* because it does not include reflected radiation; a black body is black because it does not reflect light. The distribution of the wavelengths in black body radiation is determined by temperature. Thermometers that display temperature of an object at a distance detect black body radiation and convert it to a temperature reading. Physicians commonly use this kind of thermometer to measure an infant's body temperature by placing a probe into the child's ear. The thermometer detects the radiation from the ear drum and displays the corresponding temperature.

The filament in an incandescent bulb behaves like a black body. The color of its glow is directly related to its temperature, which can be determined by measuring the intensity of the glow at various wavelengths. A filament, however, cannot be heated above its melting point; otherwise it will break, and the lamp will "burn out." Standard incandescent bulbs use filaments made from tungsten because tungsten has one of the highest melting points among metals, 3422°C [1]. That melting point allows a tungsten filament to be heated to a higher temperature than filaments made of other metals, and at that temperature, the color of the emitted light is close to white.

Eventually, however, an incandescent lamp will burn out. One of the processes that causes it to do so is the evaporation of metal atoms from the surface of the filament. These atoms leave the filament and are eventually deposited on the inside of the glass bulb, darkening it. As the atoms evaporate, the filament becomes thinner. As the filament becomes thinner, more of the electrical energy traveling through it is converted to heat, and the filament becomes hotter. Eventually, some spot will become so hot that it melts and the filament breaks. Thinning of the filament by evaporation of metal atoms can be slowed if the atoms can be returned to the filament. This process is accomplished in some lamps by putting a tiny amount of a halogen, either bromine or iodine, inside the bulb. When the metal atom comes off the filament, it reacts with the halogen to form a gaseous tungsten halide. This gas diffuses throughout the bulb, but it does not become deposited on the glass, so the glass remains clear. When the tungsten halide gas nears the hot filament, it decomposes to halogen and tungsten metal, which returns to the nearby filament, although not likely the same place where it left the filament. The addition of halogen reduces the effect of evaporation from the filament, so the filament lasts longer and the lamp can be operated at a higher temperature to produce a whiter light.

Using a video projection system to display the spectrum in this demonstration presents an opportunity to point out the difference between physical and physiological color perception. (See the section *Color Perception* in the introduction, page 4.) Consider the color yellow observed directly as the lamp is dimmed and in the projected spectrum. The color perceived from the lamp is the physical color resulting from the action of the wavelengths in the yellow region of the spectrum on the pigments in the retina of the observers' eyes. The yellow color perceived in the video-projected spectrum, on the other hand, is produced by a combination of the red, green, and blue emissions in the projection system. This combination acting on the pigments in the observers' retinas provides the physiological perception of yellow, even though there are no yellow wavelengths in the projected light.

REFERENCE

1. D. R. Lide, ed., *CRC Handbook of Chemistry and Physics,* 90th ed., CRC Press: Boca Raton, Florida (2009).

12.3

Incandescence from the Combustion of Iron and of Zirconium

Emission of light and heat when materials burn (combine with oxygen) in air are familiar phenomena. If the concentration of one or the other of the reactants is increased, the phenomena can be spectacular. Such an increase can be accomplished either by using pure oxygen or by finely dividing the material to be burned, thus providing larger surface area for the reaction.

A heated ball of steel wool is plunged into a flask of oxygen gas, and the steel bursts into flame, glowing bright yellow, throwing sparks, and ultimately melting (Procedure A). Zirconium powder is sprinkled on a heated hot plate and the metal burns, producing bright white sparks (Procedure B).

MATERIALS FOR PROCEDURE A

> tank of oxygen gas, with valve and delivery tube
>
> 5-L round-bottom flask
>
> two solid cork stoppers to fit the round-bottom flask
>
> cork ring or similar stand for round-bottom flask
>
> glass boiling beads, or sand, enough to cover the bottom of the flask to a depth of about 8 cm
>
> 35-cm piece of stiff, straight wire, about 12 gauge (a section of wire coat hanger works well)
>
> steel wool, fine (00) or extra-fine (000) grade, ball about 5 cm in diameter
>
> Meker burner

MATERIALS FOR PROCEDURE B

> 1.5 g zirconium metal: powder, 100 mesh (particle size about 150 μm)
>
> source of nitrogen gas
>
> 500-mL filter flask
>
> one-hole stopper to fit filter flask
>
> warm-water bath large enough to hold filter flask
>
> 75-mm test tube
>
> funnel to fit test tube
>
> rubber stopper to fit test tube
>
> hot plate
>
> test-tube holder

PROCEDURE A

Preparation

Add glass beads or sand to the round-bottom flask to a depth of about 8 cm. Bend a hook about 2 cm in diameter into the end of the iron wire. Pierce the cork and insert straight end of the wire through the hole, so that when the cork is seated in the mouth of the flask, the hook is slightly above the flask's equator, as illustrated in Figure 1. Remove the hook-and-stopper assembly from the flask. Form a ball of steel wool about 5 cm in diameter and attach it to the hook.

Figure 1.   Flask with sand on bottom and wire hook suspended by a cork stopper.

Fill the flask with oxygen gas. Insert the tube from the tank of oxygen into the flask so the free end of the tube is near the bottom of the flask. Open the valve on the tank and allow oxygen to flow slowly into the flask. After a couple minutes, close the oxygen valve, remove the tube, and stopper the flask with a solid stopper.

Presentation

Loosen the stopper on the flask of oxygen. Ignite the burner. Holding the end of the iron-wire hook, suspend the steel wool in the flame of the burner. The wool will glow red-hot in spots and parts will melt. When there are glowing red spots in the steel wool, quickly remove the stopper from the flask of oxygen and insert the hot wool into the flask. (To enhance the visual effect, the room may also be quickly darkened at this point.) The steel wool will burst into flame, producing bright yellowish white sparks. Some of the wool may melt, forming glowing drops that fall off the hook and drip onto the sand on the bottom of the flask.

PROCEDURE B

Preparation

Finely divided zirconium metal is often supplied either coated with a liquid or submerged in a liquid, such as water or an alcohol, to keep atmospheric oxygen from reacting with it. For this demonstration this liquid must be removed. This can be accomplished by placing about 1.5 g of the damp zirconium powder into a 500-mL filter flask and spreading it across the bottom of the flask. Stopper the flask with a one-hole stopper and connect the side arm of the flask to a source of nitrogen gas. Adjust the nitrogen to a gentle flow through the flask. Warm the bottom of the flask by placing it in a warm-water bath. Allow the nitrogen to flow through the warmed flask for 24 h. With the nitrogen still flowing through the flask, pour the dry metal powder through a funnel into a 75-mm test tube. Seal the test tube with a rubber stopper.

Turn the hot plate on high and allow it to heat for at least 10 minutes before using in the presentation.

Presentation

Caution the audience that the reaction will produce very bright light. Grasp the test tube of zirconium powder with a test-tube holder and remove the stopper. Holding the tube at arm's length, quickly sprinkle all of the zirconium powder onto the heated hot plate while the room lights are dimmed. The zirconium will burn producing heat, sparks, and very bright light. Use all of zirconium, because any remaining in the test tube may react with atmospheric oxygen and become hot enough to ignite nearby flammable material.

HAZARDS

Glass beads or sand must be used in Procedure A to prevent drops of molten iron from breaking the flask.

The flask assembly should be allowed to cool for at least 10 minutes before it is disassembled for cleaning.

Finely divided zirconium metal reacts spontaneously with oxygen in the air at room temperature, releasing heat. If the heat raises the temperature of the zirconium, the rate of the reaction increases. Therefore, once the tube of zirconium metal powder is opened, it must be used within several seconds to avoid burns to the hands. To extinguish burning zirconium, use a type D fire extinguisher to put out fire—do NOT use water, a CO_2 fire extinguisher, or a halocarbon extinguishing agent with burning zirconium.

Heat from burning zirconium can cause pitting on the surface of the hot plate.

DISPOSAL

To clean the sand or glass beads, pour the contents of the flask into a pan, remove the lumps of iron, and discard them in a solid-waste receptacle. The sand or glass beads may be saved for repeated presentations of the demonstration.

The interior of the flask should be scrubbed with soapy water using a stiff brush.

The zirconium dioxide produced can be scraped from the surface of the hot plate with a putty knife or chisel. The scrapings may be discarded in a solid-waste receptacle.

DISCUSSION

In this demonstration, two metals react with oxygen and release enough energy to heat the metals to incandescence. In Procedure A, the metal is finely divided iron in the form of steel wool that is heated in air and plunged hot into an atmosphere of pure oxygen. In Procedure B, finely divided zirconium metal is sprinkled onto a heated hot plate, and the hot zirconium particles react with atmospheric oxygen. In both cases, the reaction of metal with oxygen is highly exothermic and releases enough energy to heat the metals to incandescence. In Procedure A, the hot iron produces a yellowish incandescence. In Procedure B, the zirconium is heated even more, becoming white-hot.

Iron reacts with oxygen to form iron oxide. At room temperature and in atmospheric oxygen, this reaction is quite slow. It takes a long time for an object made of iron to completely disintegrate to the mixture of iron oxides called rust, but the rate of the reaction can increase under certain conditions. Moisture and salt both accelerate the formation of rust. A large surface-to-volume ratio of the iron also speeds the reaction, as shown by the rapid rusting of steel wool compared to that of iron nails. An increased concentration of oxygen raises the reaction rate, as does an increase in temperature. In this demonstration, these three factors, large surface-to-volume ratio, high oxygen concentration, and elevated temperature, together accelerate the reaction to dramatic effect.

The reaction of iron with oxygen is highly exothermic, whether the product is iron(II) oxide or iron(III) oxide [2]. (In this demonstration, both are formed.)

$$Fe(s) + 1/2\ O_2(g) \longrightarrow FeO(s) \qquad \Delta H^\circ = -272\ kJ\ mol^{-1}$$

$$2\ Fe(s) + 3/2\ O_2(g) \longrightarrow Fe_2O_3(s) \qquad \Delta H^\circ = -824\ kJ\ mol^{-1}$$

The energy released by the reactions increases the temperature of the steel wool to the point where it is so hot that it glows white. The energy is also sufficient to cause remaining, unreacted iron to melt (the melting point of iron is 1535°C [1]). Small pieces of burning iron are ejected from the central mass, producing a sparkler effect. Glowing liquid iron also drips from the mass at the center of the flask. If it were not for the sand or beads at the bottom of the flask, the thermal shock of having molten iron strike the glass could cause the glass to shatter. Product oxides form a smoke inside the flask, and some of it collects as a reddish-brown film on the inner surface of the flask.

Zirconium, too, combines with oxygen in a highly energetic reaction, one that is even more exothermic than the combustion of iron [2].

$$Zr(s) + O_2(g) \longrightarrow ZrO_2(s) \qquad \Delta H^\circ = -1080\ kJ\ mol^{-1}$$

The concentration of oxygen in the air is sufficient to produce an oxidation that releases enough energy to raise the temperature of finely divided zirconium and particles of product zirconium dioxide to white heat. The melting point of zirconium, 1855°C, is higher than that of iron, and that of zirconium dioxide is higher still, 2700°C [1].

REFERENCES

1. D. R. Lide, ed., *CRC Handbook of Chemistry and Physics,* 90th ed., CRC Press: Boca Raton, Florida (2009).

12.4

Chemical Reactions That Produce Light

Many chemical reactions release enough heat to increase the temperature of reactants or products to incandescence. Some exothermic reactions also emit light by other means, as well. Chapter 1 in volume 1 of this series describes many such reactions.

1.15 Burning of Magnesium
1.16 Combustion under Water
1.17 Combustion of Cellulose Nitrate (Guncotton)
1.21 Reaction of Zinc and Sulfur
1.22 Reaction of Iron and Sulfur
1.23 Reaction of Sodium Peroxide and Sulfur
1.24 Reaction of Sodium Peroxide and Aluminum
1.25 Reaction of Sodium and Chlorine
1.26 Reaction of Antimony and Chlorine
1.27 Reaction of Iron and Chlorine
1.28 Reaction of Aluminum and Bromine
1.29 Reaction of White Phosphorus and Chlorine
1.30 Reaction of Red Phosphorus and Bromine
1.33 Reaction of Potassium Chlorate and Sugar
1.34 Decomposition of Ammonium Dichromate
1.35 Reaction of Potassium Permanganate and Glycerine
1.36 Thermite Reaction
1.37 Combustion of Magnesium in Carbon Dioxide
1.38 Pyrophoric Lead
1.41 Explosions of Lycopodium and Other Powders
1.42 Explosive Reaction of Hydrogen and Oxygen
1.43 Combustion of Methane

Described in chapter 6, "Chemical Behavior of Gases," in volume 2 of this series are several reactions involving gases that also produce light.

6.7   Explosiveness of Hydrogen
6.18 Preparation and Properties of Sulfur Dioxide
6.26 Catalytic Oxidation of Ammonia
6.27 Vapor Phase Oxidations

12.5

Emission Spectra from Gas-Discharge Lamps

The characteristic emission of only discrete wavelengths (colors) of light from excited elemental atoms is a key to understanding atomic structure.

Three options—a diffraction grating (Procedure A), a compact disc (Procedure B), and a video system with either of these diffraction techniques (Procedure C)—are presented for observing the visible emission spectrum from various gas-discharge lamps. Each spectrum is composed of a unique combination of colored lines.

MATERIALS FOR PROCEDURE A

handheld, single-axis transmission diffraction gratings, one for each audience member (See Procedure A for description.)

one or more gas-discharge tubes (such as Sargent-Welch spectrum tubes)

power supply for gas-discharge tubes

MATERIALS FOR PROCEDURE B

compact discs, one for each audience member [1]

one or more gas-discharge tubes (such as Sargent-Welch spectrum tubes)

power supply for gas-discharge tubes

MATERIALS FOR PROCEDURE C

video projection system* [2]

handheld, single-axis transmission diffraction grating mounted in a slide holder, or a compact disc

one or more gas-discharge tubes (such as Sargent-Welch spectrum tubes)

power supply for gas-discharge tubes

PROCEDURE A

Preparation

This demonstration needs to be done in a room that can be darkened as completely as possible, including closing blinds, shutting doors, and so on.

Each member of the audience should have a single-axis transmission diffraction grating. This grating material is available as a thin, flexible plastic sheet and needs to be mounted in a rigid holder for convenient use. Suitable gratings are available already mounted as

---

* Various types of suitable video projection systems are described on page xxxiii. Some video projection systems are not capable of displaying a continuous spectrum but instead show a spectrum of several distinct colors. Test the system before using it to be sure that a continuous spectrum is displayed.

2-in × 2-in slides. Alternatively, bulk grating material can be used to prepare suitable gratings. A simple way to do this is to use a hole punch to make a round hole in an index card and tape a 1-cm × 2-cm piece of grating material over the hole (see Figure 1). (The audience members may also be instructed in how to make these themselves as part of the presentation.)

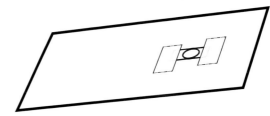

Figure 1.   Piece of grating material taped over hole punched into an index card.

Presentation

Distribute the hand-held transmission diffraction gratings mounted in slide holders to the audience.

Place the power supply for the gas-discharge lamp where the lamp is visible to all members of the audience. Insert one of the gas-discharge lamps into the power supply and turn it on. Dim the room lights. View the lamp through hand-held diffraction gratings mounted as slides. When the lamp is visible straight through the slide, the spectrum is visible to either side of the filament. (If the spectrum appears above or below the lamp, rotate the grating by 90 degrees.) The spectrum shows a set of lines of various colors depending on the gas inside the lamp.

Turn off the lamp and replace it with one containing a different gas. Turn it on, and again observe the spectrum. It will also show a set of lines of various colors, but the number and colors of the lines will be different from that of the first lamp.

Repeat this procedure with the other lamps, until the audience has observed the spectra from all of the lamps.

PROCEDURE B

Preparation and Presentation

Distribute compact discs to the audience.

Place the power supply for the gas-discharge lamp where the lamp is visible to all members of the audience. Insert one of the gas-discharge lamps into the power supply and turn it on. Dim the room lights. View the reflection of the lamp on the shiny side of a compact disc. When the reflection of the lamp is visible, its spectrum is visible to either side of the lamp. The spectrum shows a set of lines of various colors depending on the gas inside the lamp.

Turn off the lamp and replace it with one containing a different gas. Turn it on, and again observe the spectrum. It will also show a set of lines of various colors, but the number and colors of the lines will be different from that of the first lamp.

Repeat this procedure with the other lamps, until the audience has observed the spectra from all of the lamps.

PROCEDURE C

Preparation and Presentation

Display one of the lamps and its spectrum via video projection, using either transmission diffraction grating or compact disc.

When using a transmission diffraction grating, tape the grating in front of the camera lens. Focus the camera on the lamp and then turn the camera slightly to the side until the spectrum is displayed on the screen. The spectrum shows a set of lines of various colors depending on the gas inside the lamp.

When using a compact disc, display the spectrum of the lamp via video projection by mounting the disc behind and somewhat to the side of the lamp (from the perspective of the audience). Mask the lamp from the audience, so only its reflection from the disc can be seen. Focus the camera on the reflection from the disc and tip the disc sideways a bit to display the spectrum on the screen. The spectrum shows a set of lines of various colors depending on the gas inside the lamp.

Turn off the lamp and replace it with one containing a different gas. Turn it on, and again observe the spectrum. It will also show a set of lines of various colors, but the number and colors of the lines will be different from that of the first lamp.

Repeat this procedure with the other lamps, until the audience has observed the spectra from all of the lamps.

HAZARDS

The output of the power supply for the lamps is several thousand volts, so care should be exercised when it is on and powering a lamp. Always turn off the power supply before changing the lamps and any time when it is not needed.

DISPOSAL

Materials may be saved to be used for future demonstrations.

DISCUSSION

Throughout most of human history, artificial light has been produced through heat. However, at the beginning of the eighteenth century, when many investigators were studying the properties of gases and others were studying electrical phenomena, Francis Hauksbee discovered that when he placed an evacuated glass bulb containing a bit of mercury near an electrostatic generator, a glow developed inside the bulb. This is perhaps the first time that artificial light was produced without heat. As investigations of electricity advanced through the middle of the nineteenth century, it was discovered that when an electric current passed through a partially evacuated glass tube, a glow emanated from the tube. The production of light by passing electrical energy through a low-pressure gas is the basis for "neon" signs and all fluorescent lighting.

The pressure of the gas inside a gas-discharge lamp is about 6 to 30 torr (about 1% to 4% of atmospheric pressure). A high voltage, ranging from 10,000 to 40,000 volts, is applied across the metal electrodes at either end of the tube, causing electrons to be ejected from the negative electrode. These electrons collide with atoms of gas, and some of the energy of the electrons is transferred to the gas. The energy absorbed by the gas causes the valence electrons in the gas to jump temporarily to higher energy levels, and when these valence electrons return to lower energy levels, they emit energy in the form of light. The color of the light is determined by the difference between the high energy levels and the lower energy levels, and this difference is characteristic of the gas. (See the section *Emission Spectra* in the introduction, page 5.)

The pressure of gas must be low enough so that electrons ejected from the cathode can travel a significant distance before colliding with a gas molecule. However, if the gas pressure is too low, many electrons encounter no gas, and the intensity of the glow is diminished. The pressure is selected to give the brightest uniform glow. Alternating current is used with gas-discharge lamps, so that both electrodes are temporarily cathodes where the glow appears. With alternating current the glow appears throughout the gas.

Depending on its atomic composition, each gas emits certain wavelengths, producing different colors of the discharge lamps. Table 1 lists several gases available in gas discharge tubes, along with the color of the glow and the spectrum of the emitted light.

Table 1.   Color and spectra of gas discharges.

Gas	Color	Emission spectrum
Hydrogen	lavender	
Helium	white to orange	
Neon	red-orange	
Argon	violet to pale lavender-blue	
Krypton	gray off-white to green	
Xenon	gray or blue-gray to dim white	
Nitrogen	similar to argon but duller, pinker	
Oxygen	violet to lavender, dimmer than argon	
Mercury vapor	light blue, intense ultraviolet	
Sodium vapor	bright orange-yellow	

The most familiar gas discharge tubes are fluorescent lamps that contain mercury vapor and argon gas. The pressure of mercury vapor by itself is too low to produce much of a glow, so argon is added to increase the pressure. When an electron strikes an argon atom, the energy absorbed by it can be transferred to a mercury atom, and the mercury atom will then emit the energy. Most of the energy emitted by energized mercury atoms is in the ultraviolet range of the spectrum and is therefore invisible. To convert this energy to visible light, the interior of the tube is coated with a fluorescent material that absorbs the ultraviolet energy and reemits it as lower-energy visible light. The fluorescent material is usually a mixture of substances that are selected to produce light of the desired color, which in most fluorescent lamps is white.

Using a video projection system to display the spectrum in this demonstration presents an opportunity to point out the difference between physical and physiological color perception. (See the section *Displaying Small Phenomena to a Large Audience,* page xxxiii.) Consider the color yellow observed directly in a sodium vapor lamp and in the projected spectrum. The color perceived from the lamp is the physical color that results from the action of wavelengths in the yellow region of the spectrum on the pigments in the retina of the observers' eyes. The yellow color perceived in the video-projected spectrum, on the other hand, is produced by a combination of the red, green, and blue emissions in the projection system. This combination acting on the pigments in the observers' retinas provides the physiological perception of yellow, even though there are no yellow wavelengths in the projected light.

REFERENCES

1. R. C. Mebane and T. R. Rybolt, "Atomic Spectroscopy with a Compact Disc," *J. Chem. Educ.*, *69*(5), 401–402 (1992). The authors explain how to use a compact disc to view the emission spectrum from mercury-vapor street lamps.
2. F. Juergens, "Spectroscopy in Large Lecture Halls," *J. Chem. Educ.*, *65*(3), 266–267 (1988). The author describes a large diffraction grating (constructed from multiple sheets of transmission diffraction grating) that allows all members of the audience in a large lecture hall to observe emission spectra.

12.6

Colored Flames from Metal Ions

The emission of visible light from heated metal salts is responsible for the colors of fireworks. Heating metal salts in a flame also produces these colors and can be used to identify some metal ions.

When a mist of a metal-salt solution is sprayed into the flame of a Meker burner, a colored fireball appears (Procedure A) [1]. Burning a metal-salt solution in methanol produces a colored flame (Procedure B). Different metal-salt solutions produce different colors.

MATERIALS FOR PROCEDURE A

1 g of one or more of the following substances: sodium chloride, NaCl; lithium chloride, LiCl; calcium chloride, $CaCl_2$; strontium chloride, $SrCl_2$; barium chloride, $BaCl_2$; potassium chloride, KCl; copper(II) chloride, $CuCl_2$; boric acid, H_3BO_3

misting sprayer bottles, having a capacity of 100 to 250 mL, one for each of the substances to be used, plus one more

100 mL methanol, CH_3OH, for each of the substances, plus 100 mL more

Meker burner

fireproof surface, such as a sheet of cement board

MATERIALS FOR PROCEDURE B

0.2 g of one or more of the following substances: sodium chloride, NaCl; lithium chloride, LiCl; calcium chloride, $CaCl_2$; strontium chloride, $SrCl_2$; barium chloride, $BaCl_2$; potassium chloride, KCl; copper(II) chloride, $CuCl_2$; boric acid, H_3BO_3

140-mm diameter porcelain evaporating dishes, one for each of the substances

methanol, 20 mL for each of the substances

PROCEDURE A

Preparation

Prepare a solution of each of the available substances by combining about 1 g of solid with 100 mL of methanol. For most chlorides, not all of the solid will dissolve. Transfer each solution to a different spray bottle and label each bottle with the name of the salt it contains. Put 100 mL of pure methanol into another bottle and label it.

Immediately before presenting, test the bottles to make sure the nozzles produce a mist and do not leak.

Presentation

Perform this demonstration only in a well-ventilated area. Set the Meker burner on a non-flammable surface and light it. Dim the room lights. Hold the nozzle of the bottle about 25 cm from the flame, and as you pull it away, spray a fine mist of methanol into the flame to show the appearance of the flame when the methanol ignites.

With each of the chloride solutions, spray a fine mist into the flame and observe the color of the flame.

Pleasing effects can be created by spraying two or more simultaneously into the flame.

PROCEDURE B

Preparation

In each of the evaporating dishes combine about 0.2 g of each of the solids and 20 mL of methanol. Cover the dishes to prevent evaporation of the alcohol.

Presentation

Arrange the dishes on the display bench. Ignite the liquid in each dish. Dim the room lights. The flames rising from each dish will be colored by the material dissolved in the alcohol.

HAZARDS

Overspray will produce a film of flammable methanol on the surface around the burner. Care should be taken to minimize the amount of overspray by gently spraying only a fine mist into the flame. Do not spray a stream of liquid.

Methanol and metal salts can be irritating to mucous membranes. Spray so that no one inhales the mist.

DISPOSAL

The solutions may be stored for repeated presentation of the demonstration. To preserve the sprayer mechanisms, the solutions should be rinsed from them by inserting the tube into pure methanol and spraying several times into a sink. The mechanism can become clogged if the solution is not rinsed from the sprayer.

The solutions may be discarded by flushing them down the drain with water.

DISCUSSION

In a flame, each of the substances produces a characteristic color that is the combination of the various wavelengths of light emitted by the energetically excited metal ions in the flame. Table 1 displays the emission spectra of the various elements used in this demonstration, along with the colors they produce in a flame.

An interesting observation concerns the *perceived* green flame colors from barium, copper, and boron and the emission spectra in Table 1. There are emissions in the green wavelength region for barium and copper, but no green wavelength emission for boron. There are, however, prominent yellow and blue emissions from boron, and the eye *perceives* this combination as green. That is, the physiological color, green, is *constructed by the eye* from the physical colors, yellow and blue. (See the section *Color Perception* in the introduction, page 4, for a discussion.)

Table 1.   Flame color and emission spectra of the metal ions.

Gas	Color	Emission spectrum
Sodium	yellow	
Lithium	deep red	
Potassium	violet	
Calcium	red-orange	
Barium	yellowish green	
Strontium	red	
Copper	bluish green	
Boron	light green	

REFERENCE

1. K. A. Johnson and R. Schreiner, "A Dramatic Flame Test Demonstration," *J. Chem. Educ.*, *78*(5), 640–641 (2001).

12.7

Light-Emitting Diodes

Voltage and Temperature Effects

In a light-emitting diode (LED) the energy for light emission is provided by the potential (voltage) drop across the diode—the higher the energy of light emitted, the higher the required voltage. According to the Planck relationship, $E = hc/\lambda$, the energy of long-wavelength red light is less than that of short-wavelength blue light.

Red and yellow LEDs glow when connected to a 1.2-volt cell, whereas green and blue LEDs do not. When two cells are used, the green LED glows, and when three cells are used, the blue LED also glows (Procedure A). When different colored LEDs are immersed in liquid nitrogen, the color and intensity of their glow changes [1, 2] (Procedure B).

MATERIALS FOR PROCEDURE A

three 1.2-volt NiMH cells (larger D cells are more visible, but smaller AA cells also work)

battery holder to accommodate at least three of the NiMH cells (with accessible terminals to make connections to one, two, or three of the cells in series)

two small clip leads, i.e., a wire with alligator clips on both ends (one red and one black, or other different colors)

red LED with colorless lens

yellow LED with colorless lens

green LED with colorless lens

blue LED with colorless lens

Select a red LED with a 1.7-volt operating voltage, yellow and green of 2.1 volts, blue of 3.1 volts or higher. Test before using; LEDs vary in composition and not all will perform as indicated. Do not leave LEDs connected to higher voltages for more than a few seconds; the low-voltage (red) ones may burn out when connected to higher voltage. Colorless lenses are preferable to colored lenses because with colorless lenses, there is no question of the color of the light produced by the semiconductor.

MATERIALS FOR PROCEDURE B

9-volt battery, with clip

390-ohm, 1/4-watt resistor

two-socket connector to hold LED (e.g., Molex 22-01-3027 with two Molex contacts 08-05-0114)

two 10-cm lengths of insulated 22-gauge multistrand wire (preferably one with red insulation and the other with black)

two 2-cm pieces of heat-shrink tubing large enough to fit over resistor, or electrical tape

several LEDs of various colors

about 500 mL of liquid nitrogen

Styrofoam cup

PROCEDURE A

Preparation and Presentation

Insert one of the 1.2-volt cells into the battery holder.

Connect the black clip lead to the negative terminal of the battery and the red clip lead to the positive terminal, so that there is a 1.2-volt difference between the red and black leads.

Connect the free end of the black clip lead to the short lead of the red LED. Attach the free end of the red clip lead to the longer lead of the red LED. The LED will emit red light. Detach the clip leads from the LED.

Connect the free end of the black clip lead to the short lead of the yellow LED. Attach the free end of the red clip lead to the longer lead of the yellow LED. The LED will not emit light. Detach the clip leads from the LED. Repeat this with the green and blue LEDs, which will also not emit light.

Insert a second 1.2-volt cell into the battery holder, so that it is in series with the first. Attach the black clip lead to the negative terminal of the first cell and the red clip lead to the positive terminal of the second cell, so that there is a difference of 2.4 volts between the red and black leads.

Connect each of the four LEDs, one by one to the two clip leads, in the order red, yellow, green, and blue. The red LED will glow red, the yellow will emit yellow light, and the green will emit green light. The blue will not glow.

Insert a third 1.2-volt cell into the battery holder, so that it is in series with the first two. Attach the black clip lead to the negative terminal of the first cell and the red clip lead to the positive terminal of the third cell, so that there is a difference of 3.6 volts between the red and black leads.

Connect each of the four LEDs, one by one to the two clip leads, in the order red, yellow, green, and blue. All four will glow with 3.6 volts.

PROCEDURE B

Preparation

Assemble an LED circuit as illustrated in Figure 1. The circuit allows the connection of various different LEDs to a 9-volt battery in order to illuminate them. The circuit is made from a 9-volt battery clip, a 390-ohm, 1/4-watt resistor, and a connector designed to attach

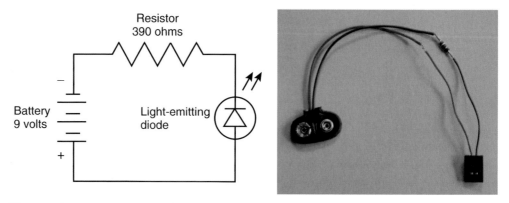

Figure 1. Diagram for LED circuit and photo of assembled circuit (for clarity, shown without the necessary insulation on the solder joints).

to a pair of pins on a circuit board. A surplus jumper cable can also be adapted. If heat-shrink tubing will be used for insulation, slip one 2-cm piece over each of the two leads of the battery clip. Solder the red cable of the battery clip to the red cable of the connector. Solder one lead of the resistor to the black cable of the battery clip and solder the other lead of the resistor to the black cable of the connector. Slide the heat-shrink tubing over the solder connections and shrink it in place; alternatively, the solder connections may be insulated by wrapping them with electrical tape.

Test the LED circuit by attaching a 9-volt battery to the battery clip and plugging an LED into the connector. The longer lead of the LED should be inserted into the connector at the red cable and the shorter lead at the black cable. The LED should glow.

Before presenting the demonstration, test each of the LEDs in the circuit at room temperature and at liquid-nitrogen temperature to see how each behaves. Fill the Styrofoam cup with liquid nitrogen. One by one, insert an LED into the connector and submerge the glowing LED into the liquid nitrogen. Most LEDs will become brighter as they cool, and the color of their glow will shift to shorter wavelengths. Some, however, will dim. For the demonstration, use those that brighten.

Presentation

Fill the Styrofoam cup with liquid nitrogen. Insert one of the LEDs into the connector and observe the color and brightness of its glow. Dim the room lights to make the glow more visible. Submerge the glowing LED into the liquid nitrogen. The glow of the LED will be visible through the cup, and as the LED cools, its glow will increase. The color of the glow may also change to a shorter wavelength. Remove the LED from the liquid nitrogen. Its brightness and color will return to the original as it returns to room temperature.

Repeat the procedure with LEDs of other colors.

HAZARDS

Liquid nitrogen (77 K = −196°C) can cause severe frostbite. Use care not to allow the liquid to contact skin and handle it with appropriate protective gloves.

DISPOSAL

Allow any extra liquid nitrogen to boil away. The circuit, LEDs, and other components may be kept for future use.

DISCUSSION

Light-emitting diodes are ubiquitous in our electronic age, producing everything from the single-color glow of a power indicator to the full-color motion of video screens in arenas and on billboards. An LED is a semiconductor device that converts electrical energy to the energy of electromagnetic radiation. (See the section *Light-Emitting Diodes and Injection Laser Diodes* in the introduction, page 49, for a discussion of the principle of these devices.) LEDs can produce visible light of many different wavelengths and therefore, colors, as well as invisible infrared and ultraviolet radiation. Although as a group LEDs produce radiation over a wide wavelength range, a single LED produces only a narrow band of radiation. Thus, to produce red light and green light, a device that uses LEDs must use at least two different LEDs, one that produces red light and a second that produces green light. The two LEDs are sometimes manufactured to appear as one LED with three leads.

Light-emitting diodes that emit different colors of light require different minimum voltages to operate. In general, blue LEDs require a higher minimum voltage than red LEDs because a photon of blue light carries more energy than a photon of red light. A typical blue LED emits light with a wavelength of around 430 nm. The energy of a photon of this light is

$$E = \frac{hc}{\lambda} = \frac{(6.6 \times 10^{-34} \text{ J s})(3.0 \times 10^8 \text{ m s}^{-1})}{430 \times 10^{-9} \text{ m}} = 4.6 \times 10^{-19} \text{ J}$$

This is the energy of one photon, which corresponds to the energy lost by an electron as it moves through the LED. This energy can be correlated to the voltage that is needed to operate the LED, because voltage corresponds to the energy possessed by a number of electrons having a charge of one coulomb. That is, 1 volt = 1 J coulomb^{-1}. A mole of electrons has a charge of 96,485 coulomb, so the voltage that is needed to produce blue light is

$$V = \frac{(4.6 \times 10^{-19} \text{ J electron}^{-1})(6.02 \times 10^{23} \text{ electron mol}^{-1})}{96485 \text{ coulomb mol}^{-1}} = 2.87 \text{ J coulomb}^{-1} = 2.87 \text{ volt}$$

Thus, a blue LED requires about 2.9 volts to emit blue light. A red LED emits light with wavelengths around 660 nm, which corresponds to a voltage of about 1.9 volts. Light-emitting diodes that emit colors of intermediate energy require intermediate voltages to operate. (The equation above can be used to determine the value of Planck's constant from the voltage required to produce light of a particular wavelength in a simple laboratory exercise using LEDs [3].)

A diode is an electronic device that conducts electricity in only one direction as long as the voltage is not too high. This property distinguishes LEDs from ordinary incandescent and fluorescent lamps, which glow when electricity flows in either direction. Therefore, to obtain light from an LED, direct current must be used, rather than the alternating current used with common lamps. An LED will glow when connected to a battery in one direction but will not glow when the connection is reversed.

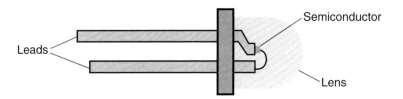

An LED is a semiconductor device. A semiconductor is a material that conducts electricity neither well, as a metal does, nor very poorly, as an insulator does. A metal conducts a current when even the tiniest voltage is applied to it, whereas an insulator conducts only when an extremely high voltage is applied and some require tens of thousands of volts. A semiconductor does not conduct when a tiny voltage is applied, but it does conduct when a moderate voltage, sometimes only a fraction of a volt, is applied.

The difference in electrical conductivity between conductors, insulators, and semiconductors can be explained by band theory, which is derived from molecular orbital theory by treating a single crystal of solid as a single, giant molecule—a macromolecular crystal.

According to molecular orbital theory, the number of molecular orbitals in a molecule is the sum of the number of atomic orbitals of its constituent atoms. The valence orbitals of the atoms are the ones most involved in bonding, and molecular orbital theory concentrates on those when describing bonding in molecules.

A qualitative view of molecular orbital theory applied to a molecule that contains two atoms, each with one atomic orbital involved in forming a bond with the other atom, shows the molecule to have two molecular orbitals. One of these molecular orbitals is of lower energy than the atomic orbitals, and the other is of higher energy. The lower-energy orbital is called a *bonding orbital,* and the higher-energy orbital is called an *antibonding orbital.* The electrons from the atomic orbitals occupy the lower-energy atomic orbitals, the bonding orbitals.

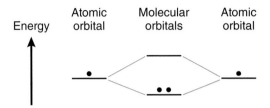

As the two bonded atoms become closer, the energy of the bonding orbital decreases, while the energy of the antibonding orbital increases, leading to a larger difference between the energy of the bonding orbitals and the energy of the antibonding orbitals.

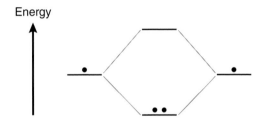

In a crystalline, macromolecular solid that contains many atoms, there are many atomic orbitals involved in forming many molecular orbitals. The atoms that are close to each other participate in molecular orbitals, forming low-energy bonding orbitals. Atoms that are farther apart participate in bonding orbitals whose energies are not quite as low. Atoms still farther apart have bonding orbitals of still slightly higher energy. As a result, in a large, crystalline solid, there are many bonding orbitals differing slightly in energy from each other. Together, these bonding orbitals form a band of closely spaced orbitals, hence the name "band theory." The situation is the same with the antibonding orbitals. In a large, crystalline solid, the antibonding orbitals occupy a band of energies. The situation is represented in this diagram.

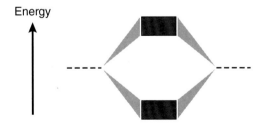

In many metals, the number of valence electrons is less than the number of valence orbitals. Therefore, the bonding orbitals in a metal are not filled. Because the energy difference between the various bonding molecular orbitals is extremely small, it is easy for the electrons to move from one orbital to another. Therefore, metals readily conduct electricity.

In nonmetals, the number of valence electrons is generally greater than the number of valence orbitals, and the extra electrons begin filling the antibonding orbitals. The occupation of many antibonding orbitals weakens the macromolecular structure, promoting the formation of discrete, smaller molecules. Consequently, most solid, crystalline nonmetals, such as iodine, are not macromolecular structures, but rather are arrays of individual molecules (I_2 in the case of iodine).

Some substances have the same number of electrons as orbitals. In a macromolecular crystal of these substances, the bonding molecular orbitals are filled, while the antibonding molecular orbitals are empty. Some of these substances are insulators, some are conductors, and some are semiconductors. Band theory is particularly useful for explaining the differences among these types of substances.

An example of a substance with an equal number of valence electrons and orbitals is carbon. Carbon atoms are relatively small, and in the densest form of carbon, diamond, the atoms are relatively close together. Therefore, the gap in energy between the bonding and antibonding molecular orbitals is large. The energies of the bonding orbitals are low, and the bonds are strong. All bonding orbitals are filled, so electrons cannot move from one to another. The strong bonds produce a solid that is very hard, and the immobile electrons produce a solid that is resistant to the flow of electric current. It takes electrons with very high energy to move from the bonding orbitals to the antibonding orbitals. Such a high-energy electron in the nearly empty antibonding orbitals can move from one antibonding orbital to another. The vacancy in the bonding orbitals can also appear to move, as electrons move from occupied bonding orbitals to the vacancy. Such a vacancy is called a *hole,* and holes move in the bonding orbitals the way electrons move in the antibonding orbitals. Thus, when a very high voltage is applied to normally nonconductive diamond, it can be made to conduct.

In the same periodic group as carbon is silicon. In solid, crystalline silicon the atoms are arranged in the same pattern as carbon atoms in diamond. However, silicon atoms are larger than carbon atoms, so their atomic orbitals do not overlap as much as those of carbon. This means that the band gap between the bonding and antibonding orbitals in silicon is smaller than that in diamond. Therefore, silicon can be made to conduct electricity by applying a smaller voltage than that required for diamond. The voltage is small enough that silicon is classified as a semiconductor.

Other network solids have structures similar to those of silicon and diamond. The element germanium, which is directly below silicon in the periodic table, has a similar structure. And so does the compound gallium arsenide, GaAs. The element gallium is just to the left of germanium on the periodic table, and arsenic is just to the right. A germanium atom contains one fewer valence electron than germanium, and an arsenic atom contains one more. On average, then, the atoms in gallium arsenide have the same number of valence electrons as germanium, and the solid attains the same structure, with a similar band gap.

The compound gallium arsenide can be prepared in such a way that there is a slight excess of gallium. Because gallium has fewer valence electrons, a few vacancies, holes, are created in the bonding orbitals. If a voltage is applied to this solid, electrons can be forced into the antibonding orbitals, as they are in silicon, forcing current to flow. Because there are some holes in the low-energy bonding orbitals, the electrons in the antibonding orbitals can lose energy by moving into these holes. This energy can be released in the form of light. The energy of the light emitted corresponds to the energy difference between the bonding and antibonding orbitals, that is, to the band gap.

The size of the band gap depends on the composition of the semiconductor. By adjusting the composition of the semiconductor, the band gap can be adjusted, and so can the energy (color) of the light emitted. The temperature of the semiconductor also affects the magnitude of the band gap. As the temperature decreases, the atoms get closer together and the band gap grows. Thus, when an LED is chilled, the energy of the light it emits increases. Therefore, the color shifts from the red end of the spectrum toward the blue—red can become orange, yellow can become green, and so forth. The emitted light also becomes brighter, because the conductivity increases as the temperature goes down. More electrons produce more light.

Not all LEDs undergo a shift toward the blue with decreasing temperature. The blue shift occurs if other factors, such as atomic arrangement, do not change. However, many solids, including semiconductors, go through subtle phase changes as the temperature changes. These phase changes correspond to different arrangements of the atoms in the solid, and different arrangements lead to different band gaps and therefore to different behavior and color. Some LEDs will show a color shift to the red, while others will stop glowing completely. Thus, it is necessary to test the LEDs to be used in this demonstration to be sure that they are free from these complicating factors.

REFERENCES

1. M. G. D. Baumann, J. C. Wright, A. B. Ellis, T. Kuech, and G. C. Lisensky, "Diode Lasers," *J. Chem. Educ., 69*(2), 89–95 (1992).
2. G. C. Lisensky, R. Penn, M. J. Geselbracht, and A. B. Ellis, "Periodic Properties in a Family of Common Semiconductors," *J. Chem. Educ., 69*(2), 151–156 (1992).
3. P. J. O'Connor and L. R. O'Connor, "Measuring Planck's Constant Using a Light Emitting Diode," *The Physics Teacher, 12*(7), 423–425 (1974).

12.8

Electrogenerated Chemiluminescence

When an electric current passes through an electrolyte solution, reactions must occur at the electrodes—oxidations at the anode and reductions at the cathode. If either (or both) of these reactions results in production of an electronically excited species, light may be emitted as it returns to the ground state. This emission resulting from passing an electric current through the solution is called *electrogenerated chemiluminescence.*

When a voltage is applied across two electrodes in an orange solution, an orange glow appears at the positive electrode (anode).

MATERIALS

0.21 g sodium dihydrogen phosphate monohydrate, $NaH_2PO_4 \cdot H_2O$

4.0 g disodium hydrogen phosphate heptahydrate, $Na_2HPO_4 \cdot 7H_2O$

100 mL distilled water

0.075 g tris(2,2′-bipyridine)ruthenium(II) chloride hexahydrate, $Ru(bpy)_3Cl_2 \cdot 6H_2O$

0.2 g ethylenediaminetetraacetic acid disodium salt dihydrate (disodium EDTA dihydrate), $Na_2C_{10}H_{14}N_2O_8 \cdot 2H_2O$

250-mL widemouthed jar with screw cap, or 250-mL beaker

two screen electrodes, platinum or high-chromium stainless steel, about 3 cm × 15 cm

3-volt battery pack (This may be assembled from two C or D cells in a holder.)

two clip leads, i.e., a wire with alligator clips on both ends

video projection display (for presentation to a large audience)*

PROCEDURE [1]

Preparation

In a 250-mL beaker, dissolve 0.21 g $NaH_2PO_4 \cdot H_2O$ and 4.0 g $Na_2HPO_4 \cdot 7H_2O$ in 100 mL of distilled water. This produces a solution that is 0.015 M in $H_2PO_4^-$ and 0.15 M in HPO_4^{2-}, which is a buffer solution having a pH of 7.85.

In the buffer solution, dissolve 0.075 g $Ru(bpy)_3Cl_2 \cdot 6H_2O$. This produces a solution in which the concentration of $Ru(bpy)_3^{2+}$ is 0.0010 M. In this solution, dissolve 0.2 g of disodium EDTA dihydrate ($Na_2C_{10}H_{14}N_2O_8 \cdot 2H_2O$). This produces a sodium EDTA concentration of 0.005 M.

Pour the electrolyte solution into a 250-mL wide-mouthed jar (or leave it in the 250-mL beaker). Use plastic

---

* Various types of suitable video projection systems are described on page xxxiii.

clamps to attach the electrode screens to opposite sides of the jar with the screens extending to the bottom of the solution.

Presentation

Display the jar (or beaker) with its mounted electrodes to the audience. With a clip lead, attach one of the electrodes to the negative terminal of the 3-volt battery pack. Dim the room lights and allow about 30 seconds for the viewers' eyes to become dark adjusted. Connect the remaining electrode to the positive terminal of the battery pack. An orange glow will appear at the edges of the positive electrode and gradually spread across the electrode. After the glow covers the entire electrode, gently swirl the electrolyte solution. The glow will disappear while the solution is swirled. Allow the solution to rest, and the glow will gradually reappear.

HAZARDS

Tris(2,2′-bipyridine)ruthenium(II) chloride hexahydrate can cause eye and skin irritation on contact, so gloves should be worn when it is handled.

Ethylenediaminetetraacetic acid disodium salt can cause eye and skin irritation on contact, so gloves should be worn when it is handled.

DISPOSAL

The electrolyte may be stored in the sealed jar for repeated presentations of the demonstration.

To dispose of the solution, leave the container open in a chemical hood until the solvent, water, has evaporated. Discard the residue in a container for organic waste.

DISCUSSION

This demonstration shows that light can be generated when an electric current passes through a solution thereby causing electron-transfer reactions at the electrodes. In order for this to happen, the reactions that occur at one of the electrodes must result in the production of an electronically excited species. In this demonstration, the electronically excited species is a complex ion containing ruthenium(II) bonded to three 2,2′-bipyridyl bidentate ligands, $Ru(bpy)_3^{2+}$.

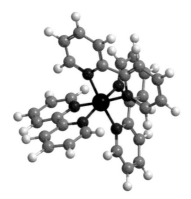

Figure 1.   Structure of tris(2,2′-bipyridine)ruthenium(II), $Ru(bpy)_3^{2+}$.

The orange glow observed in this demonstration results when an excited-state $Ru(bpy)_3^{2+}$ ion relaxes to the ground state by emitting a photon:

$$Ru(bpy)_3^{2+*} \longrightarrow Ru(bpy)_3^{2+} + h\nu \text{ (orange)} \qquad (1)$$

This, then, is a form of chemiluminescence, a process in which a chemical reaction generates an excited-state product that gives off light. The chemical reaction in this case is an electrochemical one, so electrogenerated chemiluminescence (ECL) is an accurate description of the process. That this is the source of the glow is indicated by the fact that the spectrum of the electrogenerated glow is nearly identical to the luminescence spectrum of $Ru(bpy)_3^{2+}$ that is produced when it is excited with 500 nm light [2, 1].

The excited $Ru(bpy)_3^{2+*}$ can be produced by at least three pathways [3]. The ratio of the ECL from each pathway depends on the concentrations of the reactants, the pH of the solution, and the material from which the electrodes are made. One pathway produces the excited complex in a reaction between an oxidized and a reduced form of the complex.

$$Ru(bpy)_3^{3+} + Ru(bpy)_3^{+} \longrightarrow Ru(bpy)_3^{2+*} + Ru(bpy)_3^{2+} \qquad (2)$$

The luminescent reaction takes place at the positive electrode, the anode, where oxidation is the electrode reaction. The $Ru(bpy)_3^{3+}$ can be produced electrochemically by oxidation of $Ru(bpy)_3^{2+}$ at the anode.

$$Ru(bpy)_3^{2+} \longrightarrow Ru(bpy)_3^{3+} + e^- \qquad (3)$$

However, to form $Ru(bpy)_3^{+}$ from $Ru(bpy)_3^{2+}$ requires a reduction reaction. To produce this change in the vicinity of the anode requires the use of another compound, called a *coreactant*. Some commonly used coreactants in studies of this reaction include tertiary alkyl amines, the most common of which has been tri-*n*-propyl amine, $(CH_3CH_2CH_2)_3N$ [4]. The amine is oxidized to a radical cation at the anode.

$$(CH_3CH_2CH_2)_3N \longrightarrow (CH_3CH_2CH_2)_3N^{\cdot+} + e^- \qquad (4)$$

The cation may lose a hydrogen ion, leaving a radical.

$$(CH_3CH_2CH_2)_3N^{\cdot+} \longrightarrow (CH_3CH_2CH_2)_2NC^{\cdot}HCH_2CH_3 + H^+ \qquad (5)$$

The radical reacts with $Ru(bpy)_3^{2+}$, reducing it, and forming an amine product P, that ultimately loses one of the propyl groups as an aldehyde. (The amine is a sacrificial reactant in this system.)

$$(CH_3CH_2CH_2)_2NC^{\cdot}HCH_2CH_3 + Ru(bpy)_3^{2+} \longrightarrow Ru(bpy)_3^{+} + P \qquad (6)$$

The sum of reactions (3), (4), (5), (6), and (1) produces the excited state ruthenium complex dication, $Ru(bpy)_3^{2+*}$, that emits the orange glow as it returns to the ground state.

In a second pathway to the excited-state ruthenium complex, the oxidized complex from reaction 3 reacts with the radical from reaction 5 to produce the excited state ruthenium complex and the amine degradation product, P.

$$(CH_3CH_2CH_2)_2NC^{\cdot}HCH_2CH_3 + Ru(bpy)_3^{3+} \longrightarrow Ru(bpy)_3^{2+*} + P \qquad (7)$$

In a third pathway to the excited-state ruthenium complex, the reduced complex from reaction 6 reacts with the radical cation from reaction 4 to produce the excited state ruthenium complex and reform the tri-*n*-propyl amine [3].

$$(CH_3CH_2CH_2)_3N^{\cdot+} + Ru(bpy)_3^{+} \longrightarrow (CH_3CH_2CH_2)_3N + Ru(bpy)_3^{2+*} \qquad (8)$$

Note that reaction 8 competes with reaction 5 for the fate of the amine radical cation, $(CH_3CH_2CH_2)_3N^{\cdot+}$, produced in reaction 4. This competition helps to explain the pH dependence of the ECL emission in this system.

The tertiary alkyl amine used in this demonstration is the disodium salt of EDTA.

$$NaO_2CCH_2-N-CH_2CH_2-N-CH_2CO_2Na$$

with CH_2 and CO_2H branches on each nitrogen.

Presumably this amine functions in a manner analogous to tri-*n*-propyl amine. The EDTA salt is used instead of the amine because it is readily available, very soluble in water, odorless, and generally free of hazards.

The ECL emission depends on the kind of electrodes used, because the relative ease of oxidation of the ruthenium complex and coreactant depends on the electrode material. The electrodes must be inert in the solution at the voltages used. Platinum is inert. High-chromium stainless steel also works and is much less expensive. To be inert, the stainless steel must be high in chromium, such as 18-8 grade, which is 18% chromium and 8% nickel. Steel containing less chromium is not inert, and the corrosion reaction at the anode interferes with the reactions that produce luminescence. Mesh is used so that the reactive area is large. The large surface area and small volumes within the mesh provide more reacting species formed in close proximity to one another, which results in a brighter glow. A convenient source for suitable electrodes is the fine mesh used in the heads of some electric shavers.

Analytical techniques based on ECL, many of which use the $Ru(bpy)_3^{2+}$ system, are employed in a large number of applications including clinical diagnosis, forensic investigations, environmental and pharmaceutical studies, and detection of biological warfare agents. The principle is pretty much the same in all cases: somehow associate the compound of analytical interest (the analyte) with the ECL reagent and then subject the reagent to an electric current, as in this demonstration, and quantitatively measure the emitted light, which is proportional to the amount of analyte present.

One specific example from a very large number of studies on ECL-based analysis is a highly sensitive technique for detecting DNA [5]. One method for detecting a particular single-stranded DNA base sequence (the target) is to hybridize it (form a double helix) with the complementary strand (the probe) that is then removed from the sample and brings along the target. In the study described in this reference, the target DNA and any others in the sample are chemically attached to microscopic (1 μm) plastic beads that have each been impregnated with more than a billion molecules of a water-insoluble $Ru(bpy)_3^{2+}$ salt. Conditions for the chemical attachment are chosen so that very few DNA strands are attached to any bead. The probe DNA strands are attached to another larger (10 μm) kind of bead that is magnetic. Many of the probe DNAs are attached to each magnetic bead. When the two kinds of beads are mixed in solution under appropriate conditions for DNA hybridization, the target DNA bound to the plastic beads interacts with and binds (double helically)

Figure 2.   Scanning electron micrograph of the plastic beads impregnated with $Ru(bpy)_3^{2+}$ and attached via DNA hybridization to the magnetic beads used to separate them from the hybridization solution.

to probe DNAs bound to the magnetic beads. Thus plastic beads with bound target DNA get stuck to the magnetic beads, but others without a bound target DNA do not stick. A magnet is used to remove the magnetic beads and their cargo of plastic beads (with target DNA strands) from the solution, Figure 2.

The mixture of beads is put in an organic solvent that dissolves the beads and releases the $Ru(bpy)_3^{2+}$ into the solution. Then the solution is subjected to an electric current that produces ECL, and the emission is quantitatively measured. Since each plastic bead contains more than a billion ECL-reagent molecules, there is a factor of about a billion amplification of the number of DNA target strands in the sample. In the experiments reported, the range of detection of DNA techniques was from about a femtomole (10^{-15} mol) to 10 nanomoles, a 10 million-fold range. ECL techniques are highly sensitive and selective and are relatively easy to use, which explains their wide versatility.

REFERENCES

1. C. Bohrman and M. W. Tausch, "Elektrochemolumineszenz mit unbedenklichen Chemi-calen," *Chemie in Unserer Zeit, 36*(3), 164–167 (2002).

2. N. E. Tokel and A. J. Bard, "Electrogenerated Chemiluminescence. IX. Electrochemis-try and Emission from Systems Containing Tris(2,2′-bipyridine)ruthenium(II) Dichlo-ride," *J. Amer. Chem. Soc., 94*, 2862–2863 (1972).

3. W. Miao, J.-P. Choi, and A. J. Bard, "Electrogenerated Chemiluminescence 69: The Tris(2,2′-bipyridine)ruthenium(II), $(Ru(bpy)_3^{2+})$/Tri-*n*-propylamine (TPrA) System Re-visited: A New Route Involving $TPrA^{•+}$ Cation Radicals," *J. Am. Chem. Soc., 124*, 14478–14485 (2002).

4. J. K. Leland and M. J. Powell, "Electrogenerated Chemiluminescence: An Oxidative-Reduction Type ECL Reaction Sequence Using Tripropyl Amine," *J. Electrochem. Soc., 137*, 3127–3131 (1990).

5. W. Miao and A. J. Bard, "Electrogenerated Chemiluminescence. 77. DNA Hybridiza-tion Detection at High Amplification with $[Ru(bpy)_3]^{2+}$-Containing Microspheres," *Anal. Chem., 76*, 5379–5386 (2004).

12.9

Chemiluminescence

Chemiluminescence is the release of light energy by chemical reactions. All chemiluminescent reactions that produce visible and/or ultraviolet radiation are oxidation-reduction reactions. (See the section *Chemiluminescence* in the introduction, page 61.) In addition to the demonstrations of chemiluminescence in this volume, an entire chapter devoted to chemiluminescence in volume 1 in this series contains the following demonstrations.

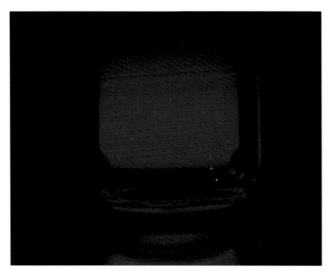

Figure 1.   The chemiluminescence of singlet molecular oxygen, Demonstration 2.1.

2.1   Singlet Molecular Oxygen
2.2   Lightsticks
2.3   Sensitized Oxalyl Chloride Chemiluminescence
2.4   Oxidations of Luminol
2.5   Luminol Chemiluminescent Clock Reactions
2.6   Two-Color Chemiluminescent Clock Reaction
2.7   Hydrogen Peroxide Oxidation of Lucigenin
2.8   Air Oxidation of White Phosphorus
2.9   Air Oxidation of Tetrakis(dimethylamino)ethylene
2.10 Chemiluminescence of Tris(2,2′-bipyridyl)ruthenium(II) Ion
2.11 Explosive Reaction of Nitric Oxide and Carbon Disulfide

12.10

Chemiluminescence from the Explosive Reaction of Nitrous Oxide and Carbon Disulfide

Many familiar oxidation reactions—for example, burning natural gas in a Bunsen burner or exploding hydrogen-oxygen mixtures—involve molecular oxygen as the oxidant. However, if the species to be oxidized is very reactive, other sources of oxygen for a flame or explosion will work just as well.

When a mixture of nitrous oxide gas and carbon disulfide vapor is ignited at the top of a large glass tube, the flame travels rapidly down the tube, producing a bright blue light and a loud noise.

MATERIALS

cylinder of nitrous oxide gas, N_2O (with valve and rubber tubing)

2.5 mL carbon disulfide, CS_2

glass tube, 58 mm in diameter and 122 cm (4 feet) long

two rubber stoppers to fit glass tube

ring stand and iron ring

pneumatic trough

syringe, 3-mL capacity or larger, with needle

PROCEDURE

Preparation

Stopper one end of the glass tube and fill the tube with water. Stopper the other end. Immerse one end of the tube in a bucket half-filled with water and remove the stopper from that end. Support the tube vertically so that it will not fall over.

Attach a piece of flexible tubing to the outlet of the nitrous oxide (N_2O) tank. Insert the free end of the tubing into the open end of the water-filled glass tube. Open the valve of the nitrous oxide tank and fill the glass tube with nitrous oxide gas by water displacement. As the water is displaced, it may need to be removed to prevent the bucket from overflowing. When the glass tube is full of gas, stopper it under water. Remove the tube from the bucket and place it upright through the iron ring on a ring stand.

Under a hood, draw 2.5 mL of carbon disulfide (CS_2) into a syringe. Gently loosen but do not remove the upper stopper of the tube. Insert the syringe needle between the stopper and the wall of the tube. Quickly inject all of the carbon disulfide into the tube and remove the syringe.

With one hand, hold the loosened stopper in place and do not seal tightly. Keeping one hand on the loosened stopper and the other on the second stopper, remove the tube from the stand and invert it several times to ensure that the carbon disulfide is vaporized and well mixed with the nitrous oxide gas. During the mixing, pressure will build inside the tube. Release the pressure by slightly opening and quickly shutting the loosened stopper. Continue inverting the tube and releasing the pressure until you no longer hear the gas escaping from the tube. Tighten the stopper and place the tube through the ring on the ring stand. The tube is now ready for the demonstration. Present the demonstration as soon as possible.

Presentation

Keep your hands and face away from the top of the tube, because the reaction is quite violent. For maximum visual effect, darken the room lights. To ignite the mixture, light a match, carefully remove the upper stopper, and drop the burning match into the tube. A bright blue flame will travel rapidly down the tube through the mixture of gases. As the flame travels down, a roaring sound is produced. With the room lights turned on, a yellow coating will be visible on the inside of the tube.

HAZARDS

Nitrous oxide is narcotic at high concentrations and should be used only in well-ventilated areas.

Carbon disulfide is extremely flammable and toxic. The explosive range is 1 to 50% (v/v) in air [1]. The vapor is extremely irritating to the eyes and malodorous.

In addition to their individual hazards, these gases form an explosive mixture, as shown by this demonstration.

DISPOSAL

After the reaction, the tube may contain unreacted gases as well as gaseous products. Therefore, it should be stoppered and then opened under a hood. After the gases have been vented, the solid remaining on the walls of the tube can be removed by scrubbing with a long-handled brush and a detergent solution. The residue is most easily removed immediately after the reaction.

DISCUSSION

This demonstration is a variation of a Demonstration 1.44, Explosive Reaction of Nitric Oxide and Carbon Disulfide, in volume 1 of this series. In the procedure described here, nitrous oxide, N_2O, is used in place of nitric oxide, NO. A significant reason for the replacement is that nitric oxide reacts rapidly with atmospheric oxygen to form extremely toxic nitrogen dioxide gas. The hazards associated with nitrous oxide are far less. Nevertheless, it is narcotic in high concentrations, hence its common name "laughing gas." It also supports combustion, as this demonstration dramatically shows.

The combustion of carbon disulfide in nitrous oxide produces a number of products. One of the products is elemental sulfur, which forms a yellow deposit on the interior of the tube. Another product is sulfur dioxide, whose odor can be detected after the reaction. The bright blue emission accompanying the reaction corresponds to a continuum from 490 nm to 310 nm. The emission has been ascribed to a triplet-singlet transition in SO_2 [2]. Generation of triplet SO_2 may follow this reaction path:

$$SO(^3\Sigma) + N_2O(^1\Sigma) \longrightarrow SO_2(^3\Sigma) + N_2(^1\Sigma)$$

The formation of SO probably results from the abstraction of O from N_2O by CS_2.

At room temperature nitrous oxide is a colorless and odorless gas. Carbon disulfide is a liquid that boils at 46°C. It is quite volatile, having a vapor pressure of 400 torr at 28°C.

REFERENCES

1. M. J. O'Neil, ed., *The Merck Index,* 14th ed., pp. 293–294, Merck: Whitehouse Station, New Jersey (2006).
2. W. Roth and T. H. Rautenberg, *J. Phys. Chem.*, *60,* 379–81 (1956).

Properties of Light

12.11

The Conversion of Light Energy to Thermal Energy

Although almost all the light that strikes a transparent substance like water is transmitted, a small amount is always absorbed, and if the water is colored, even more light is absorbed. The energy of the light that is absorbed does not disappear but instead is converted to other forms, most commonly into increased motion of the molecules, which manifests itself as an increase in temperature of the liquid.

Two beakers, one containing colored water and the other colorless, are set under a bright lamp for 15 to 30 minutes. The temperature of the colored water rises a few degrees more than that of the colorless water.

MATERIALS

100 mL water

ten drops each of red, green, and blue food coloring

two 100-mL beakers

spot lamp or work lamp, 150 watts or higher, incandescent or halogen

thermometer readable to 0.1°C (digital or temperature probes interfaced to a computer with the output projected are most convenient)

PROCEDURE

Preparation and Presentation

Pour 50 mL of water into each of two 100-mL beakers. Add 10 drops of red food coloring, 10 drops of green, and 10 drops of blue to the water in one of the beakers. Swirl the beaker to mix the colors uniformly. The colored water will be almost black. Set the two beakers next to each other on a white surface, such as on a sheet of paper.

Measure and record to 0.1°C the temperature of the liquid in each beaker. Turn on the spot lamp and adjust it so it shines directly down onto the beakers. Adjust the beakers under the lamp so they receive the same illumination.

After the beakers have been under the lamp from 15 to 30 minutes, turn off the lamp. Measure the temperature of each liquid. Determine the temperature change for each liquid. The colored liquid will have undergone a greater temperature change.

HAZARDS

There are no hazards associated with this demonstration.

DISPOSAL

Flush liquids down the drain.

DISCUSSION

Both liquids will undergo a temperature change, because both absorb radiation from the lamp. Pure water absorbs infrared radiation, which the lamp produces, but it does not absorb significant amounts of visible radiation. The colored solution absorbs infrared radiation and also most of the visible radiation that falls on it.

In the colored solution, light is absorbed by dyes from the food coloring. These dyes are molecules that have low-lying unoccupied electronic energy levels. When a photon of colored light encounters a dye molecule, and the photon's energy corresponds to the energy difference between the ground and excited electronic state of the dye, the latter can absorb the photon's energy. The dye can then release the absorbed energy in one of three ways: the energy can be emitted as a photon of visible light, the energy can cause the molecule to undergo a change (a chemical reaction), or the energy can be converted to motion in the dye molecule. (See the section *Excited Molecules* in the introduction, page 44).

Some of the energy is reemitted in all directions, which results in scattering of the light, or it is emitted at longer wavelengths as fluorescence or phosphorescence. Some of the dye molecules may undergo a chemical change, which is why many dyes gradually fade in the light. But much of the energy is converted to motion (kinetic energy of translation, vibration, and rotation) in the molecule. This increased kinetic energy in the dye molecules is transferred by collisions to other molecules, including the solvent water molecules, and results in the greater increase in temperature of the dye solution compared to pure water.

Extensions of this demonstration that could be good student projects include investigating different light (radiation) sources (for example, heat lamps, flood lights, or sunlight) and the effect of individual dyes (for example, red, green, and blue) in place of the mixture.

REFERENCES

1. H. D. Burrows, M. Graca Miguel, and A. Correia Cardoso, "Teaching Experiments and Demonstrations in Photochemistry at the Introductory Level," *European Photochemistry Association News, 29,* 39–43 (1987).
2. H. D. Burrows and A. Correia Cardoso, "Radiationless Relaxation and Red Wine," *J. Chem. Educ., 64*(12), 995 (1987).

12.12

Refraction and Diffraction

The Separation of White Light into Colors

The rainbow in the sky after a rainstorm and the play of colors reflected off a compact disc are both the result of white light separated into its component colors. This happens because different colors of light are differently refracted (rainbow) or diffracted (compact disc).

In this demonstration, a beam of white light is refracted through a flask of water. The refraction results in the separation of the white light into colors (Procedure A). Reflection of white light from closely spaced surfaces of oil on water, of a soap film, and of a compact disc results in diffraction and the separation of the white light into colors (Procedure B).

MATERIALS FOR PROCEDURE A

drop of india ink

250-mL (or larger) beaker

stirring rod

about 2 L water

2-L round-bottom flask

cork ring or other support for flask

slide projector

2-in × 2-in piece of poster board

projection screen or other large white surface, such as poster board or a wall

video projection system (optional)*

MATERIALS FOR PROCEDURE B

document camera and associated projection system

black, matte-finish sheet of paper to cover stage of document camera

glass crystallizing dish, 150-cm in diameter

about 500 mL water

drop of turpentine

bubble soap with wand

compact disc

PROCEDURE A [1]

Preparation

Put the drop of india ink into 100 mL of water in a beaker. Stir the mixture to make it uniform. Pour half of the water into a 2-L round-bottom flask. Fill the flask with water, stopper it, and set it upright on a stand.

---

*Various types of video projection systems are described on page xxxiii.

Prepare a narrow beam of white light. This can be done with a slide projector using a slide prepared by cutting a piece of poster board the size of a slide (2 in × 2 in) and punching a clean, round hole about 3 to 4 mm in diameter in the center of the square. Cover the top of the projector to reduce scattered light.

Position the projector so its beam enters the round-bottom flask roughly at the flask's equator, about halfway between the center and the edge of the flask (see Figure 1). Place a screen or white board to the side behind the projector. Adjust the positions so that, when viewed from above, the path of the refracted and reflected beams of light through the flask are visible, and that the reflected beam strikes the screen and shows a rainbow at the edge.

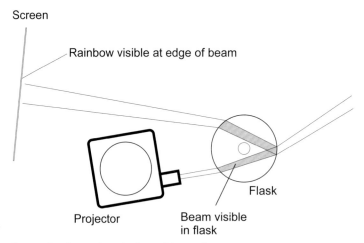

Screen

Rainbow visible at edge of beam

Flask

Projector

Beam visible in flask

Figure 1.   Set-up for Procedure A viewed from above.

For presentation to a large group, mount a video camera directly above the flask, pointing down so it captures the view of the beams passing through the flask.

Presentation

If the number of observers is small, have them gather around the flask. Otherwise, direct their attention to the video display. Turn on the slide projector and point out where the beam enters the flask. Dim the room lights. In the dark, the path of the light through the flask will be visible. The beam is refracted and changes direction as it enters the flask. At the opposite side of the flask, a portion of the beam is reflected back toward the screen. As the beam exits the flask, its direction changes again. Each time the beam is refracted, the various colors in the white light are bent to different degrees. The separated colors can be seen on the screen, at the edge of the reflected beam.

PROCEDURE B

Preparation and Presentation

Place a sheet of black, nonreflective paper on the stage of a document camera. Set the glass crystallizing dish on the paper and fill it with water to a depth of several centimeters. Turn on the document camera's top lights to brightly light the dish. Drop a single drop of turpentine onto the surface of the water. The oil will spread to a thin film, and rainbow colors will appear on the surface. Remove the dish from the projector.

Dip a bubble-blowing wand into bubble soap to form a film on the ring. Hold the ring horizontally over the black paper on the document camera. A rainbow of colors appears in the film.

Set a compact disc on the black paper with its shiny side up. Tilt the disc slowly back and forth, and a rainbow of colors will appear to move across its surface.

HAZARDS

Turpentine is flammable, so no flames should be present when it is used.

DISPOSAL

Pour water down the drain.

DISCUSSION

Refraction and diffraction are two physical processes that separate light by wavelength, and therefore, into separate colors. Refraction occurs when light crosses from one medium to another, such as from air to water. When a beam of light crosses such an interface at any angle other than perpendicular, its direction of travel changes. The degree of change (refraction) depends on the wavelength of the light, and therefore, on its color. Diffraction occurs when light is reflected from or transmitted by a material having closely spaced surfaces or lines where the material is nonreflective or nontransparent. Both phenomena are results of the wave nature of light. (See the section *The Wave Model of Light* in the introduction, page 12, for a discussion of the relationship between refraction and diffraction and the wave nature of light.)

Procedure A demonstrates the refraction of a beam of white light. The direction of travel of the beam changes as it goes from air to glass to water, and again when it goes from water to glass to air. (It also changes when it is reflected from the back surface of the flask.) The variation of refraction with wavelength is observed in the rainbow that appears at the edge of the reflected beam on the screen.

Procedure A also illustrates how a rainbow is produced in nature, where tiny droplets of water in the air refract and reflect the sunlight . In this demonstration, the flask corresponds to a large water drop. As demonstrated here, a rainbow is formed when light is reflected back toward its source, so in nature, a rainbow always appears when the sun is behind the observer. In the procedure as described, most of the beam is not reflected at the back surface of the flask, so the observed rainbow is pale. Reflection from the back surface can be increased and the observed rainbow brightened if the back inner surface of the flask is silvered. This, however, reduces the similarity with natural rainbow formation.

Diffraction occurs when light is reflected from two closely spaced surfaces. If the distance between the surfaces is close to the wavelength of the light, the reflected waves can interact either constructively or destructively. Constructive interference produces an amplification of the wave, whereas destructive interference diminishes the intensity of the reflected wave. Whether interference is constructive or destructive depends on the relationship between the magnitudes of the distance between surfaces and the length of the wave. For a given distance, some wavelengths interact constructively and others destructively. Therefore, some colors in the reflected light will be enhanced and others diminished, which produces colored light from white light.

The thin film of turpentine has a thickness similar to that of the wavelength of visible light. When light is reflected from the top surface and from the bottom surface of the film, the interaction between these reflections produces an enhancement of some colors and a diminishing of others. Because the thickness of the film varies somewhat from place to place, the reflected light varies in color. A similar process produces color in light reflected from a thin soap film, such as that of a soap bubble. With a compact disc, the reflecting surfaces are next to each other rather than above and below. The distance between the "tracks" on the disc is comparable to the wavelength of visible light, so the reflections from the tracks interfere, producing color in the reflection.

Turpentine is a fluid obtained by the distillation of resin from pine trees. It is composed mainly of terpenes, a large group of compounds biochemically produced from isoprene units, as shown in this diagram.

Turpentine is a mixture with an average molar mass of about 136, which corresponds to two isoprene units. Its boiling range is 150° to 170°C. It is used in this demonstration because it readily makes a thin film on water. The effectiveness of turpentine for this purpose was known in the seventeenth century. Robert Hooke mentions having tried the experiment on March 29, 1664, and reporting the results to the Royal Society on May 24 in a handwritten journal that was rediscovered in 2006 [2].

REFERENCES

1. R. M. Sutton, ed., "Rainbow," in *Demonstration Experiments in Physics,* p. 387, McGraw-Hill: New York (1938).
2. R. Adams and L. Jardine, "The Return of the Hooke Folio," *Notes Rec. R. Soc., 60*(3), 235–239 (2006).

12.13

Disappearing Glass
Index of Refraction

The indices of refraction of glass and air are different, but a large sheet of transparent, colorless glass is often invisible, so we occasionally bump into a glass door and flying birds collide with windows. However, we have no trouble seeing a colorless glass tumbler or beaker, because refraction by the curved surfaces makes them visible. If the medium is not air, but rather a liquid with the same index of refraction as the glass, this refraction disappears and so does the glass.

A glass rod is lowered vertically into a beaker containing two layered liquids. Where the rod passes through the upper liquid, the rod disappears (Procedure A). A liquid is poured into a smaller beaker inside a larger one until the liquid overflows and covers the smaller beaker. As the smaller beaker is covered, it disappears (Procedure B). A bundle that contains several glass rods surrounding one at the center is lowered into a liquid until the bundle is submerged. When the bundle is submerged, the surrounding rods disappear, leaving only the central one visible, and it appears to float unsupported in the liquid (Procedure C).

MATERIALS FOR PROCEDURE A

150 mL vegetable oil, such as Wesson Vegetable Oil (pure soybean oil)

several milliliters baby oil or light mineral oil, clear and colorless, with no additives (optional)

400-mL beaker

500-mL tall-form beaker

dropping pipet, about 1 mL

glass stirring rod, borosilicate (Pyrex or Kimax), about 20 cm in length

paper towels

MATERIALS FOR PROCEDURE B

2 L vegetable oil, such as Wesson Vegetable Oil (pure soybean oil)

several milliliters baby oil or light mineral oil, clear and colorless, with no additives (optional)

400-mL borosilicate beaker (Pyrex or Kimax)

2-L beaker

2-L Erlenmeyer flask

glass stirring rod, longer than the depth of larger beaker

tongs

paper towels

MATERIALS FOR PROCEDURE C

2 L vegetable oil, such as Wesson Vegetable Oil (pure soybean oil)

several milliliters baby oil or light mineral oil, clear and colorless, with no additives (optional)

2-L beaker

six borosilicate glass rods (Pyrex or Kimax), with diameter about 0.5 cm and length about 10 cm

one soft-glass (soda-lime) rod, with diameter about 0.5 cm and length about 10 cm

one glass stirring rod, borosilicate (Pyrex or Kimax), longer than the depth of larger beaker

ca. 1 meter of black carpet thread

two small rubber bands

paper towels

PROCEDURE A

Preparation

Test the oil to be sure that the borosilicate glass disappears when immersed in the oil. The oil may be adjusted to improve the effect as described in this paragraph. Pour 150 mL of vegetable oil into a 400-mL beaker. Dip the one end of the glass stirring rod into the oil, and the portion of the rod that is submerged should be nearly invisible. If the rod is still partially visible, the composition of the oil may be adjusted to decrease the visibility of the rod by adding a small amount of light mineral oil (baby oil). Add about one-half milliliter of baby oil to the vegetable oil in the beaker, gently stir the mixture until it is uniform, and observe the glass rod in the oil mixture. Continue to add small increments of baby oil until the glass can no longer be seen.

Put 150 mL of water into the 500-mL tall-form beaker. Slowly pour the oil from the 400-mL beaker onto the top of the water.

Presentation

Holding the stirring rod vertically above the beaker containing a layer of oil on top of a layer of water, slowly lower one end of the rod through the oil layer into the water. Where the rod is surrounded by oil, it is invisible. Move the rod in a slow circular motion, as though stirring the liquids. The rod appears to be in two pieces, one above the oil and the other below.

PROCEDURE B

Preparation

Test the oil to be sure that the borosilicate glass disappears when immersed in the oil. The oil may be adjusted to improve the effect as described in this paragraph. Pour 1500 mL of vegetable oil into a 2-L Erlenmeyer flask and observe while a borosilicate glass stirring rod is lowered into the oil. If the rod within the vegetable oil is visible, slowly add small increments of baby oil with stirring until the part of the rod in the liquid can no longer be seen.

Presentation

Set the smaller beaker upright inside the larger one. Slowly pour the oil down the stirring rod into the smaller beaker until the oil overflows and eventually covers the smaller beaker. The smaller beaker will be invisible (except for the graduations) when submerged in the oil. Use tongs to partially raise the smaller beaker out of the oil to show that it is still there. Return the beaker to the oil, and it will again disappear.

PROCEDURE C [1]

Preparation

Test the oil to be sure that the borosilicate glass disappears when immersed in the oil. The oil may be adjusted to improve the effect as described in this paragraph. Pour 1500 mL of vegetable oil into a 2-L Erlenmeyer flask and observe while a borosilicate glass stirring rod is lowered into the oil. If the rod within the vegetable oil is visible, slowly add a small amount (less than 1 mL) of baby oil to the vegetable oil and gently stir the mixture until it is uniform. Continue to add increments of baby oil until the part of the rod in the liquid can no longer be seen.

Near one end of the soft-glass rod, wrap thread several times around the rod until the outer diameter of the resulting thread ring is about 1 mm larger than the rod's radius. Secure the thread wrapping by with a knot. Repeat this at the other end of the rod.

Hold the six borosilicate rods around the thread-wrapped rod. Secure the bundle of seven rods together temporarily with two rubber bands, one at each end. Wrap thread several times around the center of the bundle and tie it securely. Leave one end of the thread about 30 cm long, so that the entire bundle can be lifted by this thread, and cut the other end short. Remove the two rubber bands. If necessary, adjust the position of the thread wrapping, so that when held by the free end of thread, the bundle is tipped at an angle of about 10 degrees from the horizontal.

Presentation

Pour 1500 mL of the vegetable oil into the beaker. While holding the free end of the thread, slowly lower the bundle of stirring rods into the oil. Do this slowly, with the rods tipped slightly off horizontal, so the air between the rods can escape. Continue lowering the bundle into the oil until it is completely submerged. When immersed in oil, the borosilicate rods will be invisible. Only the center rod, the soft-glass rod, will be visible, seemingly floating in the center of the ring of thread. Raise the bundle of rods to show that all are still there, and then immerse the bundle again.

HAZARDS

Use caution when handling oil-coated glassware because it is very slippery.

DISPOSAL

The oil may be saved and reused.

The slippery, oil-coated glassware should be washed with detergent and water to remove the oil.

Check local regulations for proper disposal procedure of used oil.

DISCUSSION

Clear, colorless glass immersed in water is visible because the index of refraction of glass is different from that of water. For that matter, the glass is visible in air, too, because the indices of refraction of glass and air are different. This demonstration shows what happens when glass is surrounded by another material whose index of refraction is almost the same: the glass becomes invisible. The index of refraction of borosilicate glass, which is sold under the trade names of Pyrex and Kimax, is almost the same as that of many vegetable oils such as soybean oil or canola oil. In Procedure A, a glass rod made of borosilicate glass is placed into a beaker containing a layer of oil floating on a layer of water. Where the rod passes through the oil, it is invisible, and the rod eerily seems to be divided into two pieces. In Procedure B, a small borosilicate beaker disappears completely (except for its graduations) when immersed in the oil. In Procedure C, the borosilicate rods in a bundle disappear, leaving one nonborosilicate rod in the bundle to float mysteriously in the center of a ring of thread.

We are able to see transparent objects, objects that do not absorb light, when they reflect or refract light. Reflection occurs at the surface of an object. If the surface is flat, such as a pane of window glass, we may see in the reflection an image of the surroundings. If the surface is not flat, the image will be distorted. Refraction also occurs at the surface of a transparent object, but rather than being reflected, light that strikes the surface is transmitted through the transparent object to the opposite surface, where the light emerges. When light travels through a transparent medium, its direction may have changed twice, so the objects that we observe through the transparent medium may not be exactly where they appear to be. When we look at a table with a transparent goblet resting on it, we see both light that has traveled through the goblet and light that has traveled past it. The light that traveled past it shows us the surroundings of the goblet. The light that traveled though the goblet shows us what is behind the goblet, but in a displaced and distorted fashion. We interpret the distortions as a result of light having traveled through the goblet, and this is how we "see" the goblet (disregarding for the moment that the goblet may also reflect some light).

Each time light crosses the interface between one transparent medium, such as air, and another, such as glass, refraction can occur. Refraction is a change in direction of the path of the light when it crosses the interface between two media. This change in direction occurs because the speed of light is affected by the medium through which it travels. Light travels fastest through a vacuum, somewhat more slowly through air, and even more slowly through glass. (Details of how refraction occurs are presented in the introduction, beginning on page 12.) The ratio of the speed of light in a vacuum to the speed of light in a medium is called the *index of refraction* of the medium. When light crosses the interface between two media, the degree by which its direction changes is related to the ratio of the indices of refraction of the two media. If the indices of refraction are very different, the change of direction is large. If the indices are similar, the change is small. If the indices are the same, there is no change in direction. If there is no change in direction when light travels through an object, then our view of the background behind the object is undistorted, and we have no way of detecting the object—the object disappears.

In this demonstration, a glass rod disappears where it is surrounded by vegetable oil. The demonstration exploits the fact that vegetable oil and borosilicate glass have very similar indices of refraction. Both have an index of refraction of about 1.47. Neither vegetable oil nor borosilicate glass is a pure compound; both are variable mixtures. Therefore, their indices of refraction vary somewhat. The index of refraction of vegetable oil is usually slightly higher than that of borosilicate glass. For this reason, the best results in this demonstration are obtained when the index of the oil is adjusted to match the particular glass that is to be used. The adjustment can be made by adding to the vegetable oil a different oil whose index of refraction is slightly less than that of the glass, such as a light mineral oil, which is sold as baby oil. Because of the difference in index of refraction and the miscibility of the two oils, small increments of mineral oil can be added to the vegetable oil until the borosilicate rod becomes invisible.

Borosilicate glass is noted for its very low thermal expansion, so its temperature may be changed rapidly without the risk of breakage. This makes it well suited for laboratory use. Several different formulations are manufactured and sold under various trade names, such as Pyrex and Kimax. As the name borosilicate suggests, this glass contains boron and silicates. Pyrex is made from a mixture containing about 81% silica (SiO_2), 13% boron oxide (B_2O_3), 4% sodium oxide (soda, Na_2O), 2.3% aluminum oxide (alumina, Al_2O_3), and miscellaneous traces of other compounds [2]. It has an index of refraction of 1.474 for light with a wavelength of 589 nm (the sodium D lines).

Soft glass, also called soda-lime glass, is a mixture containing about 73% silica, 14% sodium oxide (soda, Na_2O), 7% calcium oxide (lime, CaO), and perhaps small amounts of the oxides of magnesium and aluminum [3]. It is called soft glass because it has a softening temperature of only 720°C, which is lower than that of borosilicate glass, namely 820°C. The index of refraction of soft glass is about 1.52 at 589 nm.

The slight yellow tint of the vegetable oil could improve the visibility of the colorless glass submerged in it. The intensity of color depends on the length of the path that light travels through the oil. Colorless glass in the colored oil reduces the path of the light and reduces the intensity of the color where the glass is located. However, the thickness of the glass is small compared to the path through the oil, so changes in color intensity are imperceptible.

A variation of Procedure B has been used as a "magic" trick in which a broken beaker is restored to one piece. The trick starts with a large beaker filled with oil containing a smaller beaker without graduations already submerged in it. The smaller beaker is not visible. Then, pieces of a broken beaker are dropped into the large beaker. After all of the pieces have been added, tongs are used to fish the whole beaker from the oil, thus appearing to have reassembled the pieces.

All of the procedures can be adapted for use as eye-catching hallway displays. In Procedure B, fill the small, inner beaker about two-thirds full with oil and fill the larger, outer beaker to the same level, giving the appearance of a third of a beaker floating on the surface of the oil. In Procedure C, a mechanism that remotely lowers and raises the bundle of rods can be used.

Other liquids besides vegetable oils have an index of refraction near 1.47 and will make borosilicate glass disappear when it is immersed in them. Light corn syrup diluted with about 5% water is one such liquid. Historically, various mixtures of carbon disulfide, carbon tetrachloride, and benzene were used [4], but the use of these suspected carcinogens should be avoided.

Table 1.   Indices of refraction.

Material	Index of Refraction
Soybean oil	1.468 [5]
Corn oil	1.472 [5]
Water	1.333 [5]
Light mineral oil	1.46
Pyrex glass	1.474 [2]
Soda-lime glass	1.52 [3]

REFERENCES

1. R. Hipschman, "Disappearing Glass Rods," in *Exploratorium Cookbook II: A Construction Manual for Exploratorium Exhibits,* p. 104, Exploratorium Store: San Francisco, California (2002).
2. *Technical Bulletin AL-223,* Aldrich Chemical Company, Milwaukee, Wisconsin (2005).
3. *Technical Bulletin AL-224,* Aldrich Chemical Company, Milwaukee, Wisconsin (2005).
4. R. M. Sutton, ed., "Refractive Index," in *Demonstration Experiments in Physics,* 383–384, McGraw-Hill: New York (1938).
5. D. R. Lide, ed., *CRC Handbook of Chemistry and Physics,* 90th ed., CRC Press: Boca Raton, Florida (2009).

12.14

Disappearing Gel

Index of Refraction

Most transparent, colorless objects suspended in a glass of water are visible, because they differ in index of refraction from the water. If the object is itself almost all water, its refractive index is essentially the same as water and it will be invisible when immersed.

As a clear, colorless liquid is poured into a beaker filled with white, glistening crystals, the crystals disappear (Procedure A). A glass dish on an overhead projector casts a dark shadow on the screen, but when water is poured into the dish, it becomes transparent, revealing a message (Procedure B). A clear, colorless lump of gel with a nail through it is suspended by a black thread. When the lump is lowered into a clear, colorless liquid, the lump disappears and the nail appears to float in the center of the loop of thread (Procedure C).

MATERIALS FOR PROCEDURE A

dehydrated Water-Gel Crystals,* irregularly shaped, about 5 cm³

resealable plastic bag, gallon size

two 1-L tall-form beakers

MATERIALS FOR PROCEDURE B

dehydrated Water-Gel Crystals, irregularly shaped, about 2 cm³

crystallizing dish, about 15 cm diameter

resealable plastic bag, about 0.5 L size

permanent marking pen

clear, colorless transparency

overhead projector

projection screen

MATERIALS FOR PROCEDURE C

dehydrated Water Gel Crystals, large spheres, about five crystals

resealable plastic bag, about 0.5 L size

50 cm of heavy black thread, such as carpet thread

small aluminum nail, about 2 inches long

beaker, about 250 mL

pencil

---

* Water-Gel Crystals is a trade name of the Water-Gel Crystals Company for its anionic polyacrylamide superabsorbent polymer. The Chemical Abstracts Service (CAS) registry number for the substance is 71042-87-0. Similar material may be available from other sources.

PROCEDURE A

Preparation

The Water-Gel Crystals need to be fully hydrated before use in this demonstration. Hydration should be started a day before the presentation. To do this, half fill a gallon sealable plastic bag with water, and add about 5 cm^3 of dehydrated Water-Gel crystals. Seal the bag and allow it to rest over night. The crystals will swell to irregularly shaped, clear, colorless lumps.

Shortly before the presentation, fill a 1-liter tall-form beaker three-quarters full with the expanded, hydrated lumps. Fill a second beaker three-quarters full with water.

Presentation

Display the beaker containing the expanded lumps. Hold it up and turn it so the viewers can see the lumps. When brightly lit, the lumps will glisten and refract light, distorting the view through the beaker.

Pour water from the second beaker into the beaker containing the lumps. Cover the lumps with water. The lumps will disappear, and the beaker will appear to contain only water. Carefully decant the water from the beaker of lumps back into the beaker of water. The lumps will reappear where they are not covered by water. The process may be repeated several times.

PROCEDURE B

Preparation

The Water-Gel crystals need to be fully hydrated before use in this demonstration. Hydration should be started a day before the presentation. To do this, half fill a gallon sealable plastic bag with water, and add about 5 cm^3 of dehydrated Water-Gel crystals. Seal the bag and allow it to rest over night. The crystals will swell to irregularly shaped, clear, colorless lumps.

Set the 15-cm crystallizing dish on the transparency, and use a marker to draw an outline of the dish on the transparency. Inside the outline write a message, such as "Science Is Fun!" Place this transparency on the stage of an overhead projector, focus the image, and then turn the projector off.

Shortly before the presentation, half-fill the crystallizing dish with expanded Water Gel lumps, and set the dish over its outline on the overhead-projector transparency.

Presentation

Turn on the overhead projector. The screen will show a dark shadow of the dish. Slowly pour water into the dish. As the lumps in the dish are covered, the shadow on the screen will brighten, and when the lumps are completely covered with water, the message on the transparency will be clearly visible.

PROCEDURE C

Preparation

The Water-Gel crystals need to be fully hydrated before use in this demonstration. Hydration should be started a day before the presentation. To do this, half fill a quart sealable plastic bag with water, and add several dehydrated Water-Gel spheres. Seal the bag and allow it to rest over night. The spheres will swell to clear, colorless spheres with a diameter about 20 times their original.

Select one of the hydrated spheres, and carefully tie a heavy, black thread around it, leaving one end of the thread about 30 cm long and trimming the other end short. Insert a 2-in aluminum nail through the center of the sphere, perpendicular to the loop of thread.

Tape the free end of the thread to the center of a pencil. Rotate the pencil to wind the thread around it. Lay the pencil across the rim of a 600-mL beaker so that the nail and gel sphere are suspended in the beaker. Rotate the pencil to adjust the position of the nail so that it is suspended about one-third of the way up from the bottom of the beaker. Use tape to fix the thread at this position on the pencil. Add water to the beaker until it is about two-thirds full and the suspended gel is completely submerged. The gel sphere will be invisible, and the nail will appear to be floating in the center of the loop of thread.

Presentation

Display the beaker with the nail seeming to float in the center of the loop of thread. Elicit from the audience suggestions of what may be holding the nail in position. Lift the pencil and remove the gel sphere and nail from the water to show how the nail is suspended.

The beaker with the Water-Gel–sphere-and-nail assembly suspended in water may be displayed for extended periods, such as in a corridor demonstration. As water evaporates, it should be replenished.

HAZARDS

Do not put dry Water-Gel crystals down the drain. They can clog the drain when they swell from absorbed water.

DISPOSAL

The hydrated Water-Gel crystals may be stored in sealed plastic bags in a refrigerator for several weeks and reused in repeated presentations of the demonstrations. Alternatively, they can be discarded in a solid-waste receptacle.

DISCUSSION

When we see a clear and colorless solid, such as hydrated Water-Gel Crystals, surrounded by clear and colorless air, we can distinguish the crystals from the surrounding air because they reflect light and distort the light that travels through them. This distortion occurs because the hydrated crystals and the air have very different indices of refraction. (Index of refraction is described in the section *Refraction and Reflection* in the introduction, page 32.) The irregular surface of the hydrated crystal reflects light in random directions; however, the hydrated crystals disappear when immersed in water because they have an index of refraction that is virtually identical to that of water. This phenomenon is directly demonstrated in Procedure A. In Procedure B, the phenomenon is used to make transparent a container of crystals obscuring an image on an overhead projector, thereby revealing the image. In Procedure C, the invisible crystal is used to make a solid object appear suspended in the middle of a beaker of water.

We can distinguish objects from their surroundings only when they differ in color or index of refraction. When light passes from water into the hydrated crystals, its direction of travel is not altered. Neither is it altered when light travels from the crystal into water. Therefore, light that travels through the crystals in water is not distorted. Because there is no distortion, we have no way to detect the presence of the crystals—they disappear.

We are able to see transparent objects (objects that do not absorb light) when they reflect or refract light. Reflection occurs at the surface of an object. If the surface is flat, such as a pane of window glass, we may see in the reflection an image of the surroundings. If the surface is not flat, the image will be distorted. Refraction also occurs at the surface of a transparent object, but rather than being reflected, light that strikes the surface is transmitted through the transparent object to the opposite surface, where the light emerges. When light travels through a transparent medium, its direction may have changed twice, so the objects that we observe through the transparent medium may not be exactly where they appear to be. When we look at a table with a transparent goblet resting on it, we see both light that has traveled through the goblet and light that has traveled past it. The light that traveled past it shows us the surroundings of the goblet. The light that traveled though the goblet shows us what is behind the goblet, but in a displaced and distorted fashion. We interpret the distortions as a result of light having traveled through the goblet, and this is how we "see" the goblet (disregarding for the moment that the goblet may also reflect some light).

Each time light crosses the interface between one transparent medium, such as air, and another, such as glass, refraction can occur. Refraction is a change in direction of the path of the light when it crosses the interface between two media. This change in direction occurs because the speed of light is affected by the medium through which it travels. Light travels fastest through a vacuum, somewhat more slowly through air, and even more slowly through glass. (The details of how refraction occurs are presented in the introduction, beginning on page 12.) The ratio of the speed of light in a vacuum to the speed of light in a medium is called the *index of refraction* of the medium. When light crosses the interface between two media, the degree by which its direction changes is related to the ratio of the indices of refraction of the two media. If the indices of refraction are very different, the change of direction is large. If the indices are similar, the change is small. If the indices are the same, there is no change in direction. If there is no change in direction when light travels through an object, then our view of the background behind the object is undistorted, and we have no way of detecting the object—the object disappears. The effect in this demonstration is similar to that in Demonstration 12.13, Disappearing Glass, which exploits the similarity in the indices of refraction of borosilicate glass and vegetable oil.

It is not surprising that the hydrated Water-Gel Crystals have the same index of refraction as water, because the hydrated crystals are more than 99.7% water by mass. The remainder of the crystal, the material of the dry crystals, is a polymer. The polymer is a cross-linked copolymer of acrylamide with ammonium and sodium acrylates [1]. Both acrylamide and the acrylates (carboxylates) are derivatives of acrylic acid (prop-2-enoic acid).

$$CH_2{=}CH{-}\overset{\overset{\textstyle O}{\|}}{C}{-}OH \qquad CH_2{=}CH{-}\overset{\overset{\textstyle O}{\|}}{C}{-}NH_2 \qquad CH_2{=}CH{-}\overset{\overset{\textstyle O}{\|}}{C}{-}O^-\,NH_4^+$$

acrylic acid   acrylamide   ammonium acrylate

The polymer chain is a sequence of carbon atoms with amide and carboxylate groups distributed along it.

$$-CH_2{-}\underset{\underset{\textstyle CO_2NH_4}{|}}{\overset{\overset{\textstyle CONH_2}{|}}{CH}}{-}CH_2{-}\underset{\underset{\textstyle CONH_2}{|}}{CH}{-}CH_2{-}\underset{\underset{\textstyle }{|}}{\overset{\overset{\textstyle CO_2Na}{|}}{CH}}{-}CH_2{-}\underset{\underset{\textstyle CONH_2}{|}}{CH}{-}CH_2{-}\underset{\underset{\textstyle CONH_2}{|}}{CH}-$$

The amide groups and carboxylate groups are both hydrophilic, that is, both attract water molecules, and both can bind to water molecules through strong hydrogen bonding, which accounts for the ability of the polymer to hold water.

Three-dimensional structure is incorporated into the polymer by cross-linking. Cross-links tie separate polymer chains together, making the hydrated polymer rigid. Cross-linking between chains is produced by a variety of agents, such as dimethacrylates.

$$CH_2{=}C{-}\overset{\displaystyle O}{\overset{\|}{C}}{-}O{-}CH_2CH_2{-}O{-}\overset{\displaystyle O}{\overset{\|}{C}}{-}C{=}CH_2$$

$$\underset{CH_3}{|} \qquad\qquad\qquad \underset{CH_3}{|}$$

ethylene dimethacrylate

The two carbon-carbon double bonds in the linking agent allow the molecule to be incorporated into two separate polymer chains, forming a link between the chains. The Water-Gel Crystal material consists of a set of parallel polymer chains linked to each other by regularly spaced cross links. Cross linking results in a three-dimensional network, which gives three-dimensional structure to the hydrated polymer.

An aluminum nail is used in Procedure C, because it will not corrode while submerged. Iron and copper may corrode, and the corrosion of both produces colored products, brown in the case of iron, blue from copper. Colored corrosion products will make the gel visible and limit the longevity of the presentation as a corridor demonstration.

REFERENCE

1. Composition information supplied by the distributor, JRM Chemical, Inc., 4881 NEO Parkway, Cleveland, OH 44128.

12.15

Observing the Transmission Spectra of Dyes

The diffraction of a particular wavelength of light by a diffraction grating is the same no matter what the source of the light. If a wavelength of light is removed from a light beam, it will not show up in the diffracted beam.

The diffraction of monochromatic laser light is compared to the corresponding color in the transmission spectrum of white light. Then, the transmission spectra of various solutions of food coloring dyes are displayed by overhead projection and compared to the full visible spectrum.

MATERIALS

four 6-cm × 12-cm pieces of heavy paper or cardstock, e.g., black construction paper or pieces of manila file folder

single-edged razor blade or X-Acto knife

overhead projector

projection screen

linear transmission diffraction grating, 1000 to 500 lines/mm, about 15 cm × 15 cm square or larger

intact manila file folder

adhesive tape

red laser pointer (optional)

green laser pointer (optional)

dropper bottles of green, red, and blue food coloring

four Petri dishes

water

PROCEDURE [1, 2]

Preparation

Use two 15-cm × 30-cm pieces of stiff, opaque paper to mask the stage of the overhead projector, leaving only a vertical slit, about 0.5 cm wide, through which light can pass. (Black construction paper works well, and it has the advantage of darkening the stage of the projector and reducing stray light, thereby enhancing visibility of the projected image.) Lay two more pieces of masking material on top of and at right angles to the first two, leaving only a rectangular opening about 8 cm long × 0.5 cm through which light can pass. The image projected on the screen should be a narrow, vertical rectangle. The edges of the masking material should be straight and clean to produce a well-defined rectangle. Use tape to fix together the four pieces of masking material.

Prepare a diffraction-grating holder to attach the grating to the head of the overhead projector, as represented in Figure 1. Using an X-Acto knife or single-edged razor blade,

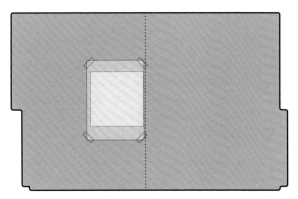

Figure 1.   Transmission grating holder assembly.

cut an opening about 10 cm square in one flap of a manila folder, near the fold. Lay a piece of single-axis transmission grating over the opening, so that its axis is perpendicular to the fold in the folder. (The axis of the grating can be determined using the overhead projector with the rectangular mask on its stage. With the rectangle of light projected onto the screen, hold the grating vertically in beam of the projector, near the projector head. When the axis is horizontal, the projected image will show colored spectra above and below the projected rectangle. When the axis of the grating is vertical, the projected image will show spectra to the right and left of the projected rectangle.) Fix the properly oriented grating to the folder with tape.

Suspend the diffraction grating vertically in the beam of the overhead projector by draping the open folder over projector head, as illustrated in Figure 2. Position the projector so that the white, undiffracted image of the slit and one of the diffracted spectra (right or left) of the slit are projected onto the screen. Turn off the projector.

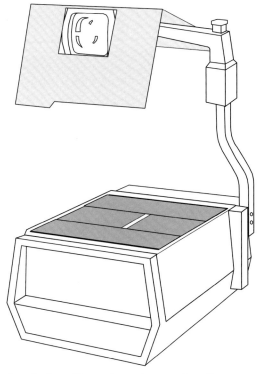

Figure 2.   Overhead projector with mask on stage and grating suspended on head.

Put enough water in each of three Petri dishes to cover the bottom of the dish. Set a dish over the slit in the masking material on the overhead projector. Turn on the overhead projector and observe the projected spectrum. Add a drop of green food coloring to the dish and mix the contents till the color is uniform. Observe the projected spectrum. The blue portion of the spectrum will be diminished. Add more drops of food coloring until the blue portion of the spectrum is completely gone. Repeat this process with the remaining Petri dishes, using a red food coloring in one and blue in the other. Use enough color in each dish to obscure a portion of the projected spectrum.

Presentation

Part 1. To illustrate how the grating separates colors (optional).

With the projector off, hold a red laser pointer perpendicular to the stage of the overhead projector and direct its beam upward through the projector's lens and through the grating. The laser beam will produce on the screen a bright, undiffracted image along with one or more dimmer diffracted images. Note the distance between the undiffracted and diffracted images.

Repeat the procedure of the previous paragraph using a green laser pointer in place of the red one. The distance between the undiffracted and diffracted images from the green laser will be smaller than those from the red laser. This observation can be used to predict that when white light passes through the diffraction grating, the blue end of the diffracted spectrum will be nearer the undiffracted, white rectangle than will be the red end.

Turn on the overhead projector and observe the diffracted spectrum from white light. The diffracted spectrum will go from blue near the undiffracted beam to red away from it.

With the projector on, repeat the diffraction of the red and green laser pointers. Holding the laser pointer perpendicular to the stage of the overhead projector near one end of the slit, align the undiffracted laser beam with the center of the undiffracted white beam. The diffracted red laser beam will line up with the red region of the diffracted spectrum, while the diffracted green laser beam will line up with the green region.

Part 2. To observe the transmission spectra of colored solutions.

Darken the room and turn on the projector. Place a Petri dish half full of water on the stage of the overhead projector. Position the dish so it covers the lower half of the slit. The edge of the dish will show in the diffracted spectrum as a dark band dividing the spectrum into two spectra. The upper spectrum is that of the light passing unrestricted through the slit, the lower spectrum that of the light passing through the glass of the Petri dish and the water in it.

Replace the dish of water with the one containing green solution. The lower spectrum will now show the light transmitted by the green solution. Only green light passes through the solution. Red, orange, and violet are missing from this spectrum.

Repeat the procedure of the previous paragraph with the dish of red solution and the dish of blue solution. Note which colors are transmitted and which absorbed by each.

The absorptions of two different solutions can be compared by covering the lower half of the slit with one Petri dish and the upper half with another dish.

HAZARDS

The hazards in this demonstration are the usual ones associated with the use of lasers. To avoid eye damage, never look directly at the beam or point it at the audience. Avoid reflections from shiny surfaces, because these could also do damage to you or the audience.

DISPOSAL

Flush the aqueous solutions down the drain or store them in capped bottles with the other materials for future demonstrations.

The grating material can be stored attached to the holder for repeated presentations of the demonstration.

DISCUSSION

Substances that are colored appear that way because they absorb some wavelengths (colors) of light and not others. Opaque objects reflect the wavelengths that are not absorbed, whereas transparent materials transmit them. In this demonstration, the white light from an overhead projector shines on various colored solutions, and the color of the solution is observed when it is projected onto the viewing screen. The spectrum of colors transmitted by the colored solution is also viewed directly and compared to the full spectrum of white light.

It is not surprising that the color of light transmitted by a dye solution and the colors visible in its transmission spectrum are similar. The transmission spectrum of a solution that appears green shows colors mainly in the green region of the spectrum. Similarly a blue solution transmits mainly in the blue region. A purple solution may transmit in both the blue and the red regions, because purple is the color we perceive when viewing a combination of red and blue light.

What is perhaps more interesting than the colors that appear in the transmission spectrum are the colors that are absent. The colors that are missing are those that actually interact with the dye. Light of the missing colors has the energy necessary to promote electrons in the dye molecules from the ground state to some excited state. Light of these energies (and these colors) are absorbed by the dye molecules and is removed from the light beam as it passes through the solution. (See the section *Absorption Spectra* in the introduction, page 6, for a discussion of absorption of light.) The absorbed energy may cause the dye to increase in temperature and may cause photochemical changes in the dye. Such photochemical changes cause dyed fabrics and paint to fade or even change color on prolonged exposure to a bright light such as sunlight.

A transmission diffraction grating such as the one used in this demonstration causes a light beam to be deflected (diffracted) by an amount that depends on the wavelength of the light. (See the section *Diffraction* in the introduction, page 14, for a discussion of diffraction gratings.) Light with long wavelengths (red light) is diffracted more than light with short wavelengths (violet light). This fact accounts for the observations in this demonstration that the red region in the spectrum is farther from the undiffracted beam than the violet region and that the red laser light is more diffracted than the green laser light. Viewers are sometimes surprised that the diffracted laser beams appear in the appropriately colored region of the spectrum of white light. The source of the light is irrelevant to the degree of diffraction; it is the diffraction grating that determines how much a particular wavelength of light will be diffracted.

A linear transmission diffraction grating is made from a transparent material, often plastic but sometimes glass, that has an array of closely spaced straight lines that are not transparent. The spacing of these lines in a particular grating is often specified in units of lines per millimeter. To produce satisfactory diffraction in this demonstration, a transmission grating with between 500 and 1000 lines/mm is appropriate. A grating with more closely spaced lines will produce a wider separation of wavelengths and colors in the diffracted spectrum. A grating with 1000 lines/mm will produce a broader spectrum than one with 500 lines/mm. A wider spectrum makes the colors of the transmitted and absorbed light easier to discern, especially when the projector needs to be close to the display screen.

The procedure in this demonstration calls for using colored solutions prepared using food colors to dye water. Of course, any colored solution may be used, such as fruit or vegetable juices and oils. All that is required is that the liquid be transparent and not cloudy. Also suitable are solutions of common laboratory chemicals, such as copper(II) sulfate (blue), nickel(II) sulfate (green), potassium permanganate (purple), sodium chromate (yellow), and manganese(II) sulfate (pink). If these solutions are used, however, appropriate precautions must be used for safe handling and disposal. For the purpose of comparing the transmitted color of a solution to the transmission spectrum, food coloring has the advantage of being readily available, relatively safe, and conveniently discarded.

REFERENCES

1. D. H. Alman and F. W. Billmeyer, Jr., "A Simple System for Demonstrations in Spectroscopy," *J. Chem. Educ., 53,* 166 (1976).
2. S. Solomon and C. Hur, "Overhead Projector Spectrum of Polymethine Dye: A Physical Chemistry Demonstration," *J. Chem. Educ., 72,* 730 (1995).

12.16

Dichroism

Transmission versus Reflection

Usually, the transmitted and reflected colors of an object are the same. Thus, the difference in transmitted and reflected colors of dichroic (two-color) glass or plastic is surprising and thought provoking.

A piece of colored glass is placed on the stage of an overhead projector. Light transmitted through the glass is a different color than light reflected from the glass.

MATERIALS

overhead projector

projection screen

piece of opaque cardboard large enough to cover the entire stage of the projector

single-edged razor blade or X-Acto knife

one or more pieces of dichroic glass*

PROCEDURE

Preparation

With a single-edged razor blade or other sharp knife, cut a hole in the center of cardboard; the hole should be smaller than the smallest piece of dichroic glass. Position the overhead projector so the audience can see both the projected image and the stage of the projector.

Presentation

Display the piece of dichroic glass to the audience so they can observe the color. With the room darkened and the overhead projector turned on, place the dichroic glass over the opening in the cardboard. The projected image will show the color of the light transmitted by the glass, which is different than the color of the glass observed earlier. While keeping the glass over the hole in the cardboard, lift the edge of the glass nearer the audience, so they can see light reflected from the bottom surface of the glass, as well as the light transmitted and projected onto the screen. The color of the reflected light is complementary to the transmitted color.

HAZARDS

There are no hazards associated with this demonstration.

DISPOSAL

Save the materials for future demonstrations.

---

* Dichroic glass is available from Coatings by Sandberg, Inc., 856 N. Commerce Street, Orange, CA 92867, www.cbs-dichroic.com.

DISCUSSION

The distinguishing characteristic of dichroic (two-colored) glass is that the color of the light it transmits is different than the color of the light it reflects. Furthermore, these two colors can shift depending on angle of view. Figure 1 shows the colors transmitted and reflected by two different samples of dichroic glass.

Modern dichroic glass has a multilayer coating of metal oxides placed on the glass. Multiple ultrathin layers of different metal oxides (gold, silver, titanium, chromium, aluminum, zirconium, magnesium, silicon) are evaporated onto the surface of the glass in a vacuum chamber. Some glasses have as many as 30 layers of these materials, but the total thickness of the coating is less than 0.001 mm (= 1 micron = 1000 nm). (Compare the structure and properties of dichroic glass with those of iridescent film, Demonstration 12.17.)

NASA developed dichroic glass filters in the 1950s and 1960s for use in satellites and space suits to protect instrumentation and astronaut's eyes from harmful radiation. The great advantage of these filters (compared to others that transmit some wavelengths and absorb others) is that most of the unwanted wavelengths are reflected, so little energy is trapped in the glass and it does not heat up.

Much of the dichroic glass you might see today is in art objects like jewelry, stained-glass windows, decorative lamps, etc. These are part of a long tradition of dichroic glass in art that dates back to at least the fourth century AD in Rome, the source of the Lycurgus Cup (Figure 2), now in the British Museum [1]. The glass in this cup contains colloidal

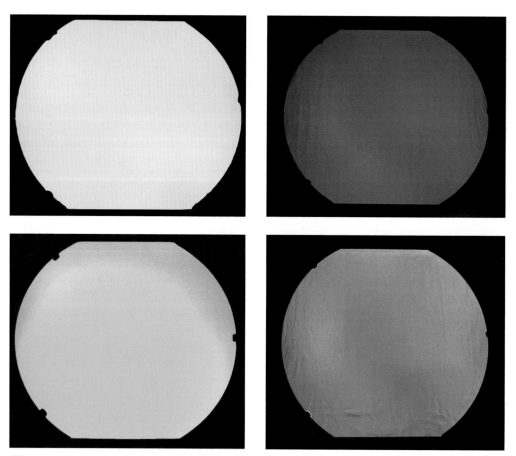

Figure 1.   The reflected (left) and transmitted (right) light from two different types of dichroic glass, yellow/violet and cyan/orange.

Figure 2.   The Lycurgus Cup viewed in reflected light (left) and transmitted light (right).

gold-silver alloy particles on the order of 50–100 nm in diameter, which are responsible for its dichroism. Such metal nanoparticles are still used to prepare dichroic materials, especially in organic polymer films [2, 3].

REFERENCES

1. I. Freestone, N. Meeks, M. Sax, and C. Higgitt, "The Lycurgus Cup: A Roman Nanotechnology," *Gold Bulletin, 40*(4), 270–277 (2007).
2. Y. Niidome, S. Urakawa, M. Kawahara, and S. Yamada, "Dichroism of Poly(vinylalcohol) Films Containing Gold Nanorods Induced by Polarized Pulsed-Laser Irradiation," *Jpn. J. Appl. Phys., 42,* 1749–1750 (2003).
3. W. Caseri, "Optically Anisotropic Metal-Polymer Nanocomposites," in *Metal-Polymer Nanocomposites,* L. Nicolais and G. Carotenuto, eds., pp. 265–285, Wiley-Interscience: New York (2005).

12.17

Iridescence from a Polymer Film

The play of colors from an iridescent polymer film is eye catching and raises interesting questions about how the nature of the interactions of light with matter and the structure of the film lead to this phenomenon.

Colored iridescent film is displayed to show its iridescence. When viewed against a white (light-reflecting) background, the film appears mostly red, but it looks mostly green when viewed against a black (light-absorbing) background.

MATERIALS

several 8.5-in × 11-in pieces of red-green iridescent film (such as used for gift wrap) [1]

cardboard transparency mounting frame, or transparent, colorless sheet protector

11-in square of white poster board

11-in square of black poster board

PROCEDURE

Preparation

Mount the pieces of red-green iridescent film in a cardboard frame, such as a transparency frame, or in a transparent, colorless sheet protector. Increasing the number of sheets used will increase the amount of reflected light but decrease the amount of transmitted light. Select the number of sheets that produces similar reflected and transmitted intensities. This number will depend on the film used.

Presentation

Display the film to the audience, tilting it at various angles so that they can observe the colors produced by the film. Hold the film immediately in front of a white background. The film will appear mostly red. Hold the black background against the film. The film will appear mostly green.

HAZARDS

There are no hazards associated with this demonstration.

DISPOSAL

The materials may be saved for repeated presentations of the demonstration, or they may be discarded in a general-purpose waste receptacle.

DISCUSSION

Iridescent plastic film is made by laminating many very thin layers of two different polymers, such as polystyrene and polymethylmethacrylate, that have different indices of refraction.

For example, that of polystyrene is 1.59, and that of polymethylmethacrylate is 1.49. At each interface, a portion of the incident light is reflected. When the light waves reflected from different interfaces are in phase, there is constructive interference between the waves, and that wavelength is enhanced. When the light waves are out of phase, those wavelengths are diminished. (See the section *Interference* in the introduction, page 16, for a discussion of constructive and destructive interference.) The light that is reflected is diminished in some wavelengths and enhanced in others, so it appears colored. The light that is transmitted is enhanced in the complementary wavelengths, so it appears the complementary color. (Compare the structure and properties of iridescent film with those of dichroic glass, Demonstration 12.16.)

The appearance of the reflected light depends on several factors. One factor is the relative refractive indices of the layers: the greater the difference in refractive index of the layers, the more light that is reflected, and so the brighter the reflection. The brightness of the reflected light also depends on the number of layers: the greater the number of layers, the more light is reflected, and the reflection appears brighter. The color of the reflected light depends on the thicknesses of the different layers.

When the film is viewed against a white background, the light observed has travelled through the film and has been reflected back from the background, completing two trips through the film. In this configuration, the observed color is the color of the transmitted light.

When the film is viewed against a black background, the light that travels through the film is absorbed by the background, so only light reflected by the film is observed. The color observed in this configuration is the color of the reflected light.

REFERENCE

1. E. H. Levine, "Polymers and the Visible Spectrum," *J. Chem. Ed.*, *69*, 122 (1992).

12.18

The Photoelectric Effect

When light of sufficiently short wavelength shines on a metal surface, electrons are ejected from the surface. The kinetic energy of these photoelectrons is a function of the wavelength of the incident light. Einstein accounted for this photoelectric effect on the basis of the Planck quantum hypothesis. His explanation was a major factor in the acceptance of the hypothesis and the advent of a quantum mechanical view of the world.

When a negatively charged metal sheet connected to an electroscope is irradiated with 365-nm ultraviolet light, the electroscope leaves do not move. When the sheet is irradiated with 254-nm radiation, the electroscope leaves gradually move.

MATERIALS

zinc sheet or piece of heavy-duty aluminum foil, about 5 cm square or larger

steel wool, extra-fine (000) grade, to polish zinc

250-mL or larger beaker, or similarly sized block of paraffin wax

electroscope, Braun or gold-leaf (Alternatively, construct a simple electroscope as described in Preparation below using a stiff wire, lightweight aluminum foil, audiotape, or Christmas-tree tinsel; a 500-mL Erlenmeyer flask; a stopper to fit the flask; and adhesive tape.)

clip lead, i.e., a wire with alligator clips on both ends

dual-wavelength (254 nm and 365 nm) ultraviolet (UV) lamp, or two separate lamps

stand to mount lamp facing down

polyvinyl chloride (PVC) plastic rod or tube, about 10 mm in diameter and 25 cm long

piece of knitted wool or fur, about 25 cm square

video projection system (for larger audiences)*

PROCEDURE [1, 2, 3]

Preparation

If zinc is to be used, clean one surface of the zinc sheet by rubbing it with fine steel wool until it is shiny. This cleaning must be done shortly before the presentation, to minimize the amount of oxide on the zinc surface.

Lay the polished zinc or aluminum foil, with the shiny side up, onto an electrically insulating support, such as an inverted glass beaker or block of wax. Attach one end of the clip lead to the zinc sheet and the other end to the probe of the electroscope. Mount the UV lamp above and close to the zinc sheet. For a large audience, a video projection system will be necessary to project an image of the electroscope for everyone to see.

If necessary, assemble a simple electroscope as represented in Figure 1. First, using sharp scissors, cut a strip of lightweight aluminum foil about 2–3 mm wide and 20 cm long.

---

* Various types of suitable video projection systems are described on page xxxiii.

Figure 1. Electroscope assembly.

Alternatively, a 20-cm piece of audiotape or Christmas-tree tinsel may be used. Bend a piece of stiff wire (about 15 cm of 16- or 18-gauge copper or a large paper clip) to make an L shape with a foot about 1 cm long and leg about 10 cm long. Make a radial cut from the edge to the center the full length of the stopper and insert the leg of the wire, so the foot will be held about 2 cm below the stopper in the flask. Carefully fold the aluminum foil strip in two across the foot of the wire and hold it in place with a small piece of adhesive tape folded over the wire. Insert the stopper in the flask with at least 9 cm of each half of the foil strip hanging free from the wire. The aluminum foil strips must be relatively long, because they are heavy compared to the very thin gold foil electrometers, and need to be quite unconstrained in order to move apart when repelled by the relatively small charges involved in this demonstration.

Test the electroscope-and-zinc-sheet assembly. Charge the PVC rod by vigorously rubbing it with the piece of wool or fur. Bring the charged PVC rod near the metal sheet. If the apparatus has been properly assembled, the leaves in the electroscope will move apart, and when the PVC rod is moved away from the metal sheet, the electroscope leaves will move together.

Presentation

Rub the PVC rod with wool. Bring the charged rod near the metal sheet and observe that the leaves of the electroscope move apart. Move the rod away and the leaves move back together. Touch the charged rod to the metal sheet to charge the electroscope. When the rod is moved away, the leaves of the electroscope will remain separated.

Turn on the 365-nm UV lamp and irradiate the zinc sheet. The electroscope leaves will not move, indicating that there is no change in the charge on the electroscope. Extinguish the 365-nm lamp and turn on the 254-nm UV lamp. The electroscope leaves will gradually move together, indicating a loss of charge. After the electroscope leaves have closed about half way, extinguish the lamp.

HAZARDS

Ultraviolet radiation is harmful to the eyes. Do not look directly at the radiation from the ultraviolet lamp, and do not direct the light toward the observers.

DISPOSAL

The materials may be saved for future demonstrations.

DISCUSSION

This demonstration shows that ultraviolet light of sufficient energy will cause a change in electrical charge of a piece of metal. The metal is first touched with a negatively charged rod, which transfers negative charge to the metal and to the electroscope that is attached to it. The electroscope registers this negative charge when its thin metal leaves move apart, because both leaves become negatively charged and like-charged objects repel one another. When the metal sheet is irradiated with ultraviolet light having a wavelength of 254 nm, the leaves of the electroscope move together, indicating that the electroscope and the metal sheet to which it is attached are losing charge. This does not happen when the metal sheet is irradiated with 365-nm light, indicating that this light has insufficient energy to eject charge from the metal.

The amount of energy required to eject electrons from the surface of a metal is expressed by the work function of the metal. The work function of zinc is 4.33 eV, which is 6.94×10^{-19} J, and that of aluminum is similar, 4.28 eV (6.86×10^{-19} J) [4]. For zinc, this energy corresponds to electromagnetic radiation with a wavelength of 286 nm, and for aluminum, a wavelength of 289 nm. Thus, 368-nm radiation does not possess enough energy to eject electrons from either metal, while 254-nm radiation does. Ejection of electrons by the shorter-wavelength ultraviolet radiation reduces the negative charge on the metal sheet. Since the sheet is connected to the electroscope by an electrically conducting wire, as electrons are ejected from the metal sheet, they move from the electroscope toward the zinc, and the electroscope leaves lose their charge. As the charge on the electroscope diminishes, the leaves repel each other less, and they move together.

Although 254-nm light has sufficient energy to eject electrons from the surface of zinc or aluminum, the process is affected by a number of factors. For example, if the metal surface is not flat and smooth, more energy is required. Ejection of electrons is more efficient from the shiny side of the aluminum foil and from a polished zinc surface. Furthermore, the work function of a metal expresses the energy required to eject electrons into a vacuum. The air surrounding the metal sheet in this demonstration also affects the energy required to eject electrons. In this demonstration, to facilitate the ejection of electrons, the zinc sheet is given a negative charge before being irradiated. The negative charge repels the ejected, negatively charged electrons, and increases the efficiency of their ejection from the metal.

When a polyvinyl chloride (PVC) rod is rubbed with wool or fur, the rod becomes negatively charged. At the same time, the wool or fur becomes positively charged. This happens because the mechanical energy of rubbing supplies energy to strip electrons from the polyamide molecules (proteins) in the wool or fur and transfer them to the polyvinyl chloride molecules in the rod. When the rod is held near the metal sheet, the deflection of the electroscope leaves increases, indicating that the electrical charge has been transferred to the electroscope. The negative charge of the PVC repels electrons in the conductive metal, and these electrons move through the conductive wire into the electroscope. When the PVC is touched to the metal sheet, electrons from the PVC transfer to the sheet and, through the wire, to the electroscope, giving them a negative charge.

When any two materials rub against each other, creating friction, electrons can be transferred from one to the other. This phenomenon of charge transfer through friction is called the *triboelectric effect.* The direction of charge transfer can be used to devise a triboelectric series, which is a list of materials in order of tendency to lose electrons. The table contains such a list, called a *triboelectric series,* in which a material will transfer electrons through friction to any material below it in the list. This series is only an approximate guide, because there are many factors that affect the transfer, such as atmospheric humidity and surface treatment of the material. The materials chosen to generate a negative charge in this demonstration were selected because PVC is near the bottom of the series, while wool and fur are above it. This assures that the charge generated on the PVC will be a negative charge.

Table 1.   A triboelectric series of selected materials.

rabbit fur
glass
human hair
nylon
wool
paper
hard rubber
polyester
polyvinyl chloride (PVC)
Teflon

If you have available an operating CRT (cathode ray tube—an older computer monitor or television set, for example), the charges on the PVC can be further demonstrated by holding the charged rod near the screen. The negatively charged PVC will repel the negatively charged electrons that are creating the image on the CRT screen and create a darker area near where the PVC is held.

REFERENCES

1. R. M. Sutton, ed., "Surface Photoelectric Effect," in *Demonstration Experiments in Physics,* pp. 488–489, McGraw-Hill: New York (1938).
2. H. F. Meiners, ed., "Photoelectric Effect," in *Physics Demonstration Experiments,* pp. 1169–1170, Ronald Press: New York (1970).
3. A. J. Beehler, "Demonstrating the Photoelectric Effect Using Household Items," *The Physics Teacher, 48*(5), 348–349 (2010).
4. D. R. Lide, ed., *CRC Handbook of Chemistry and Physics,* 90th ed., CRC Press: Boca Raton, Florida (2009).

12.19

The Tyndall Effect
Scattered Light Is Polarized

Light scattering, which causes the sky to appear blue and sunsets to be particularly colorful, also causes the scattered light to be polarized. Photographers use polarizing filters to take advantage of the polarization of scattered sunlight to produce interesting effects in their photographs.

A beam of light passing through a dilute milk solution is viewed through a polarizing filter. The brightness of the scattered light depends on the orientation of the filter.

MATERIALS

small aquarium, 2–10 gallons

water to fill aquarium

for each gallon of water, 25–50 mL milk

slide projector, with empty slide frame to allow white beam to be projected

projection screen

polarizing filter, square or circle with a diameter about two-thirds the depth of the aquarium

stirring rod

PROCEDURE

Preparation

Arrange the aquarium, slide projector, and screen as shown in Figure 1. The beam of the projector should pass through the aquarium side to side, so the audience can see the length of the beam as it passes through. The screen should be placed at the side opposite from the projector and adjusted so the spot where the beam strikes it can be viewed by the audience. Add water to the aquarium until it is nearly full.

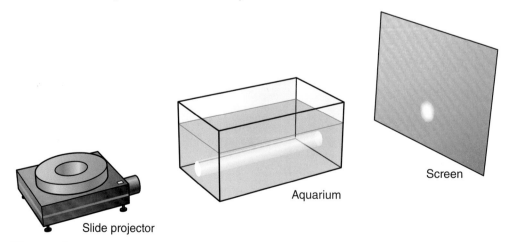

Slide projector

Aquarium

Screen

Figure 1. Arrangement of projector, aquarium, and screen.

Presentation

Turn on the projector. The beam will be virtually invisible where it passes through the water, and the spot on the screen will be bright white.

For each gallon of water in the aquarium, add 25 mL of milk to the water. Stir the contents of the aquarium to distribute the milk uniformly. The beam will become visible where it passes through the aquarium. The beam will be blue, and the spot where it strikes the screen will be orange-red.

Hold the polarizing filter against the side of the aquarium so the audience can view the beam through the filter. Slowly rotate the filter. As the filter rotates, the beam viewed through the filter will dim and brighten.

As an optional part of the presentation, a second 25 mL of milk per gallon of water may be added and the mixture stirred. The blue beam will now be more diffuse. Hold the polarizing filter to the side of the aquarium and rotate it. The scattered beam will again dim and brighten, but the difference between the dimmest and brightest will be smaller than before the second portion of milk was added.

HAZARDS

There are no particular hazards associated with this demonstration.

DISPOSAL

The waste milk-water mixture should be flushed down the drain with water, and the aquarium should be rinsed.

DISCUSSION

Colloidal suspensions scatter light strongly, producing the Tyndall effect, which is named after the British physicist John Tyndall (1820–93), who extensively studied the phenomenon [1]. The theory of how light is scattered by these suspensions is well developed [2]. Shorter wavelengths are scattered more, and thus, the scattered light is rich in the blue region of the visible spectrum.

The transmitted light is correspondingly poor in this region, so it is relatively rich in the red. According to Rayleigh's limiting law for a system dilute in particles that are small relative to the wavelength of the light, the intensity of scattered light is inversely proportional to the fourth power of its wavelength, that is, it is proportional to λ^{-4}.

Light is scattered when an electron in a particle absorbs a passing light photon, is excited to a high-energy state, and then reemits the absorbed energy in a random direction. That randomness produces the scattering of the light. Light that is scattered in a direction perpendicular to the original direction of travel is polarized, and the polarization is a result of the wave nature of light. (See the section *Polarized Light* in the introduction, page 19, for a discussion of light waves and polarization.) Light is a transverse electromagnetic wave, meaning that the oscillations in the electrical field occur perpendicularly to the direction in which the light is travelling. If the light is unpolarized, the electrical field oscillates in all directions perpendicular to the direction of travel (Figure 2). In order to absorb the energy from light, an electron must oscillate in the same direction as the oscillations in the electric field of the light. Therefore, the excited electron oscillates in a direction perpendicular to the direction of travel of the absorbed light. When the excited electron emits its energy as light, the electric field of the emitted light must oscillate in the same direction as the electron. This oscillation is perpendicular to the original direction of travel of the absorbed light. If the light is emitted in a direction that is perpendicular to the original direction of the light, this

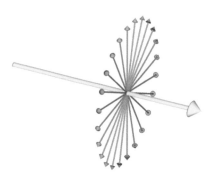

Figure 2. Direction of oscillations in electric field (blue) of a light beam (yellow), viewed obliquely.

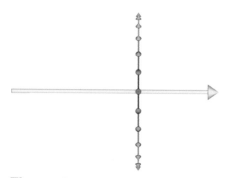

Figure 3. Direction of oscillations in electric field (blue) of a light beam (yellow), viewed from side.

light will have an electric field that oscillates only in the plane perpendicular to the original direction of travel (Figure 3). Therefore, this scattered light is plane polarized.

When the concentration of scattering particles is increased, the amount of scattering increases, as shown in the optional procedure. The scattered beam becomes more diffuse because some of the scattered light is scattered a second time. The degree of polarization of the scattered light also diminishes, as shown by the reduced difference between the brightest and dimmest positions of the polarizing filter. The degree of polarization diminishes because more of the light is scattered more than once, and at each scattering, the direction of polarization changes. As the light is scattered more, the scattered light contains light of a greater variety of polarizations.

Nature provides its own version of this demonstration every day, at sunset [3]. When viewed at an angle to the light source (the sun), the atmosphere appears blue because it scatters the shorter (blue) wavelengths of light preferentially. When viewed directly toward the light source, the atmosphere appears orange, because the longer wavelengths are not scattered as much. The latter effect is most obvious at sunset or sunrise, because the direct light of the sun passes through the greatest thickness of atmosphere at these times. It is inadvisable to attempt to look directly at the sun at any times other than sunset or sunrise, because of potential damage to the retina of the eye.

The sky is blue because blue light is more effectively scattered than is light of longer wavelengths. This does not mean, however, that no longer wavelengths are scattered. In fact, they are. When you look at the blue sky, light of all visible wavelengths enters your eyes, but because the light is relatively rich in shorter wavelengths, it appears blue. In photography, the blue color of the sky can be enhanced by using a polarizing filter. All scattered light is polarized, no matter its color. Because blue light is more likely to be scattered than red light when travelling through the atmosphere, blue light is more likely to be scattered more than once. Therefore, the blue light of the sky is less polarized than the other colors. Thus, a polarizing filter will more effectively remove the other, more polarized colors, leaving more blue light to enter the camera.

REFERENCES

1. J. Tyndall, "On the Blue Colour of the Sky, the Polarization of Skylight, and on the Polarization of Light by Cloudy Matter Generally," *Proc. Roy. Soc. (London), 17,* 223–233 (1869).
2. H. R. Kruyt, ed., *Colloid Science,* Vol. 1: *Irreversible Systems,* Elsevier: Amsterdam (1952).
3. J. Trefil, *Meditations at Sunset,* Charles Scribner's Sons: New York (1987).

12.20

Rainbow Spiral in an Optically Active Solution

Compounds, such as sugars, whose solutions rotate the plane of polarization of polarized light passing through them are called *optically active,* or *chiral.* The amount of rotation depends on the wavelength of the polarized light, so different colored light is rotated by different amounts.

When a beam of polarized light passes though a cylinder of corn syrup, the scattered light from the beam varies in color with the distance that the beam has traveled.

MATERIALS

100 mL distilled water

drop india ink

500 mL light corn syrup

500-mL glass cylinder, preferably without graduations

overhead projector with removable lens head

sheet of heavy aluminum foil or cardboard to completely cover the stage of the overhead projector

scissors

250-mL beaker

1-L beaker

glass stirring rod

piece of plastic or aluminum wrap to cover mouth of cylinder

two plane-polarizing filters

PROCEDURE [1, 2, 3]

Preparation

Measure the inside diameter of the 500-mL glass cylinder. In the center of the aluminum foil or sheet of cardboard, cut a circular hole with a diameter about 1 cm less than the inside diameter of the cylinder.

Remove the lens head from the overhead projector, so that it does not interfere with placing the cylinder in the center of the projector's stage. Place the foil or cardboard on the stage of the overhead projector with the hole at the center of the stage. This forms a mask that will allow the projector to shine a spot of light on the ceiling of the room.

Put 100 mL of distilled water into a 250-mL beaker and add one drop of india ink to the water. Stir the mixture to distribute the ink uniformly throughout the water.

Pour 10 mL of the ink-water mixture into the 1-L beaker. While stirring the mixture, slowly pour 500 mL of light corn syrup into the beaker. Stir slowly to avoid the formation of air bubbles. Continue stirring until the mixture is uniform.

Pour the corn-syrup mixture into the cylinder. Pour slowly into the center to avoid forming bubbles. Cover the mouth of the cylinder with plastic wrap or aluminum foil and allow the cylinder to stand undisturbed until any bubbles that may have formed have dissipated. This may take as long as a day or two.

Presentation

Turn on the overhead projector. A bright spot of light will appear on the ceiling above the projector. Place a sheet of polarizing film over the hole in the mask on the projector. The film will slightly dim the spot on the ceiling. Hold a second sheet of polarizing film above and parallel to the first sheet and slowly rotate the second sheet horizontally. The spot on the ceiling will alternately dim and brighten as the second polarizing sheet is rotated.

Remove the covering from the mouth of the cylinder containing the corn-syrup mixture. Set the cylinder onto the single sheet of polarizing film and center the cylinder over the hole in the mask. The spot of light projected on the ceiling is now colored. Hold the second polarizing film above the cylinder and rotate it horizontally. As the polarizing film is rotated, the color of the spot on the ceiling changes.

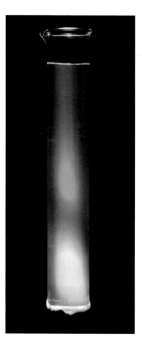

Dim the room lights and observe the scattered light as the beam travels upward through the corn-syrup mixture. The color of the scattered light varies with the distance that the beam has traveled through the mixture, giving it the appearance of a multicolored barber pole.

HAZARDS

There are no particular hazards associated with this demonstration.

DISPOSAL

The corn-syrup scattering mixture that has been prepared may be stored in a sealed bottle indefinitely. It may be stored in the cylinder if the cylinder is well sealed. To retard spoilage, add 2 mL of Consan Triple Action 20, a combination algicide-fungicide-bactericide available at most garden centers. Alternatively, any excess may be flushed down the drain with water.

DISCUSSION

In this demonstration, a beam of white, polarized light is passed through a sugar solution containing a suspension of carbon particles from india ink. When viewed from the side, light scattered by the particles from india ink appears colored, and the color varies along the path of the light through the solution.

As plane-polarized light beam travels through the sugar solution, the plane of its polarization is gradually rotated by the optically active (chiral) sugar molecules. The farther the light travels through the solution, the more its polarization plane is rotated. Furthermore, the magnitude of rotation also varies with the wavelength of the light, gradually increasing as the wavelength decreases [4]. Thus, the degree of rotation varies with the color of the light. The carbon particles from india ink scatter the light as it passes through the solution and affect the light's polarization. Light of a particular polarization is scattered more in one direction than in another. Thus, at a particular point along its path through the solution, the plane of polarized light of a particular color reaches the proper orientation to allow it to be scattered toward the viewer. The viewer will see various colors of scattered light along the path of the light through the solution. If the viewer moves around the solution, the colors will appear to move along the beam, because at different viewing angles, different colors will be scattered. Alternatively, if the viewer remains in place but the polarizing filter is rotated, the colors will also appear to move along the path of the light.

The sugar solution used in this demonstration is corn syrup, an aqueous solution containing mostly glucose, with a few other minor ingredients. The molecules of glucose are chiral, that is, the molecules are not identical to their mirror images. A solution that contains chiral molecules has the property of rotating the plane of polarized light that travels through it, with the degree of rotation depending on the distance traveled through the solution and on the wavelength of the light (see Demonstration 12.21).

India ink is a colloidal suspension of tiny particles of elemental carbon in water. The particles are of a size comparable to the wavelengths of visible light, which makes them effective scatterers of visible light. Of little importance to this demonstration but of significance to its use as an ink, india ink often also contains gum or resin that serves as a binder to hold the carbon particles together and to the paper after the ink has dried.

Another demonstration that makes use of the same cylinder containing a corn syrup–ink mixture as in this demonstration is Procedure C in Demonstration 12.24, Laser Light Is Polarized. In that demonstration, the beam of a laser pointer is directed down through the mixture. Because the laser beam contains only one color of light and the laser beam is polarized, what is observed is a beam that brightens and dims as it travels through the mixture.

REFERENCES

1. R. M. Sutton, ed., "Rotation of Plane of Polarization by Sugar Solution," in *Demonstration Experiments in Physics,* p. 425, McGraw-Hill: New York (1938).
2. A. Hultsch, "A Demonstration of Optical Activity," *The Physics Teacher, 20*(7), 476 (1982).
3. G. R. Davies, "Polarized Light Corridor Demonstration," *The Physics Teacher, 28*(7), 464–467 (1990).
4. H. Hudson, M. L. Wolfrom, and T. M. Lowry," The Rotatory Dispersion of Organic Compounds. Part XXIII. Rotatory Dispersion and Circular Dichroism of Aldehydic Sugars," *J. Chem. Soc.,* 1179–1192 (1933).

12.21

A Sugar Solution between Polarizers

A solution of sucrose, table sugar, rotates the plane of polarization of polarized light passing through it. The amount of rotation depends on the depth of the solution through which the light travels and also on the wavelength of the polarized light, so different colored light is rotated by different amounts. The beautiful effects that result from these dependencies, when the solution is viewed through crossed polarizers, are examined in Demonstration 9.51, Rotating Rainbows: A Solution in Polarized Light, in volume 3 of this series.

12.22

The Birefringence of Calcite

The speed of light passing through a transparent crystal depends on the arrangement of atoms within the crystal. In some crystals, the structure is such that the speed for one component of a light wave is different than that for the perpendicular component. (See the section *Polarized Light* in the introduction, page 19, for this model of a propagating light wave.) Crystals that have different indices of refraction in the two directions are *birefringent*.

When a mark on a surface is viewed through a calcite crystal, two images of the mark are observed. When viewed through a polarizing filter, one of the marks disappears. As the filter is rotated, this mark reappears while the other disappears.

MATERIALS

transparency film or sheet of paper

overhead projector or document camera and video projector*

projection screen

marking pen

calcite crystal, clear and colorless, with at least two parallel faces, preferably at least 2 cm thick

polarizing filter

PROCEDURE

Preparation and Presentation

Place the transparency film on the overhead projector, or place the sheet of paper on the stage of the document camera. With the marking pen, draw an X about 5 mm high on the transparency film or sheet of paper. Set the calcite crystal on top of the X. The projected image will show two Xs. Without lifting it, rotate the calcite crystal, and one X will appear to move around the other one, which remains stationary.

Hold the polarizing filter over the calcite crystal and rotate the filter. As the filter turns, one X will disappear. As the filter turns further, the X will reappear while the other disappears.

HAZARDS

Keep the calcite crystal away from acids. An acid will damage the crystal, because calcium carbonate is soluble in acids.

DISPOSAL

Materials may be kept for use in future presentations.

---

*Various types of suitable video projection systems are described on page xxxiii.

DISCUSSION

This demonstration shows that light travelling through a certain transparent material can follow two distinct paths. It also shows that the light becomes polarized and that the polarizations from the two paths are opposite. Such a material is said to be birefringent.

In a birefringent material, a single ray of light is broken into two rays when it enters the material. The direction of one of the rays is independent of the orientation of the material and is called the *ordinary ray*. The direction of the other ray depends on the orientation of the material and is called the *extraordinary ray*. In this demonstration, the fixed image is produced by ordinary rays, whereas the image that rotates around it is produced by extraordinary rays.

A birefringent crystal splits a ray of light into two because the speed of light through the crystal depends on the direction of light relative to the axes of the crystal. Light travels more quickly in one direction than in another. The speed of light through a transparent material is affected by the atoms encountered by the light as it travels. In some crystals, the arrangement of the atoms varies with direction, and the speed of light through such a crystal may also vary with direction.

When a ray of light enters a transparent material, its direction is altered (unless it strikes exactly perpendicular to the surface). A measure of how much the direction changes is expressed by the index of refraction of the material, represented by n. The index of refraction is the ratio of the speed of light in a vacuum to the speed of light in the transparent material:

$$n = \frac{c}{v_m}$$

where c is the speed of light in a vacuum, and v_m is the speed of light in the transparent material. This equation shows that the more slowly light travels in a medium, the greater the index of refraction of that medium. Because light travels at different speeds (depending on direction) through a birefringent material, a birefringent material has two different indices of refraction. Birefringent materials are also called doubly refractive.

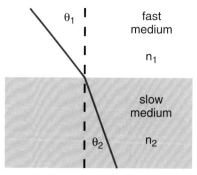

Figure 1.   Refraction of a light ray (red) crossing from one transparent medium into another.

When a ray of light travels from one transparent medium, such as air, into another transparent medium, such as calcite, the ray changes direction. This change is represented in Figure 1. Light travels faster in air than it does in calcite, so the index of refraction of air is smaller than that of calcite, $n_1 < n_2$. If the ray crosses the interface at an angle θ_1 to the perpendicular (the dashed line) and makes an angle of θ_2 in the second medium, the relationship between the angles is determined by the ratio of the indices of refraction of the two media by the equation

$$\frac{n_1}{n_2} = \frac{\sin \theta_2}{\sin \theta_1}$$

Light is an electromagnetic wave phenomenon. As a ray of light travels, electric and magnetic fields oscillate in directions perpendicular to the direction of travel. (See the section *Electromagnetic Radiation* in the introduction, page 22.) When light travels through matter, the electric field of the light interacts with electric fields created by electrons in the atoms. The stronger the interaction, the more the light is impeded, and the slower it travels. In a birefringent crystal, the interactions vary with the direction that light travels through it. The electric oscillations, all of which are perpendicular to the direction of travel, interact more strongly with the crystal in one direction than in the other direction. In the direction with greater interactions, the electric oscillations of the light are retarded more than are the oscillations in the perpendicular direction. For this reason, the ray is split into two, the ordinary ray and the extraordinary ray. Furthermore, in one ray the electric oscillations are in one direction, and in the other direction for the other ray. Thus, the two rays are polarized, and in opposite directions.

Single X viewed through calcite crystal (center). Viewed with polarizing filter on crystal (left). Viewed with polarizing filter rotated by 90 degrees (right).

The ordinary and extraordinary rays travel through a crystal of calcite at different speeds. Therefore, calcite has two different indices of refraction, an ordinary index, n_o, and an extraordinary index, n_e. The value of the index of refraction also depends on the wavelength of the light. Indices of refraction are commonly measured using the yellow light from a sodium lamp, whose wavelength is 589 nm. For this light, the value of n_o for calcite is 1.6584 and that of n_e is 1.4864 [1].

The phenomenon of birefringence was first described in a sample of calcite by the Danish scientist Erasmus Bartholin, who published a book about the phenomenon in 1669 [2]. There are many other naturally occurring birefringent crystals, including beryl, $Be_3Al_2(SiO_3)_6$; calomel, Hg_2Cl_2; ice, H_2O; quartz, SiO_2; and sodium nitrate, $NaNO_3$.

Calcite is the most stable crystal form of calcium carbonate, $CaCO_3$. Other forms are the minerals aragonite and vaterite. Aragonite changes to calcite when heated to 470°C, and vaterite is even less stable, converting to calcite at room temperature when exposed to water. Calcite crystals are trigonal-rhombohedral, although actual calcite rhombohedra are rare as natural crystals. Natural crystals show a remarkable variety of shapes including acute to obtuse rhombohedra, tabular forms, and prisms.

Calcite has a Mohs hardness of 3 and a specific gravity of 2.71 [*1*]. Pure calcite is color-less, although impurities can produce shades of gray, red, yellow, green, blue, violet, brown, or even black. Calcite ranges from transparent to opaque, depending on the purity and freedom from crystal defects. It is sometimes fluorescent or phosphorescent.

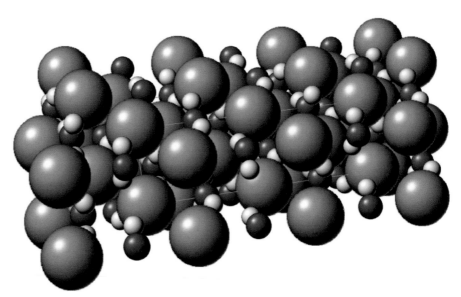

Figure 2.   Crystal structure of calcite. Blue spheres represent oxygen, gray carbon, and violet calcium ions.

REFERENCES

1. D. R. Lide, ed., *CRC Handbook of Chemistry and Physics,* 90th ed., CRC Press: Boca Raton, Florida (2009).
2. Erasmus Bartholin, *Experiments on Birefringent Icelandic Crystals,* trans. Thomas Archibald, Danish National Library of Science and Medicine: Copenhagen (1991).

12.23

A Liquid Crystal Display through a Polarizer

Liquid crystal displays (LCDs) depend on the interaction of the liquid crystals with polarized light.

The image on an LCD is alternately blocked and revealed when viewed through a rotating polarizing filter.

MATERIALS

LCD display panel, e.g., computer monitor or television

sheet of polarizing material, about 25 cm square

PROCEDURE

Preparation and Presentation

Place an LCD, such as a computer monitor or television, before the audience so that they can see the image on the screen. Hold a polarizing filter in front of the screen. Rotate the polarizing filter, and the image on the screen will darken and reappear as the filter is rotated.

HAZARDS

There are no particular hazards associated with this demonstration.

DISPOSAL

Materials may be preserved for repeated presentations of the demonstration.

DISCUSSION

This demonstration shows that the technology employed in LCD panels relies on the properties of polarized light. (See the section *Polarized Light* in the introduction, page 19, for a description of polarized light.) The liquid crystals used in an LCD panel interact with polarized light, rotating the plane of polarization. The panel also contains polarizing filters that pass light of one polarity but absorb light of the perpendicular polarity. Where the molecules of liquid crystal adjust the polarity of light to pass through the polarizing filter, the panel appears bright; where the liquid crystal molecules orient the polarity to the perpendicular, the panel appears dark. The fact that the light emitted (or reflected) by an LCD display is polarized is demonstrated by viewing the display through a polarizing filter.

Many substances can exist in more than one state. For example, water can exist as a solid (ice), liquid, or gas (water vapor). The state of water depends on its temperature. Below 0°C, water is a solid. As the temperature rises above 0°C, ice melts to liquid water. When the temperature rises above 100°C, liquid water vaporizes completely at 1 atm of pressure. Some substances can exist in states other than those of solid, liquid, and vapor. For example, cholesterol myristate (the cholesterol ester of myristic acid) is a crystalline solid below 71°C. When the solid is warmed to 71°C, it turns into a cloudy liquid. When the cloudy liquid is heated to 86°C, it becomes a clear liquid [1, 2]. Cholesterol myristate changes from the solid state to an intermediate state (cloudy liquid) at 71°C, and from the intermediate state to the liquid state at 86°C. Because the intermediate state exits between the crystalline solid state and the liquid state, it has been called the *liquid crystal state.* Molecules of substances with a liquid crystal state are generally oblong and rigid, that is, rod-shaped, as shown for cholesteryl myristate in Figure 1.

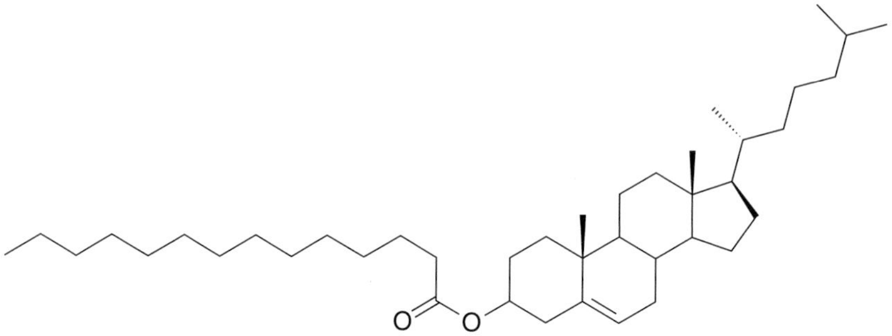

Figure 1.   Structure of cholesteryl myristate.

"Liquid crystal" also accurately describes the arrangement of molecules in this state. In the crystalline solid state, as represented in Figure 2, the molecules are arranged in a regularly repeating pattern in all directions and are held in fixed positions by intermolecular forces. As the temperature of the substance increases, its molecules vibrate more vigorously. Eventually, these vibrations overcome the forces that hold the molecules in place, and the molecules start to move. In the liquid state, this motion overcomes the intermolecular forces that maintain a crystalline state, and the molecules move into random positions, without pattern in location or orientation, as represented in Figure 3.

In materials that form liquid crystals, the intermolecular forces in the crystalline solid are not the same in all directions: in some directions the forces are weaker than in other directions. As such a material is heated, the increased molecular motion overcomes the weaker forces first, leaving the molecules bound by the stronger forces. This produces a molecular arrangement that is random in some directions and regular in others, as illustrated for one type of liquid crystal in Figure 4. The molecules are still in layers, but within each layer, they are arranged in random positions, although they remain more or less parallel to each other. Within layers, the molecules can slide around each other, and the layers can slide over one another. This molecular mobility produces the fluidity characteristic of a liquid.

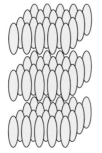

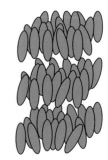

Figure 2.   Arrangement of molecules in a solid.

Figure 3.   Arrangement of molecules in a liquid.

Figure 4.   Arrangement of molecules in a liquid crystal.

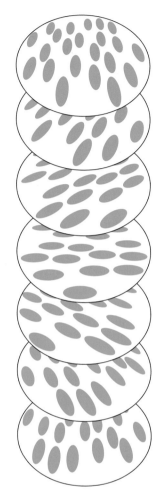

Figure 5.   Arrangement of molecules in a twisted nematic liquid crystal.

There are other molecular arrangements in liquid crystals. Many liquid crystals of tech-nological significance have the arrangement represented in Figure 5. Liquid crystals with this arrangement are called *twisted nematic* liquid crystals. In this arrangement, the long axes of the molecules lie in the plane of the layers and rotate by a small angle from one layer to the next. Twisted nematic liquid crystals are used in temperature-sensing devices that change color. They are also the most common type used in the LCDs found in calculators and watches [3]. Both of these applications depend on the way liquid crystals interact with light.

When light strikes a twisted nematic liquid crystal, some of the light is reflected. Only light with a wavelength equal to half the spacing between layers of similar orientation is reflected. Therefore, the reflected light will appear colored. As the temperature of the liquid crystal changes, the spacing between layers also changes and thus alters the wavelength of the reflected light and its observed color. Therefore, the color of the reflected light is an indi-cation of the temperature of the liquid crystal. Different substances form twisted nematic liquid crystals over different temperature ranges, and so a wide temperature range can be covered by using several substances.

When polarized light passes through a twisted nematic liquid crystal, the plane of polarization rotates, because the helical arrangement of molecules formed by the successive layers is chiral. (A helix or screw can be right- or left-handed and is not superimposable on its mirror image.). The degree of rotation depends on the number of layers of molecules the light encounters in the liquid crystal. If the axes of the molecules in the layer from which the light exits are at an angle of 90 degrees to those in the layer of entry, then the plane of polarization of the light rotates by 90 degrees. The ability of twisted nematic liquid crystals to rotate the plane of polarized light is exploited in LCDs. Figure 6 is a schematic diagram of an LCD. Ambient light enters (from the lower right) and is polarized by a polarizing filter. The polarized light passes through the front glass wall of the display, then through a transparent, electrically conductive coating on the glass, and then into the liquid crystal. The thickness of the liquid crystal is sufficient to rotate the plane of the polarized light by 90 degrees. At the back of the display, the light passes through another electrically conductive coating, glass plate, and polarizing filter. This rear polarizing filter is placed with its axis at 90 degrees to the front filter. The polarized light passes through this rear filter, because its polarization was also rotated by 90 degrees by the liquid crystal. At the back of the LCD is a mirror that reflects the light back through the cell. The light retraces its path, and an observer sees the display as relatively bright. When an electric potential is applied between the two conductive coatings, the resulting electric field affects the positions of molecules in the liquid crystal. The molecules tend to turn so they align with the electric field. When this happens, the plane of polarized light passing through the cell is no longer rotated by 90 degrees, and it cannot pass through the rear filter. Therefore, it is no longer reflected back to an observer, and the area appears dark. By the selective charging of areas of the conductive coating, patterns of dark digits or letters against a bright background are formed.

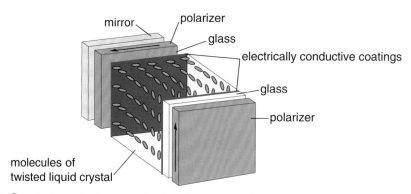

Figure 6. Schematic diagram of a liquid crystal display.

Substances that form liquid crystal structures are quite common. Approximately 0.5% of known carbon compounds have liquid crystal states. These structures are especially common in living organisms, where cell membranes are composed of molecules in a liquid crystal arrangement of parallel molecules in layers. A more mundane instance of liquid crystals is the opalescent fluid that forms in the bottom of a soap dish. Soap molecules have the appropriate oblong shape and when mixed with a little water, assume a liquid crystal arrangement.

REFERENCES

1. H. Kishimoto, T. Iwasaki, and M. Yonese, "Thermodynamic and Viscometric Studies on the Thermotropic Phase-Stability of Cholesteryl Myristate, Palmitate, Stearate, and Oleate," *Chem. Pharm. Bull., 34,* 2698–2709 (1986).
2. E. M. Barral, II, R. S. Porter, and J. F. Johnson, "Specific Heats of Nematic, Smectic, and Cholesteric Liquid Crystals by Differential Scanning Calorimetry," *J. Phys. Chem., 71,* 895–900 (1967).
3. Joseph Castellano, "Modifying Light," *American Scientist, 94,* 438–445 (2006).

12.24

Laser Light Is Polarized

The process that produces coherent, monochromatic laser emission also creates a polarized light beam.

The beam of a laser pointer is directed through a polarizing filter onto a wall. When the filter is rotated, the intensity of the spot on the wall decreases and increases (Procedure A). When the beam is aimed through an ink mixture and the scattered beam is viewed through a polarizing filter, the beam brightens and dims as the filter is rotated (Procedure B). When the beam is aimed through a mixture of corn syrup and dilute ink, the beam shows bright and dim regions (Procedure C).

MATERIALS FOR PROCEDURE A

red laser pointer

polarizing filter, about 5 cm square (or a lens from a pair of polarized sun glasses)

MATERIALS FOR PROCEDURE B

600 mL distilled water

drop india ink

500-mL glass cylinder, preferably without graduations

red laser pointer

250-mL beaker

1-L beaker

glass stirring rod

sheet of polarizing material, about 15 cm square

MATERIALS FOR PROCEDURE C

100 mL distilled water

drop india ink

500 mL light corn syrup

500-mL glass cylinder, preferably without graduations

red laser pointer

250-mL beaker

1-L beaker

glass stirring rod

piece of plastic or aluminum wrap to cover mouth of cylinder

PROCEDURE A

Preparation and Presentation

Direct the beam of the red laser pointer toward a screen or wall where the audience can see the spot. Hold a polarizing filter in the beam and rotate the filter. As the filter is rotated the spot will dim and brighten.

PROCEDURE B

Preparation

Put 100 mL of distilled water into a 250-mL beaker and add one drop of india ink to the water. Stir the mixture to distribute the ink uniformly throughout the water. Pour 10 mL of the ink-water mixture into a 1-L beaker. Add 500 mL of water to the beaker and stir the mixture to uniformly distribute the ink. Pour the diluted ink mixture into the 500-mL glass cylinder.

Presentation

Display the cylinder containing the ink mixture. Dim the lights and turn on the laser. Aim the laser beam down through the center of the cylinder. The laser beam will be visible through the side of the cylinder as it is scattered by the diluted ink. Hold a polarizing filter along the side of the cylinder, so the audience can view the scattered laser beam through the polarizer. Rotate the polarizer. As it is rotated, the scattered laser beam will dim and brighten.

If the scattered laser beam is viewed from different sides of the cylinder, the beam will appear brighter from one side and dimmer from an angle 90 degrees away.

PROCEDURE C [1]

Preparation

Put 100 mL of distilled water into a 250-mL beaker and add one drop of india ink to the water. Stir the mixture to distribute the ink uniformly throughout the water.

Pour 10 mL of the ink-water mixture into a 1-L beaker. While stirring the mixture, slowly pour 500 mL of light corn syrup into the beaker. Stir slowly to avoid the formation of air bubbles. Continue stirring until the mixture is uniform.

Pour the corn syrup mixture into the cylinder. Pour slowly into the center to avoid forming bubbles. Cover the mouth of the cylinder with plastic wrap or aluminum foil and allow the cylinder to stand undisturbed until any bubbles that may have formed have dissipated. This may take as long as a day or two.

Presentation

Remove the cover from the mouth of the cylinder containing the corn-syrup mixture. Dim the lights and turn on the laser. Aim the laser beam down through the center of the cylinder. When observed from the side, the scattered laser beam will display alternating regions of light and dark. Rotate the laser, and the bright areas will move up or down the beam. If the viewer moves around the cylinder and views it from 90 degrees away from the original viewpoint, the areas of the laser beam that were dim will now appear bright, and vice versa.

HAZARDS

When handling the laser pointer, avoid directing it or its reflections at viewers.

DISPOSAL

The ink-water mixture may be rinsed down the drain. The corn syrup/ink mixture may be saved for repeated demonstrations by sealing the cylinder with film or by transferring the mixture to a capped bottle.

DISCUSSION

Many laser beams are polarized. In procedures A and B, an ordinary polarizing filter is used to show this. The polarizing filter passes light that is polarized in the same plane as the filter, but blocks light that is polarized in a direction perpendicular to the polarization of the filter.

In Procedure A, the laser beam is aimed directly through the polarizing filter, and when the filter is rotated, the laser beam correspondingly brightens and dims. In Procedure B, the laser beam is scattered by particles from india ink, making the beam visible from the side. When the scattered beam is viewed through a polarizing filter and the filter is rotated, the scattered beam brightens and dims as the filter is turned. This procedure also shows that scattering of light causes polarization in the scattered light. If the light is already polarized, as is the laser beam, it is scattered more in one direction than in another. The brightest scattering appears at an angle 90 degrees from the dimmest.

In Procedure C, the laser beam is passed through a mixture that contains an optically active material (corn syrup) along with particles from india ink. The ink particles scatter the laser beam, and the optically active material rotates the plane of polarization of the laser beam. The rotation of the plane causes the amount of scattering at one viewing angle to vary along the length of the beam. As the viewing angle around the cylinder is changed, the intensity at a fixed distance along the beam will vary.

The output from many, though not all, lasers is polarized. For many lasers, the output is plane polarized. In the electromagnetic radiation produced by a plane-polarized laser, the electric field oscillates in a direction that is perpendicular to the direction of the laser beam. (See the section *Lasers* in the introduction, page 48.) A polarizing filter has an axis such that electromagnetic oscillations parallel to the axis pass through the filter, while the oscillations that are perpendicular to the axis are blocked. An analogy of how this happens, involving a rope and picket fence, is described in the *Polarized Light* section of the introduction on page 19.

In these demonstrations, red laser pointers seem to produce a more noticeable effect than green laser pointers. The effect is less pronounced when a green laser pointer is used, perhaps because the green beam appears so bright that even the dimmest effects produced by these procedures are still quite bright.

The sugar solution used in this demonstration is corn syrup. This syrup is an aqueous solution containing mostly glucose, with a few other minor ingredients. Glucose has *chiral* molecules, that is, molecules that are not identical to their mirror images. A solution that contains chiral molecules has the property of rotating the plane of polarized light that travels through it. The degree of rotation depends on the length of the path through the solution and on the wavelength of the light.

India ink is a colloidal suspension of tiny particles of elemental carbon in water. The particles are of a size comparable to the wavelengths of visible light, which makes them effective scatterers of visible light. Of little importance to this demonstration but of significance to its use as an ink, india ink often also contains gum or resin that serves as a binder to hold the carbon particles together and to the paper after the ink has dried.

Another demonstration that makes use of the same cylinder containing a corn-syrup-ink mixture as in this demonstration is Demonstration 12.20, Rainbow Spiral in an Optically Active Solution. In that demonstration, a beam of polarized white light is directed up through the mixture. What is observed is a beam that changes color along its length through the mixture.

REFERENCE

1. S. M. Mahurin, R. N. Compton, and R. N. Zare, "Demonstration of Optical Rotatory Dispersion of Sucrose," *J. Chem. Educ., 76,* 1234–1236 (1999).

Perception and Vision

12.25

Additive Color Mixing

All the colors we perceive are a result of the relative stimulation of the three types of cone cells in our retinas, which are sensitive to long (red), medium (green), and short (blue) wavelengths of visible light.

Red light, green light, and blue light are combined. The combined light produces different colors: red and green produce yellow, blue and green produce cyan, red and blue produce magenta, and all three together produce white. These effects are produced using three overhead projectors (Procedure A), three slide projectors (Procedure B), three LED flashlights (Procedure C), and three lightsticks (Procedure D).

MATERIALS FOR PROCEDURE A

three 12-in square sheets of cardboard, e.g., poster board

three 4-in squares of thin cardboard, e.g., manila folder

single-edged razor blade or X-Acto knife

red, green, and blue transparent filters (See Preparation for description.)

three overhead projectors (a limited version of the demonstration can be presented using two overhead projectors)

projection screen

MATERIALS FOR PROCEDURE B

three slide projectors (a limited version of the demonstration can be presented using two slide projectors)

set of three primary color slides for slide projector (See Preparation for description.)

projection screen

MATERIALS FOR PROCEDURE C

small flashlight with red LED source (wavelength around 630 nm)

small flashlight with green LED source (wavelength around 525 nm)

small flashlight with blue LED source (wavelength around 470 nm)

white, partially translucent plastic jar, with a volume of about 500 mL

MATERIALS FOR PROCEDURE D

one red, one green, and one blue lightstick, all of similar size and weight, each with a mounting hole on one end

2-in bolt, to fit through holes in lightsticks (about one-quarter inch in diameter)

about twenty washers to fit bolt

lock washer to fit bolt

nut to fit bolt

variable-speed electric drill

PROCEDURE A

Preparation

Obtain three pieces of clear, transparent, colored plastic film—one red, one green, and one blue. Each should be at least 1 in square. Best results will be obtained with a set of additive primary color transparencies, such as theater gels.* Alternatively, colored transparencies, such as report covers, can be used.

Prepare a mask for the overhead projector by cutting a 2-in square hole in the center of a piece of cardboard large enough to cover the stage of the overhead projector (about 12 in square). Do the same with two more pieces of cardboard to prepare three masks.

Make a holder for each of the three filters by cutting three 4-in squares of thin cardboard, such as manila-folder material. At the center of one edge of each square, cut a 1-in square notch that is smaller than the holes in the masks. To each of these three squares, use tape to attach one of the colored transparencies, so the notch is covered by the transparency (see diagram).

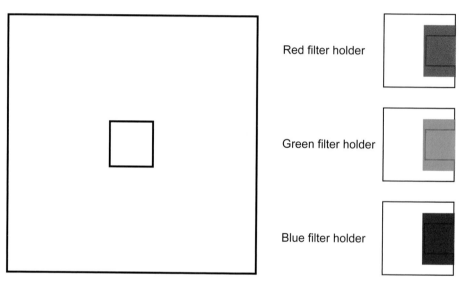

Red filter holder

Green filter holder

Blue filter holder

Mask for overhead projector (one for each projector)

Presentation

Place one of the mask boards on each stage of three overhead projectors. Turn on the projectors and darken the room lights. Cover the opening in each mask with one of the three colored filter holders, so that only filtered light is projected and all white light is blocked. Adjust the positions of the projectors so that the projected color squares overlap to form seven distinct regions: red, green, and blue alone; red plus green, red plus blue, and green plus blue; and all three together. The image on the screen will have seven distinct colors: red, green, blue, magenta (red plus blue), cyan (blue plus green), yellow (green plus red), and white (red plus blue plus green).

---

* Suitable filters are theater lighting gels: Rosco #27 medium red, #91 primary green, and #80 primary blue, available from Rosco Laboratories, Stamford, CT, www.rosco.com.

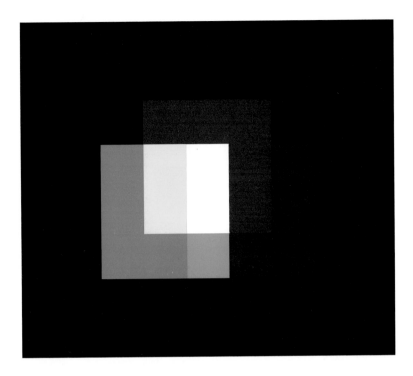

If only two overhead projectors are available, the effect of adding only two colors at a time can be displayed. The effect of adding all three primary colors, however, requires three projectors.

The difference between additive and subtractive color can be displayed by comparing the appearance of overlapping projected colors and overlapping filters on the stage of one projector. See Demonstration 12.26, Subtractive Primary Colors, for instructions.

PROCEDURE B

Preparation

A suitable set of red, green, and blue primary-color slides can be obtained from Fisher Scientific (catalog no. S63191). Alternatively, theater lighting gels can be mounted in 2-in × 2-in holders.*

Place one of the color slides into each of three separate slide projectors and position the projectors so that the images fall on the screen in an overlapping pattern with seven distinct areas: red, green, blue, red and green, red and blue, blue and green, and all three.

Presentation

Turn on the slide projectors and darken the room. Although only three colors are projected toward the screen, the image on the screen will have seven distinct colors: red, green, blue, magenta (red plus blue), cyan (blue plus green), yellow (green plus red), and white (red plus blue plus green).

---

* Suitable gels are Rosco #27 medium red, #91 primary green, and #80 primary blue, available from Rosco Laboratories, Stamford, CT, www.rosco.com.

If only two slide projectors are available, the effect of adding only two colors at a time can be displayed. The effect of adding all three primary colors, however, requires three projectors.

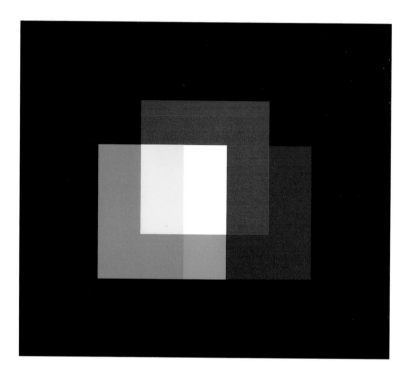

PROCEDURE C

Preparation and Presentation

Display to the audience the three colored-light flashlights, showing each of the three colors illuminated. Show also the empty white bottle. Darken the room.

Direct the light from the red flashlight into the bottle. The entire bottle will glow red. Extinguish the red light. Repeat the procedure with the green flashlight. The bottle will glow green. Extinguish the green light. Turn on both the red and the green flashlights. Direct the light into the bottle, aiming both to strike the inside of the bottle away from the audience. The audience will see the bottle glowing yellow.

Repeat the procedure of the previous paragraph with the remaining two combinations, red with blue and blue with green. Red and blue light shining into the bottle will produce a magenta glow, while blue and green will produce a glow with a cyan color.

Shine all three lights together into the bottle, and the bottle will glow white.

PROCEDURE D

Preparation

Place a washer on the 2-in bolt and insert the bolt into the hole on the end of one of the three lightsticks. (If the lightstick has a hook at the end, do not put the bolt through the hook; the hook may not be strong enough to hold the lightstick when the assembly is rotated. If the lightstick does not have a hole at the end, it should not be used in this demonstration.) Put enough washers on the bolt to be slightly less than flush with the edge of the lightstick.

Put the bolt through the hole on the second lightstick and add the same number of washers on the bolt. Add the third lightstick. Put a washer on the bolt and add the lock washer. Thread the nut onto the bolt and finger tighten it. Arrange the three lightsticks so they are 120 degrees apart, and tighten the nut securely.

Mount the bolt in the chuck of the drill. Check that the bolt-and-lightsticks assembly is securely mounted by turning on the drill to be sure that the assembly does not slip in the chuck. If the assembly slips, tighten the chuck. Turn off the drill.

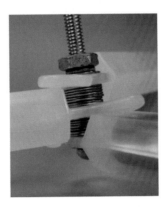

Presentation

Activate the three lightsticks. Note that one glows red, one green, and the other blue. Darken the room lights. Hold the drill with the chuck aimed directly at the audience. Turn on the drill to spin the three lightsticks slowly. Gradually increase the speed of the drill. As the speed of the drill increases, the three separate colors will blend together, forming a white disk. (The combined color may not be pure white—the perceived color depends on the relative brightness of the three lightsticks.) Do not increase the speed beyond the point at which the disk appears white, to avoid putting too much stress on the lightsticks. Turn off the drill and the three separate colors reappear. Repeat this several times.

HAZARDS

To avoid propelling lightstick fragments around the room, do not operate the drill at high speed.

DISPOSAL

The spent lightsticks may be discarded in a solid-waste receptacle. The other materials may be saved and used in future presentations.

DISCUSSION

This demonstration shows that combinations of two or three colored lights can produce colors different from the original colors. For example, a combination of red light with green light produces the appearance of yellow, even when there is no yellow light in the combination. Because the perceived yellow color is produced by adding red light and green light together, this yellow is called an *additive color*. Furthermore, because additive colors are produced by adding light, additive colors are generally brighter than the separate colors from which they are made.

The phenomenon of additive colors is a result of the way human color vision works. The retina of the human eye contains cells that are sensitive to visible light. These cells vary in their response to different wavelengths of light. In terms of this variation, there are four different types of cells. Figure 1 shows the responses for the four different types of cell. The lines labeled S, M, and L represent the responses of the three types of retinal cells that are responsible for color vision, namely the cone cells. (The fourth line, labeled R, represents the response of the rod cells, which produce vision in low-light situations, when vision is monochromatic, or black and white.) The three types of cone cells, the color-vision cells, are distinguished by their differing responses to light of varying wavelengths. One type has its maximum response at shorter wavelengths, and these cells are called the *short-wavelength cells, S* (for "short") *cells,* or *blue cells,* because short wavelengths are perceived as shades of blue. A second type has its maximum response at much longer wavelengths, and these are called the *long-wavelength cells, L cells,* or *red cells.* The third type has its maximum response at intermediate wavelengths, and these are the *mid-wavelength cells,* or *M cells,* or *green cells.*

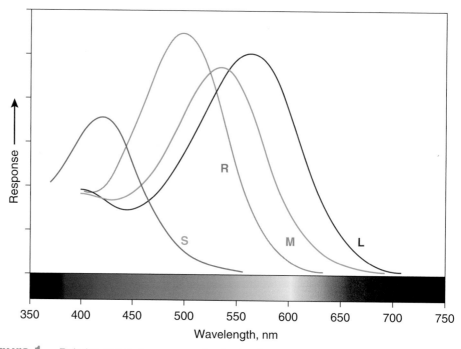

Figure 1.   Relative spectral response of retinal cells.

When light of a particular wavelength strikes the light-sensitive cells of the retina, each type of cone cell responds to a particular extent at that wavelength. The three responses are sent to the brain, where that particular combination of responses is interpreted as a particular color. For example, when light with a wavelength of 600 nm strikes the retina, it stimulates the L cells and the M cells, but not the S cells, in that region of the retina. Furthermore, the L cells are stimulated more strongly than the M cells. When these differing levels of stimulation are sent to the brain, this particular combination is interpreted as yellow. This same combination of levels can be produced in other ways that do not involve light with a wavelength of 600 nm. For example, if a combination of 550-nm light of relatively low intensity and 650-nm light at higher intensity strike the retina, the L and M cells will be stimulated to the same relative levels as when light of 600 nm strikes them. This combination of 550-nm (green) light and 650-nm (red) light will also be interpreted as yellow.

There is no unique set of primary additive colors. Any three spectrally well separated colors can be used to produce all colors of the spectrum. The red-green-blue set was first proposed by James Clerk Maxwell in 1855 and used by him in 1861 in the first demonstration of color photography [1]. An early form of color photography, the Lumiere process, used orange, green, and violet primaries [2]. The development of color television required the adoption of a standard set of primaries. It is easiest to obtain the same colors on the display screen as detected by the video camera when the screen and the camera use the same set of primary colors. When the screen of a color television or computer monitor is examined closely, perhaps using a magnifier, spots of only three colors will be observed: red, green, and blue. Although the spots do not overlap, they are too small to be optically resolved at normal viewing distances, and light from adjacent spots on the screen strike the same area of the retina, producing a sensation of the additive color.

To view the effects of adding different colors of light, it is necessary to use separate light sources, one for each color. This is why three projectors are used in procedures A and B. Placing two or three of the color filters in one projector does not produce added light colors, but subtracted colors instead. Each filter removes some wavelengths of light and transmits others. Figure 2 shows the transmission spectra of three primary-color filters. After passing through two or more filters, the transmitted light is what is left after each filter has subtracted its portion of the spectrum. The transmitted light is the subtractive color produced by the combination of filters, rather than the additive color produced by combining light transmitted by each filter. Demonstration 12.26 deals with subtractive primary colors and their relationship to the additive primaries.

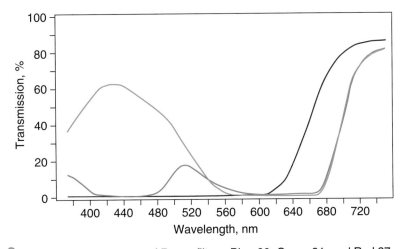

Figure 2.   Transmission spectra of Rosco filters, Blue 80, Green 91, and Red 27.

When observing colors produced by combining colored light, the most vivid results are obtained when the surroundings are dark. This is why the observations should be made in a darkened room, and why masks are used with the overhead projectors to eliminate white light from the projector. This is also why commercial displays of televisions are usually dimly lit, to enhance the colors of the images on the screens. That perceived color is strongly influenced by surroundings is also demonstrated by the Land effect (see Demonstration 12.30).

The primary transmission filters used in procedures A and B each produces light containing a fairly broad range of wavelengths. The light produced by the three LED flashlights used in Procedure C contains much narrower ranges of wavelengths, as shown in

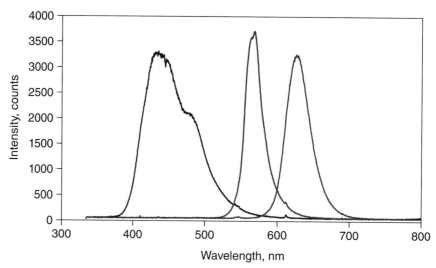

Figure 3.   Emission spectra of blue, green, and red light-emitting diodes.

Figure 3. The light produced by the colored lightsticks, as shown in Figure 4, contains a range of wavelengths wider than that from LEDs but narrower than that produced by the transmission filters.

The color observed when the three colored lightsticks are rotated is highly dependent on the relative brightness of the three lightsticks. Green lightsticks are generally much brighter than red or blue lightsticks, so the color observed by the combination of the glow from these three has a somewhat yellow-green cast rather than being pure white.

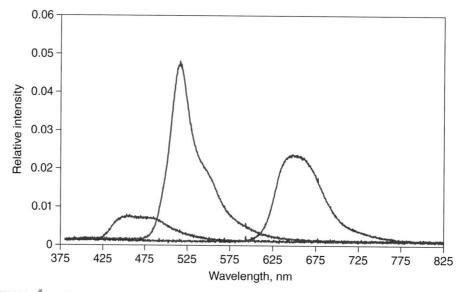

Figure 4.   Emisson spectra of typical blue, green, and red lightsticks.

REFERENCES

1. I. J. Tolstoy, *Clerk Maxwell: A Biography,* University of Chicago Press: Chicago (1982).
2. M. L. Heidingsfeld, "The New Lumiere Process of Color Photography," *The Lancet-Clinic, 99,* 26 (27 June 1908).

12.26

Subtractive Primary Colors

The three colors of inks used in your color printer are cyan, magenta, and yellow. The light reflected from these inks is missing wavelengths absorbed by the ink, that is, wavelengths are subtracted from the incident white light. Combinations of these inks leave wavelengths in the reflected light that stimulate the three types of cone cells in our retinas that are sensitive to long (red), medium (green), and short (blue) wavelengths of visible light.

Three transparent films—one cyan, one magenta, and one yellow—are placed on the stage of an overhead projector. Where the films overlap, other colors are produced, and these colors are the additive primaries: red, green, and blue.

MATERIALS

12-in square sheet of cardboard, e.g., poster board

three 4-in squares of thin cardboard, e.g., manila folder

single-edged razor blade or X-Acto knife

cyan, magenta, and yellow transparent filters (See Preparation for description.)

overhead projector

projection screen

PROCEDURE

Preparation

Obtain three pieces of clear, transparent, colored plastic film—one yellow, one cyan, and one magenta. Each should be at least 1 in square. Best results will be obtained with a set of subtractive primary color transparencies, such as theater gels.* Alternatively, similarly colored transparencies such as report covers can also be used.

Prepare a mask for the overhead projector by cutting a 2-in square hole in the center of a piece of cardboard large enough to cover the stage of the overhead projector (about 12 in square).

Make a holder for each of the three filters by cutting three 4-in squares of thin cardboard, such as manila-folder material. At the center of one edge of each square, cut a 1-in square notch smaller than the holes in the masks. To each of these three squares, use tape to attach one of the colored transparencies, so the notch is covered by the transparency (see diagram).

Presentation

Place the mask board on the stage of the overhead projector. Turn on the projector and darken the room lights. Cover part of the opening in the mask with the yellow filter, so that about half of the opening is covered by the filter and half is uncovered. The image on the screen will show a patch of white light and a patch of yellow light. Cover the uncovered area

---

* Suitable filters are theater gels: Roscolux #4390 CalColor 90 Cyan, #4790 CalColor 90 Magenta, and #4590 CalColor 90 Yellow, available from Rosco Laboratories. Available from Sargent-Welch is a set of primary subtractive color filters, catalog no. WL3664.

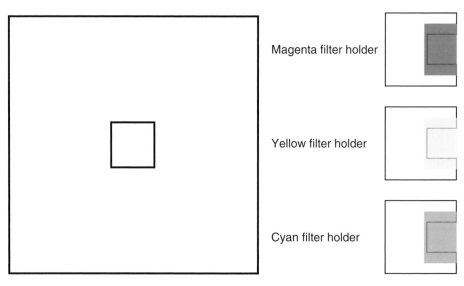

Magenta filter holder

Yellow filter holder

Cyan filter holder

Mask for overhead projector

of the mask with the cyan transparency, so that only yellow and cyan appear on the screen. Adjust the filters so that they overlap over part of the opening. On the screen will be three colored areas: yellow, cyan, and green.

Repeat the process in the previous paragraph with the remaining two combinations of filters: yellow and magenta to form red, and magenta and cyan to form blue. Place all three filters over the opening in the mask, so that there are areas covered by only one filter, areas covered by the three combinations of two filters, and an area covered by all three. The image on the screen will have seven regions: cyan, yellow, magenta, red, green, blue, and gray.

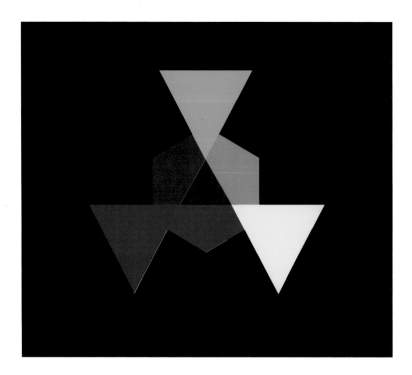

HAZARDS

There are no particular hazards associated with this demonstration.

DISPOSAL

The filters may be stored for repeated presentations of the demonstration.

DISCUSSION

This demonstration shows the effect of removing various ranges of wavelengths of light on the perceived color of the light. Each colored filter transmits light containing a limited range of wavelengths because it absorbs some wavelengths but not others. When two filters are overlapped, each removes its set of wavelengths, and the light transmitted by the two combined contains fewer wavelengths than that transmitted by each separately. Therefore, the color of the light transmitted by a combination of filters is different than that transmitted by either alone. Because the color transmitted by a combination of filters is produced by removing wavelengths from what is transmitted, the transmitted color is called a *subtractive color*. Because subtractive colors result from removing wavelengths and thereby intensity, subtractive colors are darker and duller than the colors used to produce them. This is in contrast to colors produced by adding different wavelengths together, which produce additive colors. Additive colors are presented in Demonstration 12.25.

Any filters of differing colors can be used to show that the light transmitted by a combination of filters has a different color than either of the filters alone. However, filters with the colors cyan, magenta, and yellow are particularly useful because they correspond to the standard subtractive primary colors used in modern color process printing. The transmission spectra of three subtractive primary filters are shown in Figure 1. These colors are used because their pairwise combinations yield red, green, and blue, which are the additive primary colors used in color televisions and monitors. Using cyan, magenta, and yellow as subtractive colors simplifies color matching to displays that use red, green, and blue as additive primaries.

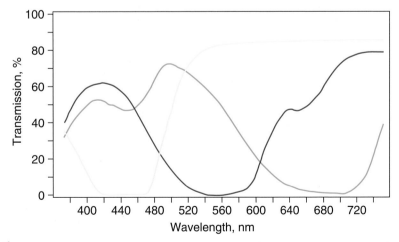

Figure 1.   Transmission spectra of subtractive primary filters cyan, magenta, and yellow.

Combinations of colors other than cyan, magenta, and yellow can serve as subtractive primaries. The pigments available to artists before the era of synthetic dyes and pigments, which began in the mid-nineteenth century, were limited to natural materials. Artists learned to combine this limited set of colors to produce a broad range of colors.

12.27

The Perception of Brightness Is Relative

Our eyes are sensitive to very small differences in the brightness of adjacent areas in a scene and not good at perceiving absolute brightness. Our perception of the brightness of a particular area depends on the relative brightness of its surroundings. Our perception is also influenced by what we expect (from experience) to see.

Figures are displayed that have gray areas that are exactly the same shade of gray but appear to be different in brightness because of their surroundings.

MATERIALS FOR PROCEDURE A

two copies of the image from Figure 1

single-edged razor blade or X-Acto knife

sheet of uniformly colored paper the same size as the image copies

document camera (optional; for display to a large audience)

MATERIALS FOR PROCEDURE B

copy of the image from Figure 2

sheet of opaque paper the same size as the image

single-edged razor blade or X-Acto knife

document camera (optional; for display to a large audience)

projected Photoshop image of Figure 2 from a computer (another option)

PROCEDURE A

Preparation

On two separate sheets of paper, print as large as possible two copies of the image of Figure 1 [*1*]. Use a single-edged razor blade or X-Acto knife to cut out and remove the two smaller gray squares from one of the sheets. For a small audience, the sheets may be displayed directly. For a large audience, a document camera can be used.

Presentation

Display the sheet with the complete image of Figure 1. The two smaller gray squares are the same shade of gray. However, the square surrounded by the darker background appears to be brighter than the square surrounded by the lighter background.

To convince the audience that the two smaller squares are the same, display the sheet of uniformly colored paper. Place the sheet with square holes over the colored paper, so that the colored paper is visible through the cut-outs. The colored square visible through the opening in the darker square appears brighter than the colored square visible through the

Figure 1.

other opening. This may be repeated with papers of other colors, and a similar effect will be observed.

PROCEDURE B

Preparation

Print a copy of the image of Figure 2 [2]. In a sheet of opaque paper of the same size, cut an opening over the "square" in the center of the checkerboard and a second opening over the second "square" from the right in the row at the front of the checkerboard. (The relevant squares are those marked by an *x* in Figure 3.) These openings should expose only the gray areas within these checkerboard squares.

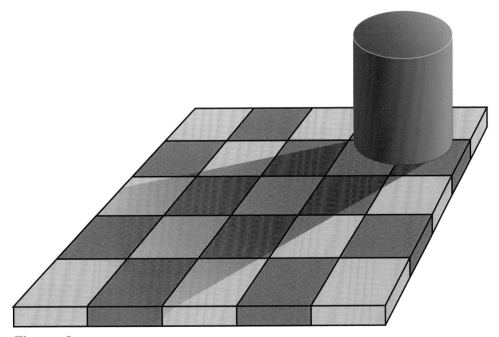

Figure 2.

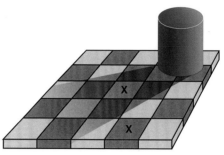

Figure 3.

Presentation

Display the image of Figure 2. Although they do not appear so, the square in the center of the checkerboard and the second square from the right in the front row of the checkerboard are exactly the same shade of gray. To show this, place the paper with openings over these squares on top of image from Figure 2. When only the two patches of gray are visible, surrounded by the uniform cover sheet, they look the same.

A complementary presentation is to project a Photoshop image of Figure 2 from a computer and then use the dropper tool to show that the two squares have identical color profiles in either the RGB or CMYK scales. This might not be as convincing for some, because the visual illusion remains. Others might find it satisfying to see that the gray shades are identical, since the proof does not depend upon their vision, which has proved not to be a reliable gauge of the true amount of light coming from the various areas of the image.

HAZARDS

There are no hazards associated with this demonstration.

DISPOSAL

Save the hardcopies of the figures and cut-out sheets and the computer files to use again.

DISCUSSION

How does the checkerboard illusion work? This discussion is an adaptation of an explanation by Edward H. Adelson [3]. The visual system needs to determine the color of objects in the world. In this case, the problem is to determine the gray shade of the squares on the checkerboard. Just measuring the light coming from a surface (the luminance) is not enough. A cast shadow will dim a surface, so that a white surface in shadow may be reflecting less light than a black surface in full light. Your visual system uses several clues that you have learned (subconsciously) from experience to determine where the shadows are and how to compensate for them, in order to determine the shade of gray "paint" that belongs to a surface. "Paint" is the color or shade of gray that you "see" in your mind, not the physical color or the physiological color of the image on your retina.

The first clue is based on local contrast. In shadow or not, a surface (checkerboard square here) that is lighter than its neighboring surfaces (squares) is probably lighter than average, and vice versa. In Figure 2, the light square in shadow is surrounded by darker squares. Thus, even though the square is physically dark, it is light when compared to its neighbors. The dark squares outside the shadow, conversely, are surrounded by lighter squares, so they look dark by comparison.

A second clue is based on the fact that shadows often have soft edges, while the boundaries of "painted" objects or surfaces (like the squares) often have sharp edges. Your visual system tends to ignore gradual changes in light level, so that it can determine the color of the surfaces without being misled by shadows. In the figure, the shadow looks like a shadow, both because it is fuzzy and because the shadow-casting object (the cylinder) is visible.

The "paintness" of the squares (that is, your interpretation of their color or shade of gray) is aided by the form of the *X-junctions* formed by four abutting squares. This type of junction is usually a signal to your visual processing system that all the edges should be interpreted as changes in surface color rather than in terms of shadows or lighting.

The result of the demonstration based on Figure 1 has a similar explanation. A gray (or colored) surface within a dark surround appears lighter (brighter) relative to an identical surface within a lighter surround. In this case, the surfaces of interest are both lighter than the surrounds, but the *contrast* is so much larger within the dark surround that its "paintness" (your interpretation) appears lighter. The demonstration shows that your interpretation of the relative brightness extends to colors as well as shades of gray.

Note that all of the clues above have their origin with the retinal receptive fields that are so good at detecting edges (changes in contrast) in your retinal images, supplemented by how you have learned to interpret the great majority of the images you receive every waking moment. (See the section *Perception* in the introduction, page 78, for a description of retinal receptive fields.) As with many optical illusions, the effects in these demonstrations really illustrate the success, rather than the failure, of your visual system. The visual system is not very good at being a physical light meter, but that is not its purpose. Its important task is to separate the information from the image into meaningful components, and thereby you perceive the nature of the objects in view, and, as much as possible, place them in context with previous experiences with similar objects.

When you view the masked and unmasked checkerboard demonstration, your perception of the shades of gray in the designated squares is different because the surroundings are different in the two cases. What you perceive is strongly influenced by the context of the input and your past experience with similar contexts. Your eye-brain perception in the unmasked case involves integration of familiar contexts—checkerboards, light and shadow, and so on. When these contexts are removed by masking all but the designated squares with a uniform surrounding, your perception is changed and you sense the identical shades of gray. You can take advantage of this phenomenon when viewing and interpreting a chemical demonstration by trying to isolate and focus on just those aspects of the system that are relevant to understanding the basic concept(s) involved. This is more difficult than in the case of the perceptual demonstration that is set up to make the isolation easy. Help from the presenter to direct your attention to particular aspects of the system is the most important guide you can have to begin this "masking." From some points of view, however, the inability to completely mask the more complicated system is fortunate, because intrusion of the "surroundings" can lead to consideration and extension of the concepts or introduction of new ones that enhance the experience and make it more fruitful. Presentation, therefore, is a balancing act between the masked and unmasked demonstration and depends a good deal on what perception is desired and/or desirable.

REFERENCES

1. E. H. Adelson, "Lightness Perception and Lightness Illusions," in *The New Cognitive Neurosciences,* pp. 339–351, 2nd ed., M. Gazzaniga, ed., MIT Press: Cambridge, Massachusetts (2000).

2. Checkerboard figure adapted from one by Prof. Edward H. Adelson, Department of Brain and Cognitive Sciences, Massachusetts Institute of Technology.

3. E. H. Adelson, "On Seeing Stuff: The Perception of Materials by Humans and Machines," in B. E. Rogowitz and T. N. Pappas, eds., *Proceedings of the SPIE: Human Vision and Electronic Imaging VI, 4299,* pp. 1–12 (2001).

12.28

The Hermann-Grid Illusion

You can see things that aren't there.

When viewing an array of black squares separated by white lines, blurry gray dots appear at the intersections of the white lines. A similar effect occurs when viewing white squares separated by black lines. One explanation of this illusion is based on a simple visualization of the receptive field architecture of the retina and the relative sizes of those fields at different locations on the retina. Experiments with variations on the traditional grid provide evidence that the explanation is more complex and requires accounting for all parts of the brain visualization system, not simply the retinal part.

MATERIALS

array of black squares separated by white lines (this may be projected on a screen or printed on individual cards), Figure 1.

array of white squares separated by black lines (this may be projected on a screen or printed on individual cards), Figure 2.

PROCEDURE

Preparation and Presentation

Display the array of black squares separated by white lines. Instruct the audience to look at an intersection of white lines near the center of the array. In their peripheral vision, they will see blurry gray dots at the other intersections of white lines at a distance from the one they have focused on. When they try to look directly at one of the gray dots, it disappears. The illusion can be observed at just about any scale and distance, from that in Figure 1 to an image projected on a screen. The ratio of the width of the white lines to the size of the black squares can also be varied. The illusion disappears when the ratio is unity; the usual ratio is about 1 to 3, as in the figures.

Repeat with the array of white squares separated by black lines. In this case, the blurry gray dots will appear at the intersections of the black lines.

HAZARDS

There are no hazards associated with this demonstration.

DISPOSAL

Save the hardcopies of the grids and the computer file to use again.

DISCUSSION

This is an optical illusion discovered in the nineteenth century and first published in 1870 by a German physiologist, Ludimar Hermann [1]. In 1960 an explanation for the illusion,

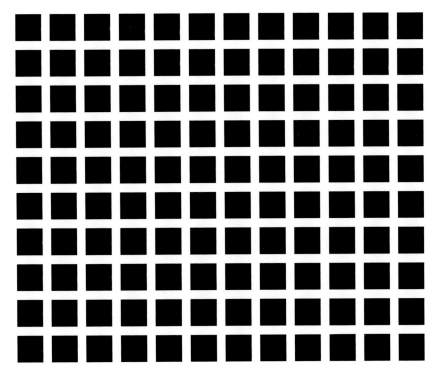

Figure 1. Classic Hermann grid.

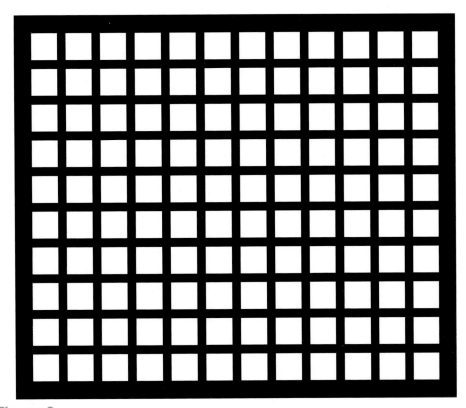

Figure 2. Inverse Hermann grid.

based on the way retinal receptive fields respond to light, was proposed by a German neurologist, Günter Baumgartner [2]. (See the section *Perception* in the introduction, page 78, for a description of retinal receptive fields.) The explanation originally proposed by Hermann [3], which was based on simultaneous contrast (see Demonstration 12.27), is remarkably similar to that based on receptive fields, although this organization of the retina was not discovered until the middle of the twentieth century. In the following discussion, we will first describe Baumgartner's explanation and then present the evidence that it is inadequate to be the whole story.

The retina contains overlapping receptive fields, each composed of two regions, a center region and a surrounding region. These fields are known as *center-surround fields,* of which there are two types, *center-on* and *center-off.* Each center-on receptive field, Figure 3, has a central region of rods and/or cones that, when illuminated, activates the ganglion cell associated with this field and a surrounding region that, when illuminated, deactivates the cell. (Center-off fields operate in the reverse manner, being activated when the center is dark and the surround is illuminated.) The intensity of the signal sent to the brain by the retinal ganglion cell depends on the relative degree of illumination of these two regions.

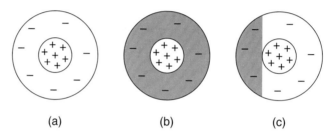

(a)   (b)   (c)

Figure 3.   Idealized representation of a center-on retinal receptive field.

When the entire light-sensitive area is uniformly illuminated, Figure 3a, the surrounding deactivating region suppresses the central activating region, and the signal sent by the ganglion cell is, at most, weak. When the central region is illuminated but the surrounding area is dark, Figure 3b, the signal sent is maximal. When the central area is illuminated, but a part of the surrounding area is dark, Figure 3c, the intensity of the response is intermediate.

Consider now what happens in your peripheral vision when you focus on one of the intersections of the black-squares Hermann grid. Suppose that the center-surround receptive fields outside the central retina are larger than those in the center and are likely to receive light from a larger area of the grid. In this case, the center-surround receptive fields that signal the white of the intersections (Figure 4b) are more suppressed by the four bits of light in the suppressing surround area (7(+) + 4(−) = 3(+)), than are those that signal the white lines between intersections (Figure 4a), which have only two bits of white light in the suppressing surround area (7(+) + 2(−) = 5(+)). The ganglion cells that signal whiteness at the intersections are more suppressed than those signaling the white lines between the intersections, and therefore less active. So the intersections look less white, that is, darker.

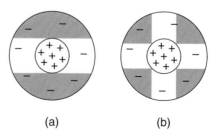

(a)   (b)

Figure 4.   Center-on peripheral retinal receptive fields getting their input from different parts of the black-square Hermann grid.

When you try to focus on one of the gray dots at these intersections, the dot disappears because the illumination from the intersection is now focused on your fovea and the receptive fields in the fovea are very small. Many adjacent fields are now entirely illuminated by the light from the intersection (with no contribution from the dark areas), so the associated ganglion cells signal uniform whiteness. Thus, this illusion provides evidence that the peripheral receptive fields are a good deal larger than those in the center of the retina. (Baumgartner used information from experiments with the grid to estimate the size of these fields.)

Check your understanding by formulating an explanation for the appearance of the blurry gray dots at the intersections of the black lines in the array of white squares separated by black lines.

Although Baumgartner's explanation for the Hermann grid illusion seems satisfying and provides a persuasive example of the receptive field model of retinal structure, a minor variation of our demonstration indicates that it is not this simple. Look again at the black-squares Hermann grid and rotate it about 45 degrees, so the white lines run diagonally and the squares are "diamonds." For almost all observers, the illusion becomes much less pronounced and, for many, it disappears completely. This is not predicted by the argument associated with Figure 2, since the angle at which the white line or lines cross the receptive fields should make no difference.

Several experiments using variations on the usual Hermann grid have been designed to test Baumgartner's explanation. The results for grids like the one in Figure 5 are particularly compelling [4]. In this grid with sinusoidal lines, the illusion disappears completely, even though the coverage of the receptive fields would be essentially unchanged from that in the usual straight-line grid.

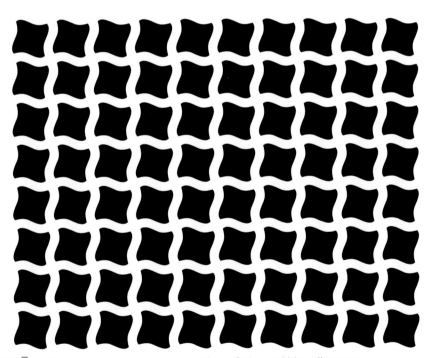

Figure 5.   Sinusoidal grid developed by János Geier and his colleagues.

Our present view of visual processing includes evidence that signals from horizontal, vertical, and slanted (oblique) lines or objects are sensed separately in the lateral geniculate nuclei and primary visual cortex of the brain before the information is combined and processed to provide the perception of what is seen. The origin of the oriented signals to these parts of the brain may be that retinal receptive fields are ellipsoidal and oriented, not essentially round, as our idealized pictures suggest. We appear to have more sensitivity to horizontal and vertical signals, but oblique signals can interfere with or alter them, and the processing in the visual cortex may be responsible for the disappearance of the Hermann grid illusion when oblique lines become part of the pattern (as in the rotated grid or the sinusoidal grid) [5].

REFERENCES

1. L. Hermann, "Eine Erscheinung simultanen Contrastes," *Pflügers Arch. gesamte Physiol., 3,* 13–15 (1870).
2. G. Baumgartner, "Indirekte Größenbestimmung der rezeptiven Felder der Retina beim Menschen mittels der Hermannschen Gittertäuschung," *Pflügers Arch. gesamte Physiol., 272,* 21–22 (1960).
3. L. Spillman, "The Herman Grid Illusion: A Tool for Studying Human Perceptive Field Organization," *Perception, 23,* 691–708 (1994). See page 692 for Hermann's explanation, in English translation, of the illusion.
4. J. Geier, L. Bernath, M. Hudak, and L. Sera, "Straightness as the Main Factor of the Hermann Grid Illusion," *Perception, 37,* 651–665 (2008).
5. P. H. Schiller and C. E. Carvey, "The Hermann Grid Illusion Revisited," *Perception, 34,* 1375–1397 (2005).

GENERAL REFERENCE

M. Livingstone, *Vision and Art: The Biology of Seeing,* Harry N. Abrams: New York (2002).

12.29

Finding the Blind Spot

Where the optic nerve exits the back of the eye there is a blind spot.

When a card containing a dot or a broken line is moved slowly from arm's length toward the face, the dot or break will disappear then reappear. This demonstrates the existence of the retinal blind spot, as well as how the brain compensates for lack of information from this area. Simple distance measurements during the demonstration can be used to find the approximate location and size of the blind spot on the retina.

MATERIALS

two 3-in × 5-in index cards for each observer

pen or pencil for each observer, if they are to prepare the cards themselves

meter sticks (optional for semiquantitative location and size of the blind spot on the retina)

PROCEDURE

Preparation

For each participant prepare two 3-in × 5-in index cards, as illustrated below. The participants can prepare these themselves if they have the cards and a pen or pencil. Place one of the cards horizontally and draw an X about 2 cm from the left edge halfway down the card. Draw a heavy dot about the same distance from the right edge and also halfway down the card. On the other card, draw a similar X on the left side, but on the right side draw a heavy vertical line with gap of about 1 cm.

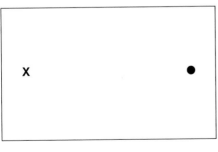

Card 1

Card 2

Presentation

In your right hand, hold the card horizontally at arm's length, with the X to the left and the dot to the right. Cover your left eye. With your right eye, look directly at the X. In your peripheral vision, you will be able to see the dot. While still looking at the X, gradually bring the card closer to your eye. As the card gets closer, at some point the dot will disappear from your peripheral vision. Continue to bring the card closer, and eventually the dot will reappear.

Repeat this procedure with the card containing the X and the broken vertical line. At arm's length, you will see in your peripheral vision the two lines with the gap between them. As you move the card closer, eventually the gap will disappear, and in your peripheral vision you will see only a single vertical line. As you move the card closer still, the gap will reappear.

The same procedures will show the blind spot in the left eye if the cards are rotated 180 degrees and the right eye is covered.

Optional: Work in pairs. While one partner does the experiment, the other should measure and record the distance from the card to the experimenter's eye when the dot (or gap) just disappears and when it reappears. Also measure and record the distance between the X and dot (or gap). Exchange roles and repeat the experiment and measurements. Use these measurements and the analysis below to find the location of the blind spot (with respect to the fovea) in each eye and its size.

HAZARDS

There are no hazards associated with this demonstration.

DISPOSAL

Cards may be collected for reuse, or the audience may be allowed to keep them to test their friends and families.

DISCUSSION

The analysis of this activity to locate the retinal blind spot and discover how the brain (of which the retina is a part) interprets the information is presented at the end of the introduction, pages 82–83. Here we will focus on the analysis of the optional distance measurements to determine the location and size of the blind spot.

A schematic diagram of the right eye focusing on the X with the dot in the peripheral vision is shown in Figure 1. For simplicity, only one light ray from each object is included.

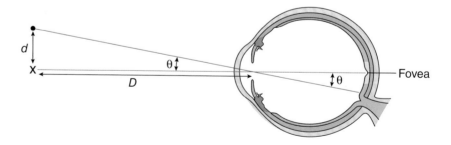

The observer is focusing on the X, so the light ray from this object falls on the fovea, the area of highest resolution at the center of the back of the eye. The distance of the object from the eye (the iris where the rays come together) is labeled D. The distance between the X and the dot is labeled d. The light ray from the dot falls away from the fovea and is shown here just about where the blind spot begins. The angle between the two light rays, θ, outside the eye is the same as the angle between the rays inside the eye.

Simple trigonometry determines the value of θ:

$$\tan \theta = \frac{d}{D}$$

$$\therefore \quad \theta = \tan^{-1}\left(\frac{d}{D}\right)$$

For one observer, when the X and dot are about 6 cm apart ($d = 6$ cm), the dot disappears at about 23 cm and reappears at about 17 cm. When the X and dot are about 10 cm apart, the dot disappears at about 45 cm and reappears at about 28 cm. For the first set of data, the angles when the dot disappears and reappears are

$$\therefore \quad \theta(\text{disappear}) = \tan^{-1}\left(\frac{6}{23}\right) = 15 \text{ degrees}$$

$$\therefore \quad \theta(\text{reappear}) = \tan^{-1}\left(\frac{6}{17}\right) = 19 \text{ degrees}$$

The corresponding values for the second set of data are 13 degrees and 20 degrees. Given the difficulty of measuring these distances accurately, the agreement of the two sets of data is quite good.

As a check on these conclusions about the position and size of the retinal blind spot, in terms of the angular distance away from the fovea, look at Figure 58 in the introduction, page 66. The blind spot is located between about 15° and 20° toward the nasal side of the retina. The figure, representing the right eye, shows that this demonstration will find the blind spot on the nasal side of the fovea. The data analyzed here also show that the demonstration locates the blind spot and gives its size just where Figure 58 indicates it should.

GENERAL REFERENCE

Almost all physiology textbooks that discuss the eye suggest some variation of this demonstration to show readers how to find their retinal blind spot.

12.30

The Land Effect

The vivid colors of the world we perceive are the result of combinations of long, medium, and short wavelengths of visible light striking retinal receptors sensitive to these wavelengths. (See Demonstration 12.25, Additive Color Mixing.) However, physiology can partially trump physics and produce the perception of an array of colors when a single color and two specially prepared gray-scale images are combined.

Two black-and-white images of the same scene are projected and aligned on a screen. When a red transmission filter is placed in the beam from one of the projectors, the projected image contains a range of colors (Procedure A). When a black-and-white image on a transparency is overlaid on a red-tinted monochrome printed image, the combined image contains a variety of colors (Procedure B).

MATERIALS FOR PROCEDURE A

full-color digital image

computer with image-editing software

two sheets of transparency film

printer to print black on transparency film

two 13-in squares of poster board

single-edged razor blade or X-Acto knife

two overhead projectors

projection screen

red and green transparent sheets, such as plastic report covers

MATERIALS FOR PROCEDURE B

full-color digital image

computer with image-editing software

sheet of transparency film

color printer to print on paper and on transparency film

PROCEDURE A

Preparation

Starting with a color digital photograph, generate two different black-and-white transparencies as follows.

Open the digital image file with photo-editing software, for example, Photoshop. If it is not already, convert the image to RGB color mode. Separate the three color channels—red, green, and blue—into three separate images. Discard the blue channel. Convert the red and green channels to gray-scale images, if they are not so already. Save the gray-scale red and green channels as new files.

Full-color image.

Red channel.   Green channel.   Blue channel.

Print each of the two gray-scale images in the same size on transparency film.

In the center of each piece of poster board cut a rectangular opening slightly smaller than the images printed on transparency film. Tape one of the two transparencies to each of the pieces of poster board so the image completely covers the openings.

Arrange two overhead projectors next to each other so that their projections overlap on a screen.

Presentation

Set one transparency on each of the two projector stages. Adjust the positions and focus of each projector so that the projected images are the same size and in focus. Carefully align the two images so that they overlap into one image. Darken the room. The image is without color. Hold a piece of transparent red film over the lens in the head of the projector that contains the transparency created from the red channel. The image on the screen will have not only red areas, but green and other colors, too, although they will not be as vivid as those in the original image. Remove the red film, and the image returns to black and white.

Hold a piece of transparent green film over the lens in the head of the projector that contains the transparency created from the green channel. The image on the screen will have not only green areas, but red and other colors, too. Remove the red film, and the image returns to black and white.

PROCEDURE B

Preparation

Starting with a color digital photograph, generate a black-and-white transparency and a red-and-white print as follows.

Open the digital image file with photo-editing software, for example, Photoshop. If it is not already, convert the image to RGB color mode. Separate the three color channels—red, green, and blue—into three separate images. Discard the blue channel. Convert the red and green channels to gray-scale images, if they are not so already. Save the gray-scale red and green channels as new files.

Open the green-channel image. Use the photo-editing software to convert the image to RGB. Add a new layer, and fill this layer with red. Lighten the black-and-white image with the red layer, combining them into one. Put a wide black border around the red-and-white image, and print it.

Green channel.   Green channel lightened with red overlay.

On transparency film, print the gray-scale image of the red-channel image in the same size as the green-channel image.

Presentation

Display the black-bordered red-and-white image. Carefully align the black-and-white transparency over the red-and-white image. A full range of colors will appear in the image when the two are aligned.

HAZARDS

There are no particular hazards associated with this demonstration.

DISPOSAL

The transparencies may be saved for repeated presentations of the demonstration. Alternatively, they may be discarded in a solid-waste receptacle.

DISCUSSION

This demonstration shows that a full-color image can be produced from a black-and-white image and a single color. The effect can be produced additively, by combining projected images, as in Procedure A, or subtractively by overlaying a transparency on an opaque image, as in Procedure B. The color images produced in this way are not as vivid as those produced using three primary colors.

The method of producing full-color images used in this demonstration was discovered by Edwin Land, the inventor of the Polaroid instant camera. In developing a method of instant color photography, Land experimented to discover various ways in which the impression of full color can be created. It was already known that full-color images could be created using either the additive primaries of red, green, and blue light or the subtractive primaries of cyan, magenta, and yellow pigments. He attempted to find a different method that could simplify the process of instant color photography. Because the phenomenon was discovered and first reported by Land, it is called the Land effect in his honor [*1, 2*]. Although he did not find a method to produce vivid color images using fewer than three primaries, he did discover that there are more factors involved in color vision than the relative intensities of three primaries.

Human perception of color depends on a variety of factors. One of these is the intensity of the various wavelengths of light that enter the eye. Light that contains high intensities of wavelengths around 600 nm and lower intensities of other wavelengths will be perceived as yellow. Light that contains wavelengths of high intensity at 650 nm, lower intensity at 550 nm, and no other wavelengths will also be perceived as yellow. However, perceived color is not determined solely by the relative intensities of wavelengths of light. Light containing exactly the same combination and intensities of light coming from two different parts of a scene can be perceived as different if their surroundings are different. (See Demonstration 12.27, The Perception of Brightness Is Relative, for an illustration of this.) Land discovered that the surroundings are so important in the perception of color, that we can see colors even if their corresponding wavelengths or combinations of primary wavelengths are not present.

An image perceived as black-and-white, or shades of gray, is composed of varying intensities of white light. A black-and-white image viewed by reflected sunlight or room light, or by projection with a white light, contains light with a full spectrum of wavelengths. It is perceived in shades of gray because the distribution of wavelengths is the same from all portions of the image, although the total intensity varies. If the balance of wavelengths is altered somewhat by filtering out a narrow band of wavelengths, then the image will appear colored. However, the image does not appear to have a single, uniform tint, but a range of different colors. The human vision system is so biased to see colors, that when the balance of wavelengths is changed, different areas of the image will appear to have different colors, as a result of their different surroundings. The visual system assigns colors according to the relative balance of short, medium, and long wavelengths across the whole visual field, so that light of a particular wavelength has no inherent color in itself but may be perceived as any color depending on the range of wavelengths in the surrounding visual field.

The human visual system evolved to perceive colors in this way, because it is advantageous to survival to perceive objects as having a constant color, independent of their illumination. As the sunlight on an apple varies with the movement of leaves, the light it reflects changes. However, finding that apple is much easier if its color does not change. The constancy of human color vision versus variation in the composition of reflected light sometimes leads to surprises for amateur photographers. The colors recorded by a camera under sunlight, incandescent light, and fluorescent light can vary considerably, even though the photographer sees all three in the same way.

REFERENCE

1. E. H. Land, "Experiments in Color Vision," *Scientific American, 200*(5), 84–99 (May 1959).
2. E. H. Land, "The Retinex Theory of Color Vision," *Scientific American, 237*(6), 108–128 (December 1977).

12.31

Saturation of the Retina

Afterimage

A flash of very bright light can blind you for a few seconds because a large number of your photoreceptor molecules have reacted and it takes time for new ones to form. A less extensive, but more dramatic, effect can be caused by staring at an image for long enough to partially saturate (deactivate) some photoreceptors.

An image consisting of a green rectangle and a magenta rectangle is viewed for 30 seconds. When the image is replaced with one having a white rectangle in place of the magenta one, the replaced area appears a more vivid green than the original green rectangle (Procedure A). A color image is viewed for about 30 seconds. When it is replaced with a blank image, or the eyes are closed, an image appears in reverse color (Procedure B).

MATERIALS FOR PROCEDURE A

large reproduction of the images in Figures 1 and 2

projection system for image, e.g., computer video projector or overhead projector

projection screen

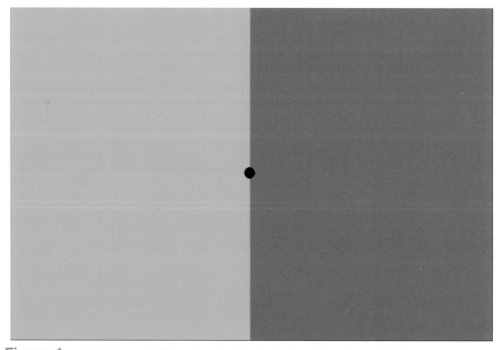

Figure 1.   Initial image for Procedure A saturated-color-vision demonstration.

Figure 2.   Second image for Procedure A saturated-color-vision demonstration.

MATERIALS FOR PROCEDURE B

large reproduction of the image in Figure 3 (See Procedure B for description.)

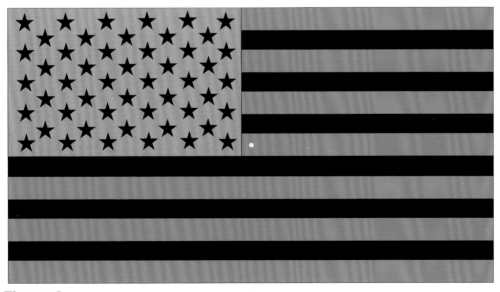

Figure 3.   American flag in reverse color, with center mark.

PROCEDURE A

Preparation

Prepare the images in Figures 1 and 2 to be displayed to the audience. This can be done by projecting them onto a screen using video projection or an overhead projector or by preparing large poster versions.

Presentation

Display the image in Figure 1 to the audience, and instruct them to stare at the dot in the center of the image. After they have viewed the image for about 30 seconds, quickly replace it with the image in Figure 2 (or, if side-by-side images are displayed, instruct the viewers to shift their gaze to the dot at the center of Figure 2). When this is done, the right side of the image will appear to be an even more intense green than that on the left.

PROCEDURE B

Preparation

Prepare the image in Figure 3 to be displayed to the audience. It may be printed on a transparency to be projected using on overhead projector. It may be projected via a video projector from a document camera, video camera, or computer camera, or the image may be transferred to a computer and displayed on a large-screen monitor. Alternatively, it may be reproduced in a large format such as a poster.

Presentation

Display the image in Figure 3 to the audience and instruct them to stare at the plus sign in the center. After they have stared for 30 seconds, quickly replace the image with a white or black view (or tell the audience members to close their eyes). A latent image in reverse colors (red, white, and blue) will appear to replace the original one.

HAZARDS

There are no hazards associated with these demonstrations.

DISPOSAL

The materials may be saved for future demonstrations.

DISCUSSION

What you see depends not only on what you are looking at but also on what you *have been* looking at. This demonstration shows that you can "see" an image, even when the light entering your eyes does not correspond to that image, as long as you have been recently viewing a related image. Your vision system has a short-term memory, although it is reversed.

The image in Figure 1 contains only two colors (other than the small dot used to help position the eyes). The two colors are complementary, and that is important to producing the effect. Because the colors are complementary, they stimulate opposite effects in the cone cells of the retina. That is what is meant by complementary colors. The green-sensing cone cells in the retina are more active in the green area of the image, whereas the red- and blue-sensing cells are more active in the magenta area.

As you stare at the image, the cone cells in your retina respond to the light that falls on them. Although you may not notice it, the signals from the cone cells gradually diminish as the light-sensitive pigments are used up faster than they are regenerated. However, only the pigments that are absorbing light are depleted [1]. On one side of the image, the pigments that absorb green are consumed. On the other side, the pigments that absorb magenta (red and blue) are consumed. Because magenta is the complement of green, the pigments that respond to green are still in ample supply. Then, suddenly, the magenta area is replaced by a white area, which will stimulate all of the color-sensing cone cells. However, because the pigments that absorb magenta are depleted, and there are ample pigments for green, the white area can stimulate only the green-sensing pigments, and only a green signal goes to the brain. Furthermore, because the green side of the image has depleted the green pigments in that area, the signal to the brain from the green side is weak. Therefore, the white side of the image appears even greener than the green side. This appearance will persist for several seconds, while the cone cells slowly recover.

A similar effect, but using more colors, is achieved in Procedure B. Here, a color image is used to deplete particular areas of the retina in those colors. When the image is replaced by white, the white can stimulate only the complementary pigments, because the others have been depleted, and a reverse-color image is perceived. If the original image is a reverse-color image (such as the flag in Figure 3), then the white field will evoke a true-color image, called an *afterimage*. The afterimage can also be perceived if, instead of replacing the image with a white field, the eyes are closed. This effect occurs because the retinal receptors are always sending signals to the brain, even when they are not receiving light stimuli. These "rest" signals are stronger from the cells that have not been saturated, so you perceive a color complementary to the one that saturated the receptors. In this case, the afterimage appears when the stimulus to the other receptors is turned off. For some observers, the stimulated afterimage against the white background is more vivid than that against the black or closed eyes.

Simple images work best in this demonstration, because even though you try to keep your eyes focused on the plus sign or the dot, it is impossible to do this precisely. When staring at the dot in Figure 1, you will probably observe that the border between the green and magenta areas becomes fuzzy. This happens because your eyes move reflexively ever so slightly to avoid the very retinal saturation that this demonstration relies upon. Because your eyes move, borders become slightly blurred. If there are many such borders, such as in a complex landscape, the color reversal produced by retinal saturation will be less distinct.

REFERENCE

1. G. S. Brindley, "The Discrimination of After-Images," *J. Physiol., 147,* 194–203 (1959).

12.32

The Persistence of Vision

Although we live in a world of blinking lights, we are usually not aware of them. This is because the chemical processes required to produce an image on the retina take some time, so the image persists for a short time after its stimulus is over.

In a darkened room, a red-orange-emitting neon voltage-tester light is swept rapidly in an arc. Alternating dashes of light and intervening dark gaps are observed (Procedure A). A blurry image is projected onto a dark background in a darkened room. When a white bar is waved several feet in front of the background, the image appears in focus where the bar is waved (Procedure B).

MATERIALS FOR PROCEDURE A

neon-light voltage tester for electrical circuits (the least expensive one available)

extension cord (or, if necessary, a lamp cord with plug) long enough to reach from an electric outlet to the presentation area

solder, soldering iron, and electrical tape (if necessary)

MATERIALS FOR PROCEDURE B

photographic slide of well-known person or place

slide projector

projection screen

poster board

white wooden stick about 1 m long (a meterstick painted white)

PROCEDURE A

Preparation

If the voltage tester has a plug, plug it into an extension cord. Many voltage testers have two electrical leads with prongs at the end to test various kinds of outlets. To prepare these testers, cut off the prong assemblies, strip about 1 cm of insulation from each lead wire and also about 1 cm of the insulation from each of the two wires of the lamp cord. Attach the voltage tester to the lamp cord—twist the wires together, solder them, and wrap with electrical tape to insulate and protect the connection.

Presentation

Plug the extension cord (or lamp cord) into a live electrical outlet and darken the room. Hold the voltage tester so the lighted lamp is visible to the audience and point out that it appears to be emitting light continuously. Swing your arm in an arc and the audience will observe a series of light dashes with intervening shorter dark gaps as the light flashes on and off. The series of light dashes will travel with your arm. Get the audience to report the number of dashes they see at any one instant (just at the end of the arc is easiest). They will probably report five or six.

PROCEDURE B [1]

Preparation

Place a piece of white poster board three or four feet in front of the projection screen. Project the image of a well-known person or location onto the board and adjust the focus so that the image is in focus on the board. Remove the board. The image will be out of focus and blurry on the screen. Remove the screen, or cover it with a black cloth, so that the projected blurry image is as dim as possible. Turn off the projector without moving it.

Presentation

Darken the room lights, and turn on the projector. Standing to the side, rapidly wave the white stick up and down in the plane where the poster board had been placed. The image will appear in focus where the stick is being waved. Wave the stick at various rates. When the stick is waved slowly, the image disappears.

HAZARDS

There are no hazards associated with this demonstration.

DISPOSAL

Materials may be saved and used in future presentations.

DISCUSSION

All the chemical and transport processes in the retina that occur after a light stimulus occurs take time, so an image does not disappear instantly. Physiological experiments show that it takes about 1/25 of a second (about 40 milliseconds [ms]) from initiation to disappearance of an image on the retina. This phenomenon is often called the *persistence of vision*.

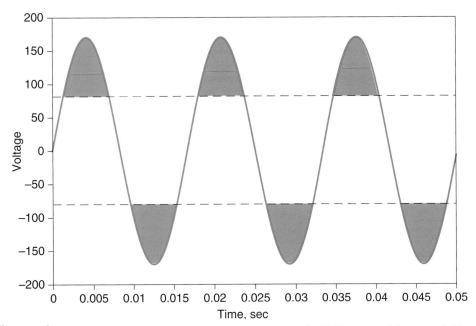

Figure 1.   Variation in voltage with time for typical 110-volt, 60 Hz power (blue curve). The dashed lines indicate the threshold voltage beyond which a typical neon lamp will glow, and the orange areas indicate when the neon lamp glows.

A neon indicator light requires several volts of potential to excite the gas to emit its red-orange light. When alternating current (the usual 60 cycle-per-second line current in North America, for example) is used to excite the emission, there are periods of light emission when the potential is high enough, Figure 1, and dark periods as the potential falls to zero when it changes sign in the middle of the cycle. Thus, there are two periods of light emission during each complete 16.7 ms (= [1 sec]/60) cycle. The light periods are about 8 ms apart (center to center). If you and the indicator light are both stationary, the light will appear to be on continuously, because the image of the on part of each cycle remains during the off part of the cycle and is then renewed at the same place on your retina during subsequent cycles.

If either you or the light is moving, so that subsequent images from the on parts of the cycles fall on different parts of your retina, you will be aware of the off (dark) periods, when no light falls on your retina and only the persistent light images remain. If the light is moved in a continuous path, say an arc, as in this demonstration, you see lighted dashes with intervening dark spaces. About five or six (depending on the intensity of the light and the sensitivity of your retina) of the light dashes are present at the same time, and the array moves along the arc keeping up with the moving light. The leading light dash in this array is the result of the emission that has just occurred. The last one is the result of the emission that occurred about 40 ms earlier, the persistence of vision period. Since the dashes are about 8 ms apart, there should be about five or six of them still on your retina after 40 ms, which is observed in Procedure A.

A similar phenomenon involving a series of images at different locations on the retina, all within the persistence-of-vision time period, is responsible for the observations in Procedure B. In this case, your retina is presented with a continuous series of overlapping images (reflected from the white stick) that together make up the projected picture. If the white stick is moved fast enough, the first, last, and all intermediate images will be present on your retina during the persistence-of-vision period, and you will perceive the entire picture as though it had all been reflected at the same time. If the stick is moved too slowly, the earlier images will disappear before the sweep is completed and only a part of the picture will be observed.

Very early in the history of motion pictures, in the nineteenth and early twentieth centuries, persistence of vision was invoked as the basis for the sense of motion when a series of still images are presented to the retina very quickly. The idea was that the persistence of one image while another, only very slightly different one, was presented, and then another, and another, created the motion on the retina. The problem with this explanation was that, at best, what would "pile up" on the retina was a series of static images. a concept that brings to mind Marcel Duchamp's painting, *Nude Descending a Staircase,* which obviously requires mental processing to "see" the motion [2].

Indeed, since the beginning of the twentieth century, extensive physiological and psychological research has shown that the motion in motion pictures and on television arises in the brain, not on the retina. In the most simplistic way of looking at this, the brain has to "fill in" the gaps between the still images to provide a sense of smooth motion. This is reminiscent of other perceptual phenomena, such as the brain filling in the information missing from the blind spot, Demonstration 12.29. Thus, when you relax watching a movie or your favorite television program, your brain is working diligently in the background to create what appears to be the same kind of movement you experience when viewing an actual occurrence.

REFERENCES

1. J. C. Sprott, *Physics Demonstrations: A Sourcebook for Teachers of Physics,* pp. 258–259, University of Wisconsin Press: Madison (2004).
2. J. Anderson and B. Anderson, "The Myth of Persistence of Vision Revisited," *Journal of Film and Video, 45*(1), 3–12 (1993). See this article for the Duchamp analogy and for the history and references to the research on the perception of motion related to movies and television.

12.33

The Imprecision of Peripheral Vision

What you see out of the corner of your eye can be an important warning, like an approaching vehicle, but the image is not very sharp, and you have to turn your head to look more directly in its direction to see it more clearly.

An image containing colored letters arranged in rings around a center dot is displayed. When vision is focused on the center dot, only the letters in the ring nearest the dot are recognizable, and the colors of the outer ring of letters are indistinct.

MATERIALS

large reproductions of the images in Figures 1 and 2 (See Procedure for description.)

PROCEDURE

Preparation

Prepare the images in Figures 1 and 2 to be displayed to the audience. They may be printed on transparencies to be projected using on overhead projector. They may be projected via a video projector from a document camera, video camera, or computer camera. The images may be transferred to a computer and displayed on a large-screen monitor.

Figure 1.

E
Y
L
V
K Q Z
X M
B
A G
S I U F • R H O F
H K
S
D T
C W N
A P J
R

Figure 2.

Presentation

Display the image in Figure 1 to the audience and instruct them to stare at the black dot in the center. As they focus on the dot, ask them if they can discern the identities of any of the letters. The letters closest to the dot will be recognizable, but those further away will be less distinct. Those in the outermost ring will not be identifiable at all.

When vision is focused on the center dot, the colors of the letters in the outermost rings will also be indistinct. Some letters will appear to have the colors of other letters, and the colors may seem to change as vision remains on the center dot.

Display the image in Figure 2 to the audience and instruct them to stare at the black dot in the center. As they focus on the dot, ask them if they can discern the identities of any of the letters. In this instance, all of the letters will be identifiable, because their sizes have been adjusted to compensate for the imprecision of peripheral vision.

HAZARDS

There are no particular hazards associated with this demonstration.

DISPOSAL

The copies or transparencies may be saved for repeated presentations of the demonstration

DISCUSSION

This demonstration shows that our peripheral vision is much less precise than that in the center of our visual field. While we stare at the dot at the center of Figure 1, which is an array of variously colored letters, we can easily determine both the identity and the color of the letters near the dot. However, the identities and colors of the letters become harder to determine as their distances from the dot increase. In Figure 2, the letters increase in size away from the dot in approximate degree corresponding to the decrease in precision of peripheral vision. In this figure, we can identify the letters and their colors.

Human vision is most precise near the center of the visual field. This is the case because at the center of the retina is an area of densely packed cone cells called the *fovea*. The fovea has about 30,000 cone cells packed into an area of about 0.3 mm^2, a density of about 100,000 cells per square millimeter [1]. The density of cone cells is only about 5,000 per mm^2 in the outer areas of the retina. The high density of cone cells in the fovea contributes to the higher precision of central vision over peripheral vision. Furthermore, the cone cells in the fovea are narrower than those in the peripheral retina, which also contributes to finer detail in the center of the visual field.

The imprecision of peripheral vision is what gives many impressionist paintings an illusion of motion. Misaligned brush strokes produces spatial imprecision. When we glance at such a painting, our peripheral vision gives us a vague idea of where the spots of color are located. Our brains combine these spots into a complete picture. Only if we examine the spots of paint carefully with our high-acuity foveal vision, do we see clearly that the strokes are misaligned. The illusion of motion is created because our visual system completes the picture differently with each glance [2].

Figure 3.   Claude Monet's *La Rue Montargueil* (1878), an example of an impressionistic painting that exploits the imprecision of peripheral vision to produce the illusion of motion.

REFERENCES

1. S. J. Williamson and H. Z. Cummins, *Light and Color in Nature and Art,* pp. 320–322, Wiley: Hoboken, New Jersey (1983).
2. M. Livingstone, *Vision and Art: The Biology of Seeing,* pp. 68–77, Harry N. Abrams: New York (2002).

12.34

The Pulfrich Phenomenon
Perception of Motion

There is a binocular component to our perception of motion. If the signal from one eye gets to the brain before the other, then the brain can be fooled into thinking there is motion where there is none. The absorption of several photons is required to send a message from a retinal photoreceptor to the brain. Thus, the brighter the light that strikes the retina (within the normal range of intensities), the faster will be its response. If one eye receives more light from an object than the other eye, the signal from the more brightly illuminated eye will be detected earlier.

When viewed with one eye covered by a dark transparent filter, a pendulum swinging from side to side will appear to follow an elliptical path moving in front of and behind the plane of its actual motion.

MATERIALS

white ball

string

large steel nuts or washers (See Preparation for description.)

stand with clamp

desk lamp

pair of sunglasses (one pair for each observer), or a lens-sized piece of transparent, deeply colored, stiff plastic film (one for each member of the audience)

PROCEDURE

Preparation

Make a pendulum by suspending the white ball on the stand with string. The pendulum should be able to swing without dwindling noticeably for at least 15 seconds. If the pendulum does not swing steadily, increase the mass of the pendulum by threading steel nuts or washers onto the string and resting them on the ball.

Place the stand so that the pendulum can swing in a direction perpendicular from the viewing line. Set the lamp so that it illuminates the swinging ball without obscuring it.

Presentation

Turn on the lamp and start the pendulum swinging perpendicular to the line of sight. With both eyes, observe the ball swing side to side. While observing, cover one eye with the dark, transparent filter or one lens of the pair of sunglasses. The ball will now seem to follow in an elliptical path. Cover the other eye with the filter. Now the ball will seem to follow an elliptical path, but in the opposite direction.

HAZARDS

There are no particular hazards with this demonstration.

DISPOSAL

The materials may be stored for repeated presentations of the demonstration.

DISCUSSION

Perception of motion can be affected by the relative brightness of an object as perceived by the two eyes. When the brightness of the object is the same in both eyes, a pendulum moving back and forth in a plane appears to be moving just so. However, when the same pendulum is viewed with a neutral-density filter in front of one of the eyes, the pendulum will appear to be moving in an elliptical orbit. This phenomenon of perception was first described by the German physicist Carl Pulfrich in 1922 [*1*].

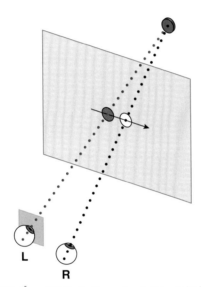

Figure 1.   White ball moving left to right in plane.

The apparent orbital motion of a pendulum actually moving in a plane results from the variation of visual response time with light intensity. The retina responds more quickly to brighter light. When the illumination on the retina decreases by a factor of ten, there is a 15-ms delay in retinal response [*2*]. When both eyes receive the same intensity from the pendulum, both eyes send signals to the brain at the same time. However, if the intensity of light to the left eye is reduced by a filter in front of the eye, then the signal from the left eye is delayed and arrives at the brain later than the signal from the right eye. This situation is represented in Figure 1. Because the signal from the left eye is delayed, the signal that reaches the brain suggests that the left eye is looking somewhat to the left of where the pendulum is seen with the right eye. That is, to the left eye, the ball appears to be at the position represented by the red dot in Figure 1. The brain uses the positions of the eyes to determine the distance to the pendulum. Because the left eye seems to be looking to the left of where the right eye is looking, the intersection of the viewing directions of the two eyes puts the apparent location of the pendulum behind the plane in which it is swinging. The apparent position of the ball is represented by the blue dot in Figure 1.

When the pendulum is moving from left to right (Figure 2), the delayed image of the pendulum from the left eye makes the pendulum appear to the right of where it is seen by the right eye. The brain interprets this to mean that the pendulum is in front of the plane in which the pendulum is actually moving, as represented by the blue dot in the figure. So, as the ball moves left to right, it appears to be behind the plane, and when it moves from right to left it appears to be in front of the plane. Together, these effects give the pendulum the appearance of moving in an elliptical orbit.

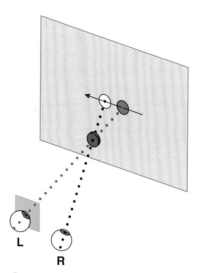

Figure 2.   White ball moving right to left in plane.

The Pulfrich effect has been used to produce three-dimensional effects on broadcast television. The half-time show during Super Bowl XXIII in 1989 and a specially produced commercial for Diet Coke were telecast exploiting this effect. In the commercial, objects moving in one direction appeared to be nearer to the viewer (actually in front of the television screen) and when moving in the other direction, appeared to be farther from the viewer (behind the television screen). Forty million pairs of paper-framed 3-D viewing glasses were distributed by Coca-Cola USA for the event.

REFERENCES

1. C. Pulfrich, "Die Stereoskopie im Dienste der isochromen und heterochromen Photometrie," *Die Naturwissenschaften, 10,* 553–564, 569–574, 596–601, 714–722, 735–743, and 751–761 (1922).
2. J. M. Williams and A. Lit, "Luminance-Dependent Visual Latency for the Hess Effect, the Pulfrich Effect, and Simple Reaction Time," *Vision Res., 23*(2), 171–179 (1983).

Photoemission: Fluorescence and Phosphorescence

12.35

Photoluminescence

Many substances, for example, fluorescent minerals and plant pigments, absorb light energy and reemit some of it at lower energy, longer wavelengths.

Several items are displayed and then viewed under black light. Under black light, the items glow various colors (Procedure A). A cylinder containing liquid is placed on the stage of an overhead projector that has been masked to project only a small spot. The spot projected through the solution is green, and where the beam passes through the liquid, a red glow is apparent (Procedure B).

MATERIALS FOR PROCEDURE A

several fluorescent materials (See Preparation for description.)

black light (long-wave ultraviolet lamp)

MATERIALS FOR PROCEDURE B

10 g frozen spinach leaves

30 mL acetone, $(CH_3)_2CO$

230 mL hexane, C_6H_{14}

30 mL saturated aqueous sodium chloride, NaCl (To prepare, combine 12 g of NaCl with 30 mL of water, stir the mixture for 15 minutes, and decant the resulting solution.)

60 mL water

1 g anhydrous magnesium sulfate, $MgSO_4$

small mortar and pestle

long-stemmed glass funnel

wad of glass wool, 1 cm in diameter

125-mL separatory funnel

50-mL Erlenmeyer flask, with stopper

at least 12-cm square piece of aluminum foil

scissors

250-mL glass cylinder

overhead projector

PROCEDURE A

Preparation

Gather a selection of materials that emit visible light when exposed to black light. Some common materials that do this include tonic water, Mountain Dew beverage, laundry detergents that contain brighteners, cloth that has been laundered in such a detergent, unsaturated vegetable oil, and white copy paper. These and other materials can be located by shining a black light around an otherwise darkened room and looking for objects that glow brightly. Other suitable materials include many minerals, such as calcite, corundum, agate, apatite, fluorite, halite, quartz, and opal. These and other minerals can be obtained in kits for this purpose.

In addition to fluorescent materials, phosphorescent materials may also be used to illustrate the difference between fluorescence and phosphorescence. Photoactive glow-in-the-dark decorations, toys, and paints are all suitable.

Presentation

Display the materials to the observers. Turn on the black light and dim the room lights. Expose each of the samples to the black light and observe the emission produced.

If phosphorescent material is available, place both it and a fluorescent material under the black light. Extinguish the black light. The emission from the fluorescent material will disappear immediately, while that from the phosphorescent material will gradually fade.

PROCEDURE B

Preparation

Squeeze the water from frozen spinach and put 10 g of it into a mortar. Add 25 mL of acetone to the spinach and grind the mixture with a pestle until the spinach has lost most of its color to the acetone.

Put a plug of glass or cotton wool in the stem of a funnel, and filter the spinach-acetone mixture through the plug into a 125-mL separatory funnel. Rinse the solid in the funnel with an additional 5 mL of acetone.

Add 30 mL of hexane and 30 mL of saturated aqueous sodium chloride to the separatory funnel. Shake the mixture vigorously, occasionally pausing and venting the funnel. Drain off the aqueous layer.

Add 30 mL of water to the funnel, shake the mixture, and drain off the water. Repeat this a second time. Pour the hexane solution into a 50-mL Erlenmeyer flask and add 1 g of anhydrous magnesium sulfate. Filter this mixture through a plug of glass or cotton wool into a clean, dry Erlenmeyer flask. This is the chlorophyll solution to be used later. Stopper the flask and keep it in the dark until it is to be used.

Cut a piece of aluminum foil large enough to cover the stage of an overhead projector. Measure the inside diameter of a 250-mL glass cylinder. Cut in the center of the aluminum foil a circular hole with a diameter 0.5 cm smaller than that of the cylinder. Put the aluminum foil onto the overhead projector with the hole at the center of the projector's stage.

Procedure

Put 5 mL of the chlorophyll solution into the 250-mL cylinder. Add 200 mL of hexane and stir the mixture to make it uniform. Set the cylinder over the hole in the aluminum foil on the stage of the overhead projector. Note the green color of the projected image. Darken the room, and the green of the projected image will be more readily visible. Also visible will be a red glow where the beam of the projector travels upward through the chlorophyll solution.

HAZARDS

Acetone and hexane are highly flammable and should be kept away from open flames. Vapors from both can irritate mucous membranes and can be narcotic when inhaled. They should be used only with adequate ventilation.

Black light (long-wavelength ultraviolet light) can damage one's eyes. Shield the black-light source from the audience and never look directly at the source yourself.

DISPOSAL

The hexane solution of chlorophyll should be discarded in a receptacle for organic waste. The aqueous solution should be flushed down the drain with water.

DISCUSSION

Certain substances, when exposed to black light, emit visible light. Human vision is not sensitive to black light, and we cannot see it, which is why it is called black. However, it is like visible light in the sense that it is energy. The energy of black light is higher than that of visible light, and this is why black light is also called ultraviolet (beyond violet) light. Many substances absorb black light, and the energy of absorbed light can be transformed into heat, can cause a chemical change in the absorbing material, or can be emitted as light. The emitted light is frequently of lower energy than the absorbed—in fact, it cannot be of higher energy. This lower-energy light can be visible, and the visible emitted light produces the glow that is observed from the materials used in this demonstration.

When a substance absorbs visible or ultraviolet light, molecules in the substance are transformed from a low-energy state to a high-energy state. Molecules in a high-energy state are not stable, and in some cases, will undergo a chemical transformation; this transformation can change their properties, such as color. This is what happens when a piece of colored cloth is bleached by sunlight. The molecules in the cloth that are responsible for its color absorb energy from sunlight, and that energy causes them to change into molecules that do not produce color. Other examples of chemical transformations caused by the absorption of light are presented in the demonstrations in the Photochemistry section, Demonstrations 12.42 through 12.54.

Molecules in a high-energy state may also lose their energy by transferring it bit by bit to other molecules through collisions with them. This has the effect of distributing the absorbed energy throughout the absorbing material, causing its molecules to vibrate more intensely. This increase in molecular vibrations is responsible for an increase in the temperature of the material. The net effect is the conversion of light energy to heat energy. This phenomenon is displayed in Demonstration 12.11, The Conversion of Light Energy to Thermal Energy.

Molecules in a high-energy state can also lose energy by emitting that energy in the form of light, that is, electromagnetic radiation. The energy of the light emitted by a molecule in a high-energy state may be the same as that of the absorbed light, but usually it is of lower energy. Some of the absorbed energy may be converted to heat before the molecule emits light, so the light emitted from this molecule must have less energy than the absorbed light. The conversion of absorbed energy to heat takes time, so the emission of light occurs some time after the absorption, although that time may be very short. Emission that occurs very shortly after absorption is called *fluorescence*. Emission that occurs at a longer time after absorption is called *phosphorescence*. Fluorescent emission occurs about 10^{-9} to 10^{-5} seconds after absorption, whereas phosphorescent emission occurs from 10^{-4} seconds to several hours after absorption.

The temporal distinction between fluorescence and phosphorescence is an indication that the two phenomena are the results of different processes. During absorption of visible or ultraviolet radiation by a molecule, the energy of an electron in the molecule increases. The electron moves from a lower-energy molecular orbital to one with higher energy. During any process involving the absorption or emission of light, however, the spin of the electron does not change. If the electron does not change spin, in a very short time it can return to its original state by the emission of light, resulting in fluorescence. While the electron is in a high-energy state, however, its spin can change by the loss of a small amount of energy, say, through collision with another molecule or by a change in the vibration within the molecule. If the spin of the electron changes, it is no longer able to return to its original state before absorption. (See the section *Electron Spin* in the introduction, page 43.) Because of this, the molecule can remain in a high-energy state for an extended period of time before it eventually decays by the emission of light. This delayed emission is phosphorescence.

In nearly all cases, some energy is lost from the absorbed light before it is emitted as fluorescence or phosphorescence. For this reason, the emitted light is of lower energy than the absorbed light. Visible emission from a material irradiated by black light is of lower energy, but this is only part of the reason black light is used in this demonstration. The other reason it is used is because it is not visible, only the fluorescence or phosphorescence are visible. It is possible to produce fluorescence or phosphorescence from visible light. For example, blue light can produce emission of yellow or red light. Such a phenomenon is responsible for the extraordinarily bright yellow plastic that is sometimes used in the handles of tools to make them stand out from their surroundings, or in the blue plastic file tags that have a reddish glow in bright light, which makes them stand out in a collection of files.

That visible light can stimulate visible fluorescence is also demonstrated in Procedure B, with a solution of chlorophyll. Chlorophyll is the light-absorbing molecule in the leaves of plants that harvests the energy of sunlight for use in photosynthesis. Chlorophyll absorbs red and blue light, but not green. Plant leaves reflect green light, which is why they appear green. In Procedure B, a solution of chlorophyll extracted from green leaves is exposed to white light. The unabsorbed light is transmitted by the solution, and this transmitted light appears green. When viewed from the side, perpendicular to the path of light through the solution, a red fluorescence is visible. The energy from absorbed blue light is emitted as red light. In a living plant, this energy would instead be captured and used to drive the biochemical reactions responsible for the plant's growth.

Most substances are not fluorescent. Few chemically pure minerals are fluorescent. Minerals that are fluorescent usually contain impurities that are necessary for fluorescence. These impurities are called *activators*. The identity and amount of activator determine the color, intensity, and duration of the emission. The most common material used in glow-in-the-dark novelties is zinc sulfide, ZnS, containing less than 1% copper chloride, $CuCl_2$. This material has been used since the 1930s and produces a yellow-green phosphorescence that can last an hour or so. Newer materials that glow with different colors, more brightly, and for longer times have been developed. Recently discovered phosphors based on strontium aluminate, $SrAl_2O_4$, with various activators, such as rare-earth compounds, emit blue, green, or red phosphorescence that lasts several hours.

12.36

The Halide Quenching of Quinine Fluorescence

Following the fate of the ultraviolet-light energy that is absorbed by a compound in solution is easier, if at least some of the energy is reemitted as visible light. The effects on the emission when solution conditions are changed provide clues.

Five beakers of liquid emit a blue glow under a black light. Varying amounts of salt (sodium chloride) are dissolved in the liquids, and the higher the salt concentration, the dimmer the glow (Procedure A). Chloride, bromide, and iodide ions at the same concentration in the liquid dim the glow to different degrees (Procedure B).

MATERIALS FOR PROCEDURE A

250 mL tonic water (a carbonated, sweetened solution of quinine in water)

2.0 g sodium chloride, NaCl

five 250-mL beakers

black light

four disposable weighing dishes

stirring rod

MATERIALS FOR PROCEDURE B

200 mL tonic water (a carbonated, sweetened solution of quinine in water)

0.10 g sodium chloride, NaCl

0.18 g sodium bromide, NaBr

0.28 g potassium iodide, KI

four 250-mL beakers

black light

three disposable weighing dishes

stirring rod

PROCEDURE A

Preparation

Pour 50 mL of tonic water into each of the five 250-mL beakers. Arrange the five 250-mL beakers in a row under a black light.

Into a disposable weighing dish, weigh 0.1 g NaCl. Into a second dish, weigh 0.3 g NaCl, into a third weigh 0.6 g NaCl, and into a fourth weigh 1.0 g NaCl. Arrange the dishes in order of increasing mass near the second through fifth beakers.

Presentation

Darken the room lights and turn on the black light. The tonic water in the beakers will emit a blue fluorescence.

Add 0.1 g of NaCl to the second beaker in the row and stir the mixture to dissolve the salt. The glow will be diminished.

Add 0.3 g of NaCl to the third beaker and stir the mixture to dissolve the salt. The glow will be dimmer than in the other beakers.

Add 0.6 g NaCl to the fourth beaker and 1.0 g NaCl to the fifth beaker and stir the mixtures to dissolve the salt. The higher the concentration of salt in the tonic water, the dimmer the glow appears.

PROCEDURE B

Preparation

Pour 50 mL of tonic water into each of the four 250-mL beakers. Arrange the beakers in a row under a black light.

Into a disposable weighing dish, weigh 0.10 g NaCl. Into a second dish, weigh 0.18 g NaBr, and into a third weigh 0.28 g KI. Arrange the dishes in the order NaCl, NaBr, and NaI near the second through fourth beakers.

Presentation

Darken the room lights and turn on the black light. The tonic water in the beakers will emit a blue fluorescence.

Add 0.10 g of NaCl to the second beaker in the row and stir the mixture to dissolve the salt. The glow will be diminished.

Add 0.18 g of NaBr to the third beaker and stir the mixture to dissolve the salt. The glow will be dimmer than in the other beakers.

Add 0.28 g KI to the fourth beaker and stir the mixture to dissolve the salt. The glow will be dimmer than in the other beakers. Although the concentrations of halide ions are the same in the three beakers, the glow in the one containing iodide ions is dimmer than that in the solution containing bromide ions, which is dimmer than the one containing chloride ions.

HAZARDS

Black light (long-wavelength ultraviolet light) can damage one's eyes. Shield the black-light source from the audience and never look directly at the source yourself.

DISPOSAL

All solutions may be flushed down the drain with water.

DISCUSSION

Quinine, Figure 1, is extracted from the bark of various species of *Cinchona,* trees that are native to the eastern slope of the Andes Mountains in South America but are now largely cultivated in Indonesia. Beginning in the seventeenth century, Europeans used the dried, ground bark as a medicine to treat malaria. The active ingredient in the bark was isolated in the early nineteenth century and made treatment more consistent. Although more effective drugs have been produced, the malarial parasite has become resistant to some of them and quinine treatment is still in use. A problem with quinine therapy is that the drug does not

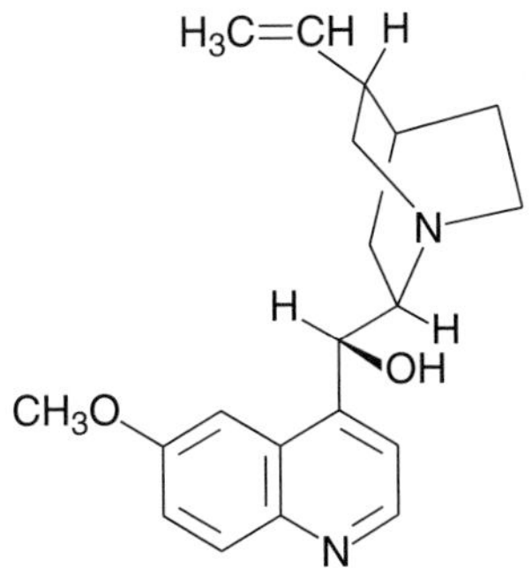

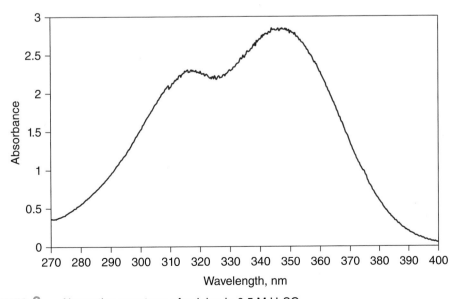

Figure 1.   Molecular structure of quinine.

kill the parasite, it only suppresses its action, so the drug must continue to be taken after the original malarial symptoms have disappeared.

Another problem with quinine as a drug is that it is extremely bitter. One story (legend) about the origin of tonic water as a drink is that the colonial British in India who were taking quinine to quell their malarial symptoms would mix it with gin, sweetened water, and lemon or lime to mask much of the bitterness. Thus were born both tonic water, sweetened and flavored (often carbonated) solutions of quinine, and the gin-and-tonic mixed drink. In the United States, the Food and Drug Administration limits the quinine content in tonic water to 83 ppm (83 mg L^{-1} by mass), about a 0.25-mM solution, because quinine can have rather severe side effects. The therapeutic oral dose of quinine is about 580 mg taken two or three times daily, so you would need to drink 14 to 21 liters of tonic water daily to get a therapeutic dose.

The relevant property of quinine for these demonstrations is its visible fluorescence when it is excited in solution with near-ultraviolet radiation. Solutions of quinine, for example, tonic water, are clear and colorless, because quinine absorbs in the ultraviolet but not the visible (Figure 2). The fluorescence emission from the excited quinine (Figure 3) peaks at about 450 nm in the blue region of the visible spectrum [1]. Solutions of quinine in dilute acid, especially sulfuric, are often used as standards in emission studies of other molecules. This is because its fluorescence quantum yield is very nearly constant, 0.56 ± 0.02 (varying slightly with the actual solution conditions) at all excitation wavelengths in the near ultraviolet [2].

Figure 2.   Absorption spectrum of quinine in 0.5 M H_2SO_4.

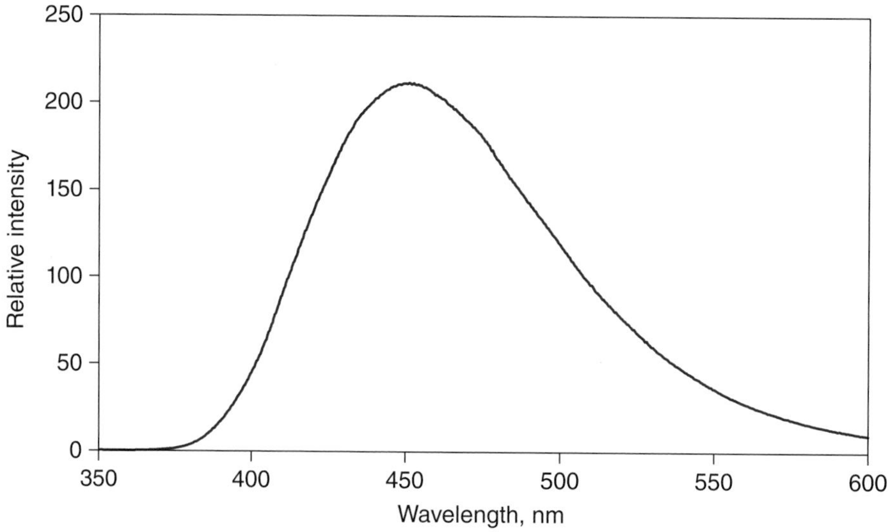

Figure 3.   Emission spectrum of quinine in 0.5 M H_2SO_4 excited at 310 nm.

Recall from the introduction, page 46, that the fluorescence quantum yield is the ratio of the number of photons emitted (from the singlet excited state) to the number absorbed. The Jablonski diagram shows that there are radiative and nonradiative pathways for the excited singlet state to lose its energy (Figure 36 in the introduction). Another way to think about quantum yield is as a ratio of the rate at which the excited state (of quinine in these demonstrations) radiates energy to the total rate at which the excited state energy is lost by both radiative and nonradiative processes.

$$\text{radiative rate} = k_r[Q^*]$$

$$\text{nonradiative rate} = k_{nr}[Q^*]$$

$$\text{quantum yield} = \Phi_0 = \frac{k_r[Q^*]}{k_r[Q^*] + k_{nr}[Q^*]} = \frac{k_r}{k_r + k_{nr}}$$

In these equations, k_r and k_{nr} are, respectively, the rate constants for radiative and all nonradiative processes for decay of the excited state, and Φ_0 is the quantum yield for fluorescence emission in a solution containing no added species that can quench the emission from Q^*, which represents the excited quinine molecule [2].

One means of quenching the emission of one species (the donor) is by adding another (the acceptor) to which energy is transferred (see introduction, page 52). Another pathway for quenching is the addition of a species (the quencher) to enhance nonradiative pathways for the loss of energy by the excited species. This is the case for quinine fluorescence quenching by halide ions. In our calculations, we can account for this pathway with an additional factor—the rate of loss of excited molecules via their collision with halide ions—and incorporate it into the equation for the quinine fluorescence quantum yield in the presence of halide ions, X^-:

$$\text{quenching rate} = k_q[Q^*][X^-]$$

$$\text{quantum yield} = \Phi = \frac{k_r[Q^*]}{k_r[Q^*] + k_{nr}[Q^*] + k_q[Q^*][X^-]} = \frac{k_r}{k_r + k_{nr} + k_q[X^-]}$$

Compare the quantum yields in the absence, Φ_0, and presence, Φ, of quencher.

$$\frac{\Phi_0}{\Phi} = \frac{k_r}{k_r + k_{nr}} \left/ \frac{k_r}{k_r + k_{nr} + k_q[X^-]} \right. = \frac{k_r + k_{nr} + k_q[X^-]}{k_r + k_{nr}} = 1 + \frac{k_q}{k_r + k_{nr}}[X^-]$$

If experiments are carried out under identical conditions (intensity and wavelength of exciting light, concentration of absorbing molecules, temperature, etc.) without and with added quencher (halide ions), we can replace the ratio of quantum yields with the ratio of intensities of emission in the absence, I_0, and presence, I, of the quencher.

$$\frac{I_0}{I} = 1 + \frac{k_q}{k_r + k_{nr}}[X^-]$$

This is the Stern-Volmer equation and predicts for this reaction pathway that a plot of I_0/I versus the concentration of quencher, $[X^-]$ in this case, should be a straight line with an intercept at unity and a slope equal to $k_q/(k_r + k_{nr})$. Our demonstration, Procedure A, shows that the intensity of fluorescence emission from tonic water (quinine) decreases as the concentration of chloride, $[Cl^-]$, in the beakers increases: 0.00 M, 0.03 M, 0.1 M, 0.2 M, and 0.3 M. Thus, although the observations are not quantitative, you observe that I_0/I increases as $[Cl^-]$ increases, as the Stern-Volmer equation predicts.

In the absence of quencher, the lifetime of fluorescence emission, τ_0 (the time required for the emission to fall to $1/e$ of its initial value when excitation is turned off), is given by $1/(k_r + k_{nr})$. Thus, you will sometimes see the Stern-Volmer equation written as

$$\frac{I_0}{I} = 1 + k_q\tau_0[X^-] \qquad .$$

Experimental measurements of fluorescence lifetimes generally require equipment that can capture emission decays that last only a few nanoseconds. The lifetime for quinine fluorescence decay, for example, is about 19 ns [3].

The likely mechanism for halide-ion quenching is *spin-orbit coupling* that increases the rate of intersystem crossing from the excited singlet to the triplet state of the quinine (pathway 4 in Figure 36) [4]. The transition involves changing (flipping) the spin of one of the electrons in the excited singlet state to give the triplet state. Electron spin is associated with a magnetic moment and flipping the spin means changing the orientation of its magnetic moment (like flipping a magnet to exchange its north and south poles). This flipping can be brought about by interaction of the electron spin magnetic moment with the magnetic fields created by the motions of the electrons in the molecule (that is, their orbital motion, which explains the origin of the term "spin-*orbit* coupling") or in nearby atoms, molecules, or ions. The faster electrons are moving, the larger the magnetic field associated with their motion and the greater the probability that the coupling will flip the electron spin. A major factor determining the speed of electron motion in an atom is its nuclear charge—the larger the nuclear charge, the faster the electron moves. Thus, atoms with high nuclear charge enhance spin-orbit coupling. This is often called the *heavy-atom effect,* because heavier atoms (higher atomic number) have a higher nuclear charge. You observe the heavy-atom effect in Procedure B, in which the concentrations of the halide ions are 0.03 M in all three solutions but the quenching (loss of emission) increases in the order of atomic number: $Cl^- < Br^- < I^-$. For these halogen atoms, the approximate spin-orbit coupling constants for quantum mechanical calculations are 8, 29, and 58 kJ mol^{-1}, respectively.

REFERENCES

1. H. Du, R. A. Fuh, J. Li, A. Corkan, and J. S. Lindsey, "PhotochemCAD: A Computer-Aided Design and Research Tool in Photochemistry," *Photochem. Photobiol., 68,* 141–142 (1998).
2. D. F. Eaton, "Reference Materials for Fluorescence Measurement," *Pure Appl. Chem., 60,* 1107–1114 (1988).
3. D. A. Barrow and B. R. Lentz, *Chem. Phys. Lett., 104,* 163 (1984).
4. N. J. Turro, V. Ramamurthy, and J. C. Scaiano, pp. 149–156, *Principles of Molecular Photochemistry: An Introduction,* University Science Books: Sausalito, California (2009).

12.37

Differentiation of Fluorescence and Phosphorescence

Fluorescence and phosphorescence are light emissions from different photoexcited states of molecules. (See the section *Excited Molecules* in the introduction, page 44.) They can be distinguished by the length of time that it takes for the emission to fade after the excitation source is extinguished and, for the same compound, by the different emission wavelengths of fluorescence and phosphorescence.

When a glycerol solution at room temperature is irradiated with long-wavelength ultraviolet light (black light) in a darkened room, a blue emission is observed. The emission disappears immediately when the light is extinguished. Cooling a second, identical solution to dry ice temperature and similarly irradiating it yields an emission that is not pure blue but grayish. When the light is extinguished, the emission changes immediately to a yellow-orange that lasts for one or two seconds.

MATERIALS

60 mL glycerol (glycerin), $C_3H_8O_3$

0.30 g 4-amino-5-hydroxy-2,7-naphthalenedisulfonic acid monosodium salt hydrate (H-acid salt), $C_{10}H_8NO_7S_2Na \cdot H_2O$

100-mL beaker

two 20-mm × 150-mm test tubes

test-tube holder that displays contents of the tubes

600-mL beaker (or a small, widemouthed Dewar flask)

300 mL isopropanol or acetone

about 300 g crushed dry ice

long-wavelength ultraviolet light (black light)

PROCEDURE

Preparation

Dissolve 0.30 g of 4-amino-5-hydroxy-2,7-naphthalenedisulfonic acid monosodium salt hydrate (H-acid salt) in 60 mL of glycerol in a 100-mL beaker. Divide the solution equally between two 20-mm × 150-mm test tubes.

Immediately before the presentation, prepare a dry-ice–acetone (or isopropanol) bath by adding crushed dry ice to a 600-mL beaker or small widemouthed Dewar flask about half full of acetone or isopropanol.

Presentation

Place one of the test tubes of H-acid solution in a holder rack that leaves the solution visible to the audience. Place the second test tube of H-acid solution in the dry-ice bath for about 10 minutes. Remove the cold solution from the dry-ice bath and place it in the holder rack next to the room temperature sample. Darken the room and irradiate both tubes with a black light. Observe the blue fluorescence from the room-temperature sample and the grayish emission from the cold sample. Extinguish the black light, and observe the yellow-orange phosphorescence from the cold sample, which lasts 1–2 seconds.

HAZARDS

H acid is irritating to the eyes, the respiratory system, and the skin. In case of contact with eyes, rinse immediately with plenty of water and seek medical advice. Immediately take off all contaminated clothing. After contact with skin, wash immediately with plenty of water. Do not empty into drains. Never add water to this product. Take precautionary measures against static discharges. Wear suitable protective clothing (safety goggles, gloves, and lab apron or coat) when preparing the solution.

Acetone (or isopropanol) is flammable. Extinguish all flames before preparing and using the dry-ice baths.

Black light (long-wavelength ultraviolet light) can damage one's eyes. Shield the black-light source from the audience and never look directly at the source yourself.

DISPOSAL

The test tubes may be stoppered and stored for repeated use.

If they are to be discarded, H acid and its container must be disposed of in a safe way according to local ordinances.

The dry-ice bath liquid may be saved for future uses like this. It will be contaminated with a little water condensed directly from the air or on the dry ice.

DISCUSSION

H acid (Figure 1) is a high-production-volume chemical, with more than 1 million pounds synthesized annually in the United States. It is used in the production of numerous synthetic dyes.

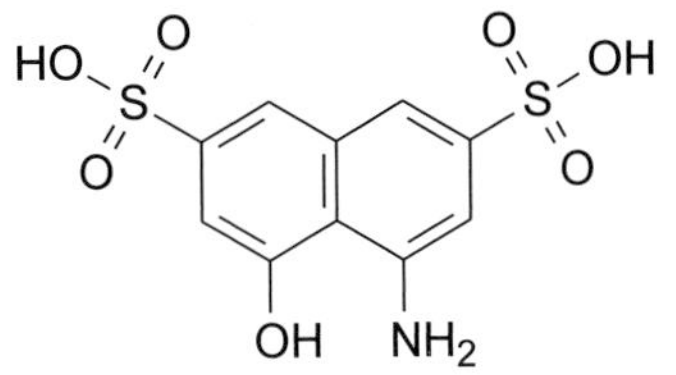

Figure 1.   Structure of H acid. In the H-acid salt, a sodium ion replaces the H on one of the sulfonyl groups.

The blue fluorescence and yellow-orange phosphorescence from H acid demonstrate that the longer-lived triplet excited state is lower in energy than the singlet excited state, as shown in the Jablonski diagram (see the introduction, page 44). The phosphorescence emission, which occurs as the triplet molecules return to the ground state, is at a longer wavelength (lower energy) than the fluorescence emission, which occurs when singlet excited molecules return to the ground state.

The triplet-to-singlet transition is slow because it is "forbidden" by quantum mechanical rules regarding the change of electron spin in this transition (see the introduction, page 45). The long lifetime of the triplet state means that nonradiative processes for return to the ground state can compete favorably with phosphorescence emission and quench the emission. Even the low concentration of molecular oxygen dissolved in solutions of phosphorescent molecules is an effective quencher in liquid solutions, where diffusion is rapid. Freezing the solution (or greatly increasing its viscosity or suspending the phosphorescent species in a solid matrix [1, 2]) slows diffusion and allows emission to better compete with quenching, as shown in this demonstration. (Fluorescence emission can also be quenched, but doing so requires higher concentrations of quencher to compete with the short fluorescent lifetimes. See Demonstration 12.36, Halide Quenching of Quinine Fluorescence.)

In 1944, in their classic study of phosphorescence, Lewis and Kasha [3] took advantage of the longer lifetime of the phosphorescent (triplet) state to isolate the phosphorescence emission from 89 compounds and record their spectra without interference from overlapping fluorescence. Their technique, the rotating-can phosphoroscope (Figure 2), was very simple, and indeed they say, "The methods of observing and measuring phosphorescence spectra are so simple that almost any laboratory has sufficient equipment."

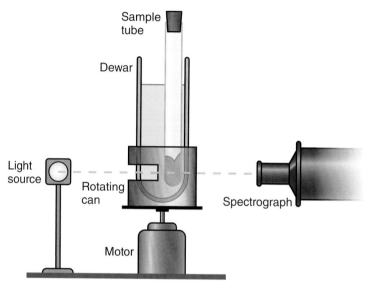

Figure 2. Rotating-can phosphoroscope.

The rotating can (rotated by the motor) in this apparatus has a section cut out about a third of the way around the can so that the sample is alternately illuminated and viewed by the spectrograph. The illumination excites any fluorescence and phosphorescence emission when the opening in the can faces the light source. As the can turns, the light is blocked and the fluorescence immediately decays, leaving only the longer-lived phosphorescence emission to be viewed when the opening faces the spectrograph.

When rotated at 1800 revolutions per minute, the emission is observed 30 times per second for about 10 ms each time. The spectra were recorded photographically (electronic methods had yet to be invented) and required exposures (run times for the apparatus) that varied from several seconds for strong emissions to more than a day for very weak emissions. Phosphorescence is generally of low intensity, even if the total number of photons emitted is substantial, because the photons are emitted over a long time, so the number per unit time is low.

Lewis and Kasha hypothesized that the phosphorescent emissive state was the triplet, modeled the ground and excited states, and with a great deal of data, tested their predictions in several experimental ways. They stated their conclusions, which have been confirmed by many further experiments in the decades since, as "Perhaps some of our conclusions may have to be modified in the light of further experiment and theory, but we are convinced that the framework is sound, and that the chemical evidence definitely proves that in all the molecules we have studied the phosphorescent state is the triplet state."

REFERENCES

1. G. N. Lewis, D. Lipkin, and T. T. Magel, "Reversible Photochemical Processes in Rigid Media: A Study of the Phosphorescent State," *J. Am. Chem. Soc., 63*, 3005–3018 (1941).
2. F. E. E. Germann, G. L. Lee, F. T. Brown, and L. M. Guthrie, "Preparation of Plates of Boric Acid Glass for Use in the Study of Luminescence," *J. Opt. Soc., 44*(6), 496–498 (1954).
3. G. N. Lewis and M. Kasha, "Phosphorescence and the Triplet State," *J. Am. Chem. Soc., 66*, 2100–2116 (1944).

12.38

Phosphorescence Excitation
Energy and Color Relationship

Glow-in-the-dark objects, like emergency signs and decorations for a child's room, must be activated by visible light before they can emit. The wavelengths of light required for excitation help to explain the mechanism of the glow.

When a beam of blue light moves across a phosphorescent panel, the panel glows where the blue excitation light illuminated it. Red light does not produce a phosphorescent glow.

MATERIALS

two phosphorescent panels large enough to be easily seen by the audience (Phosphorescent vinyl sheets should be glued to a backing of stiff cardboard, foam core, or poster board to create panels that can be easily displayed.)

stand to hold panels, such as an easel, typist's stand, or music stand

light-proof sleeve, such as an artist's portfolio, to keep panels in the dark

several penlights that produce light of different colors, including red and blue, e.g., LED pocket lights*

PROCEDURE

Preparation

Place two similar phosphorescent panels into a sleeve to protect them from light exposure. Keep the panels in the dark long enough for any residual phosphorescence to decay. This may vary from several minutes to hours, depending on the phosphorescent material used for the panels.

Presentation

Remove one of the phosphorescent panels from the sleeve and set it on a stand. Expose it for at least 15 seconds to the lights in a lighted room. Darken the lights to show the yellow-green phosphorescent glow of the panel.

With the lights still darkened, remove the second panel from the sleeve and place it on the stand. Turn on the blue penlight and use it to draw a line or figure on the phosphorescent panel. Where the blue light struck the panel, the panel will glow yellow-green.

Now use a red penlight to draw on the panel. The audience will see that you are moving the red light over the panel, but the red light will produce no phosphorescent glow.

Repeat using other colors of penlights, such as yellow, green, and violet, to draw on the panel. The yellow light will not produce a phosphorescent glow. The green light will produce a yellow-green phosphorescence that is not as bright or long lasting as that from the blue light. The violet light will produce a yellow-green phosphorescence that is brighter and longer lasting than that from the blue light [1].

---

* Phosphorescent vinyl sheets and LED lights in several colors are available from Educational Innovations (www.teachersource.com) as well as other suppliers.

HAZARDS

There are no hazards associated with this demonstration.

DISPOSAL

Materials may be saved and used in future presentations.

DISCUSSION

Ground-state electrons in many phosphorescent materials are excited to higher-energy levels when exposed to visible light, and they then emit light over an extended period of time, as they return to the ground state when the excitation light is turned off. Not all wavelengths in the visible spectrum have enough energy to cause this excitation, and this demonstration illustrates the relationship between energy and frequency (or wavelength) of light as postulated by Planck.

A common phosphorescent material is copper-doped zinc sulfide, which is usually embedded in a vinyl plastic sheet or other matrix, as in this demonstration. (See the section *Phosphorescence* in the introduction, page 46, for a discussion of emission from this phosphorescent material.) The energy required for exciting the electrons comes from the photons generated by the light sources that are used to write on the panels in this demonstration. The Planck relationship,

$$E = h\nu = h\frac{c}{\lambda}$$

shows that the energy of a photon of light is proportional to the frequency of the light, or alternatively, inversely proportional to its wavelength. Photons of blue (short-wavelength) light have higher energy than do photons of red (long-wavelength) light. (And consistent with these energetics, only one battery is required to light the red LED, but two are required for the blue.) For example, a 470-nm blue photon has an energy of

$$E = \frac{hc}{\lambda} = \frac{(6.6 \times 10^{-34} \text{ J s})(3.0 \times 10^8 \text{ m s}^{-1})}{470 \times 10^{-9} \text{ m}} = 4.2 \times 10^{-19} \text{ J} \approx 250 \text{ kJ mol}^{-1}$$

and a 630-nm red photon has an energy of

$$E = \frac{hc}{\lambda} = \frac{(6.6 \times 10^{-34} \text{ J s})(3.0 \times 10^8 \text{ m s}^{-1})}{630 \times 10^{-9} \text{ m}} = 3.1 \times 10^{-19} \text{ J} \approx 190 \text{ kJ mol}^{-1}$$

Photons from the red light do not provide enough energy to excite the electrons in these phosphorescent panels. This makes sense, because the emission from the panels is yellow-green, so the emitting energy level must have an energy that corresponds to this color, which is at a shorter (more energetic) wavelength than red light. Thus, red light cannot provide enough energy to excite the electrons to this level. Photons from blue light can, and do, provide the required energy, since the light is at a shorter wavelength (higher energy) than the yellow-green emission. The semiquantitative conclusion is that the required excitation energy lies somewhere between approximately 190 and about 250 kJ mol^{-1}.

This range can be narrowed by using lights with colors that lie between red and blue in the visible spectrum. In this demonstration, photons of yellow light (about 592 nm) do not have enough energy (and so have less energy than the yellow-green emission), but photons of green light (about 525 nm) do have enough (and thus have more energy than the yellow-green emission). This puts the required energy somewhere in the range of approximately 200–230 kJ mol^{-1}.

The absorption by copper-doped zinc sulfide phosphorescent material begins in the green region of the visible spectrum (which is consistent with the green-light excitation of the phosphorescence) and increases to a maximum in the near-ultraviolet at about 365 nm. Since the absorption of light by the phosphor increases from green to blue to violet, this sequence of colored lights produces increasing numbers of excited electrons. The number of excited electrons also depends on the intensity of the sources, which are about the same in this demonstration. Thus, among these three, green light produces the fewest excited electrons, and the glow is dimmest and of shortest duration because few emitters are produced and take little time to be used up. Violet light produces the most excited electrons, and the glow is brightest and longest lasting because many emitters are produced and take more time to be used up.

REFERENCES

1. A demonstration similar to this was included in the presentation "Using LEDs to Demonstrate Properties of Light and the Interaction of Light and Matter," S. A. S. Hershberger, A. M. Sarquis, and L. M. Hogue, Nineteenth Biennial Conference on Chemical Education, July 30–August 3, 2006, Purdue University, West Lafayette, Indiana.

12.39

Quenching Phosphorescence with Light

Glow-in-the-dark objects get dimmer with time and finally stop glowing unless rephoto-excited. The natural decay of the emission can be accelerated by light of appropriate energy.

A phosphorescent panel is irradiated by incandescent light. When the light is extinguished, the panel glows brightly. A transparent yellow film and an opaque object are held against the panel, and the panel is irradiated a second time. When the light is extinguished and the film and opaque object removed, the glow from the panel is dimmest where the yellow film had been held, brighter where the opaque object had been placed, and brightest where nothing was between the light and the panel. Where the beam of a red laser pointer is passed across the glowing screen, it glows more brightly at first and then gets dimmer.

MATERIALS

a phosphorescent panel large enough to be easily seen by the audience (The phosphorescent vinyl sheet should glued to a backing of stiff cardboard, foam core, or poster board to create a panel that can be easily displayed.)*

yellow transparency, such as from a report cover

incandescent desk lamp

red laser pointer

infrared remote control from an electronic device, such as a television set or DVD player

green laser pointer (optional)

blue laser pointer (optional)

PROCEDURE

Preparation and Presentation

Display the yellow transparency. Turn on the incandescent desk lamp and hold the transparency so the audience can see the yellow color of the transmitted light. Turn off the desk lamp.

Display the phosphorescent panel in the lighted room. Darken the lights and show the phosphorescent glow of the panel. Turn on the desk lamp, illuminate the phosphorescent panel, and then extinguish the lamp. The phosphorescent panel will glow brightly.

With the room still darkened, use a hand with outstretched fingers to hold the yellow transparency against the glowing phosphorescent panel. Turn on the desk lamp and illuminate the panel. Some of the light from the lamp should strike the uncovered panel, some should be filtered through the yellow transparency, and some should be blocked by the hand. Turn off the desk lamp, and remove the transparency and hand. The panel will glow brightly

---

* Phosphorescent vinyl sheets are available from Educational Innovations (www.teachersource.com) as well as other suppliers.

where the light from the lamp was unobstructed, less brightly where the light was obscured by the hand, and most dimly where the light was filtered by the yellow transparency.

With the room still darkened, turn on the desk lamp, illuminate the phosphorescent panel, and then extinguish the lamp. While the phosphorescent panel is glowing, aim the beam of a red laser pointer at the glowing panel. Move the beam across the panel. Where the beam has struck the panel, the panel will glow more brightly. The laser can be used to draw a figure, such as a circle, on the panel. The figure that was drawn will fade more rapidly than the rest of the panel, and within a minute, the figure will appear darker than the rest of the panel. After the glow from the panel has disappeared, and it is dark, again aim the beam of the red laser pointer at the panel. Now, the laser does not cause the panel to glow.

With the room still darkened, turn on the desk lamp, illuminate the phosphorescent panel, and then extinguish the lamp. While the phosphorescent panel is glowing, point an infrared remote control at the panel, holding it directly against the panel. Hold down a button on the remote for several seconds, and then move the remote away. The spot where the invisible infrared beam from the remote struck the panel will glow more brightly and then fade until it is darker than the rest of the panel. With some remote controls, if the control is moved quickly over the surface of the glowing panel, a series of glowing dots will form, because the infrared beam produced by the remote is flashing on and off.

If they are available, it can be shown that a green laser pointer has no effect on the glowing panel, whereas a blue laser pointer produces a bright glow that remains brighter than the rest of the panel.

HAZARDS

When handling the laser pointer, avoid directing it or its reflections at viewers.

DISPOSAL

Materials may be saved and used in future presentations.

DISCUSSION

A phosphorescent material emitting a green glow can be made to glow more brightly if it is irradiated with light of lower energy—yellow, red, or infrared—than that emitted by the phosphorescent material. At first, it may seem counterintuitive that lower-energy radiation can produce higher-energy radiation. However, the enhanced glow comes at the price of a reduced duration of the glow. The energy from red light is not converted to higher-energy green light. Instead, the red light increases the rate of emission of green light. The amount of energy available for emission by the phosphorescent material is not changed; it is used up more quickly, so the glow is brighter but of shorter duration.

The energy emitted by many phosphorescent materials comes from the absorption of visible light by the phosphorescent material. The energy of the light absorbed must be greater than that emitted by the phosphorescent material. This means that only wavelengths shorter than that of the emitted light can cause phosphorescence. When the phosphorescent material is exposed to room light, the panel absorbs only some of the wavelengths of the white light. The color of most commonly available phosphorescent materials is yellowish, which indicates that it absorbs some violet light. Violet light is the most energetic of visible light and can supply the energy for green phosphorescence.

When visible light is absorbed, electrons in the absorbing material are excited from the ground state to higher-energy levels. When the excited electrons return to the ground state, they may emit light. The return of the excited electrons to the ground state happens slowly in phosphorescent materials, so the emission occurs over an extended period of time.

Phosphorescence emission is slow, because in phosphorescent materials there is a relatively stable excited state, called a *metastable state*. This excited state is relatively stable, because an electron in this metastable state cannot return directly to the ground state. Instead, the electron must gain a small amount of energy to enter a slightly different excited state, from which it can easily return to the ground state. The excitation of an electron from the metastable state to an emissive state is called *activation*. At room temperature, thermal energy brings about the activation. The thermal activation can be speeded up by heating the phosphorescent material, as shown in Demonstration 12.40, or by low-energy radiation, as shown in this demonstration.

A common phosphorescent material is copper-doped zinc sulfide, which is usually embedded in a vinyl plastic sheet or other matrix, as in this demonstration. (See the section *Phosphorescence* in the introduction, page 46, for a discussion of emission from this phosphorescent material.) The energy required for initial electron excitation is furnished here by photons from the incandescent lamp. The excitation of an electron leaves a vacancy for an electron at lower energy, and this vacancy creates a positive center or hole. The separated electrons and holes constitute a metastable state. A small activation energy is required to bring the excited electrons and holes together, which results in long-lived emission as thermal energy slowly provides the required activation energy.

Other forms of energy can promote this recombination and emission of light. Light that does not have enough energy to promote electrons from the ground state into the excited state(s) that leads to the metastable state can still move these electrons from the metastable state to the emissive state. The greenish-yellow light of the phosphorescent glow corresponds in energy to the emissive state. Light of energy greater than the yellow-green light can populate the metastable state with electrons. Light of lower energy, the yellow light in this demonstration, cannot. Yellow, red, and even infra-red light does, however, have enough energy to promote the electrons from the metastable state to the emissive state. For this reason, a glowing panel glows even more brightly where it is irradiated by these low-energy lights. (The brighter glow is difficult to see because it is masked by the exciting light from the incandescent lamp.). When the exciting light from the incandescent lamp is extinguished, there are fewer metastable-state electrons left in the area that was irradiated by the longer wavelength light. Because the low-energy light has depleted the number of electrons in the metastable state, there are fewer metastable-state electrons and the glow is dimmed in this area.

In the area where an opaque object (the hand) prevents the exciting light from the incandescent lamp from reaching the panel, the normal thermal decay of the phosphorescence continues without any enhancement of the slow rate of loss of electrons from the metastable state. Thus, the glow from this area is dimmer than from the area that has been freshly illuminated by the incandescent lamp, but brighter than the area where the yellow light has promoted a more rapid loss of metastable-state electrons.

12.40

Quenching Phosphorescence with Thermal Energy

Glow-in-the-dark objects get dimmer with time and finally stop glowing unless rephoto-excited. The natural rate of emission and decay can be accelerated by warming the object.

A spot on a glowing phosphorescent panel is heated in the dark with a hair dryer. Where the panel is heated, the glow is brighter than the rest of the panel. After heating is stopped and the panel has returned to room temperature, the glow at the spot where the panel had been heated is dimmer than the rest of the panel.

MATERIALS

a phosphorescent panel large enough to be easily seen by the audience (The phosphorescent vinyl sheet should be glued to a backing of stiff cardboard, foam core, or poster board to create a panel that can be easily displayed.)*

handheld hair dryer

PROCEDURE

Preparation and Presentation

Plug in the hair dryer and have it handy. Expose the phosphorescent panel to room lights.

Darken the room. The phosphorescent panel will glow. Allow the panel to glow for about 30 seconds, during which time its brightness will diminish.

Before most of the glow has disappeared, turn on the hair dryer and blow hot air onto the center of the phosphorescent panel. Where the hot air strikes the panel, the glow will be brighter than in the surrounding, cooler areas.

After heating a spot on the panel for about 15 seconds, turn off the hair dryer, and allow the panel to cool. When the temperature of the panel has become uniform, the glow at the spot that had been heated will be dimmer than the glow in the surrounding areas that had not been heated.

HAZARDS

There are no hazards associated with this demonstration.

DISPOSAL

Materials may be saved and used in future presentations.

---

* Phosphorescent vinyl sheets are available from Educational Innovations (www.teachersource.com) as well as other suppliers.

DISCUSSION

This demonstration shows that when the temperature of a glowing phosphorescent material is increased, the intensity of its glow also increases. Furthermore, when a phosphorescent material that has been heated returns to its original temperature, its glow is less than that of material that has not been heated. These effects are a result of an increase in the rate of emission as temperature increases. When the rate of emission increases, the glow appears brighter. The increased rate also reduces the amount of emitting material more quickly, so when the heated material returns to its original temperature, it contains less of the emitting material, thus its glow is not as bright as material that has not been heated.

Phosphorescence occurs when certain materials absorb energy from visible or ultraviolet light and then emit this energy as visible light after the absorption has stopped. When a phosphorescent material is exposed to visible or ultraviolet light, electrons in the lowest-energy state (the ground state) in the material are excited to higher-energy levels. These higher-energy electrons eventually return to the ground state, and in the process, visible light is emitted.

A common phosphorescent material is copper-doped zinc sulfide, which is usually embedded in a vinyl plastic sheet or other matrix, as in this demonstration. (See the section *Phosphorescence* in the introduction, page 46, for a discussion of emission from this phosphorescent material.)This material absorbs energy from photons of visible room light, which elevate the energy of electrons from their ground states to higher-energy states. The excitation process also leaves electron vacancies (holes) in the lower-energy states. Some of the higher-energy states are metastable, that is, electrons in these states do not immediately return to the ground state. A small activation energy is required to bring the excited electrons and the holes together to emit light. Thermal energy supplies this activation energy and slowly brings about the required pairing of excited electrons and holes. The slow combination results in emission over a period of time, which accounts for the long-lived glow from the panels.

Raising the temperature of the panel provides more thermal energy to promote electrons from the metastable to the emissive state and thus increases the emission, causing a brighter glow. When the source of extra thermal energy is removed and the panel returns to room temperature, there are fewer metastable-state electrons left in the area that was heated. Because there are fewer metastable-state electrons, the glow is dimmed in this area.

12.41

The Fluorescence of Molecular Iodine Vapor

For some substances, visible light can produce fluorescence. At room temperature, solid molecular iodine sublimes to yield a pale violet gas, and the color of the vapor indicates that it absorbs green light. The absorbed light can cause the molecules to emit at lower energy (longer wavelengths) than green light.

The beam of a green laser pointer is directed through a flask containing iodine vapor. The beam is visible as a yellow glow in the vapor.

MATERIALS FOR PROCEDURE A

0.5 g iodine, I_2

2-L round-bottom flask

one-hole rubber stopper to fit flask

glass stopcock

vacuum tubing

plastic or rubber gloves

cork ring for 2-L round-bottom flask

water aspirator or vacuum pump

laser pointer that produces 532-nm (green) light

MATERIALS FOR PROCEDURE B

0.5 g iodine, I_2

2-L round-bottom flask

one-hole rubber stopper to fit flask

glass stopcock

vacuum tubing

plastic or rubber gloves

cork ring for 2-L round-bottom flask

water aspirator or vacuum pump

either
  slide projector
  2-in × 2-in piece of poster board
  hole punch
or
  overhead projector
  12-in × 12-in piece of aluminum foil
  scissors
  16-in × 16-in piece of black felt
  4-in × 4-in piece of black felt

PROCEDURE A [1]

Preparation

Prepare a 2-L, round-bottom flask containing only solid iodine, its vapor, and a small amount of air. Insert one arm of the stopcock through the hole of a one-hole rubber stopper that fits the mouth of the 2-L flask. Wearing protective gloves, put about 0.5 g solid I_2 into the flask. Seal the flask with the rubber stopper. Use vacuum tubing to attach the free arm of the stopcock to a vacuum pump or water aspirator. With the stopcock open, turn on the pump or aspirator. Allow the pump or aspirator to run until it has removed most of the air from the flask, as signaled by a change in the sound of the pump or aspirator. (Do not allow the pump or aspirator to run so long that there is an obvious change in the amount of solid iodine in the flask.) Close the stopcock. Detach the vacuum tubing from the stopcock, and turn off the pump or aspirator. Allow the flask to rest for at least 15 minutes to assure that iodine vapor uniformly fills the flask. The vapor of iodine will impart a pale violet color to the interior of the flask.

Presentation

Display the flask of iodine vapor resting on a cork ring. Darken the room. Aiming the laser pointer horizontally and perpendicular to the line of sight from the audience, direct the beam through the flask. Be careful to avoid directing the beam or its reflection toward the audience. Viewed from the side, the laser beam is visible as a yellow beam where it passes through the iodine vapor. With some laser pointers, the yellow beam will gradually fade and brighten, while with others, the intensity will remain constant.

PROCEDURE B [2, 3]

Preparation

Prepare a 2-L, round-bottom flask containing only solid iodine, its vapor, and a small amount of air. Insert one arm of the stopcock through the hole of a one-hole rubber stopper that fits the mouth of the 2-L flask. Wearing protective gloves, put about 0.5 g solid I_2 into the flask. Seal the flask with the rubber stopper. Use vacuum tubing to attach the free arm of the stopcock to a vacuum pump or water aspirator. With the stopcock open, turn on the pump or aspirator. Allow the pump or aspirator to run until it has removed most of the air from the flask, as signaled by a change in the sound of the pump or aspirator. (Do not allow the pump or aspirator to run so long that there is an obvious change in the amount of solid iodine in the flask.) Close the stopcock. Detach the vacuum tubing from the stopcock, and turn off the pump or aspirator. Allow the flask to rest for at least 15 minutes to assure that iodine vapor uniformly fills the flask. The vapor of iodine will impart a pale violet color to the interior of the flask.

Prepare a narrow beam of white light. This can be done with a slide projector using a slide prepared by cutting a piece of poster board the size of a slide (2 in × 2 in) and punching in its center a clean, round hole about 3 to 4 mm in diameter. Cover the top of the projector with a piece of black felt to reduce scattered light. Position the flask of iodine vapor on a cork ring so that the beam of the projector passes through the flask.

Alternatively, a narrow beam of white light can also be produced using an overhead projector. In the center of a sheet of aluminum foil large enough to completely cover the stage of an overhead projector, cut a circular hole 2 to 3 cm in diameter. Place the foil on the stage of the overhead projector with the hole at the center of the stage. On the aluminum foil, set a cork ring centered around the hole. Cut a round hole about 2 cm in diameter in the center of the larger piece of black felt. Place the felt over the cork ring with the hole centered over the hole in the aluminum foil, and drape the felt over the top of the projector. Set the 2-liter flask of iodine vapor onto the cork ring. Turn on the projector and make sure the beam of the projector passes through the hole in the foil and the hole in the felt and up through the flask. Place a 10-cm × 10-cm piece of black felt on top of the flask where the beam of the projector exits the flask. The purpose of the felt is to reduce the amount of scattered and reflected light.

Presentation

Display the flask of iodine vapor and the source of white light (slide projector or overhead projector). Turn on the projector and position the flask in the beam from the projector. Darken the room. Allow a minute or two for the eyes of the viewers to become dark adjusted. Viewed from the side, the beam is visible as a yellow-green beam where it passes through the iodine vapor.

HAZARDS

Because iodine, a moderately strong oxidizing agent, vaporizes readily at room temperature to yield toxic fumes, it should be handled only with adequate ventilation. Wear gloves while handling iodine.

Avoid directing the laser beam toward the audience; the beam is capable of causing damage if it shines directly on the retina.

DISPOSAL

The sealed flask of iodine may be stored for repeated presentations of the demonstration. Alternatively, the flask may be opened and the iodine dissolved in about 50 mL of 1 M KI (potassium iodide) solution, and the resulting solution may be flushed down the drain with water.

DISCUSSION

Generally, a beam of light is not visible when it passes through a nonscattering medium, such as a gas. For example, to view a laser beam as it passes through air, it is necessary to introduce into the air a medium, such as smoke or fog, that will scatter the light. (See the section *Light Scattering* in the introduction, page 17.) A laser beam is not visible in air because the molecules that usually compose air do not interact significantly with the light. However, iodine molecules absorb a portion of a beam of white light or the beam of a green laser. Iodine vapor is purple, indicating that it absorbs yellow-green visible light. At least some of the light energy absorbed by the iodine vapor is emitted as fluorescence, which makes visible the path of the light beam through the vapor.

The flask containing iodine vapor must be at least partially evacuated. When the total pressure of the gas is a significant fraction of atmospheric pressure, the fluorescence from iodine vapor is very dim and difficult to see. The fluorescence is dim, because the energy absorbed by iodine molecules is transferred to other molecules when the energized iodine molecules collide with them. As the pressure of the gas inside the flask is reduced, the frequency of collision between molecules decreases, and the time between collisions increases. As the time between collisions increases, more iodine molecules lose energy by emission than by collision. As more molecules emit, the emission becomes brighter. A water aspirator can reduce the pressure inside the flask to the vapor pressure of water, about 15 torr at 18°C. At this pressure, the emission from iodine is easily visible. A vacuum pump can reduce the pressure even more, which will result in an even brighter glow. The minimum pressure inside the flask will be the vapor pressure of iodine, which is about 0.3 torr at 25°C [4].

The iodine emission resulting from excitation by a green laser pointer may alternately fade and brighten. This fluctuation occurs because the wavelength of the light produced by the laser varies slightly as the temperature of the laser changes during operation [1]. The variability in the wavelength has an effect on the brightness of the emission, because the absorption of iodine depends strongly on the wavelength. Figure 1 is a visible absorption spectrum of iodine vapor and shows the large variations in absorption with small changes in wavelength. As the laser warms, the size of its lasing cavity may vary, which in turn changes the wavelength of light that the laser emits, and hence the amount of energy the iodine vapor can absorb. As the amount of energy absorbed changes, so does the amount that can be emitted, and the emission varies in intensity.

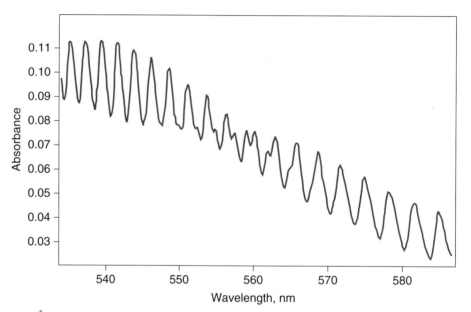

Figure 1.   Visible absorption spectrum of iodine vapor.

The emission produced from white light is more difficult to see than that produced by the green laser pointer. The visibility of the emission is enhanced by reducing stray light from the white-light source (slide projector or overhead projector) and by minimizing the amount of light scattered when it passes through the glass of the flask. Scattered light can be reduced by placing a light absorber, such as black felt, where the light beam exits the flask. Even with careful attention to reducing stray and scattered light, the emission from white light can be difficult to see. The brighter the source, the more visible will be the emission, so high-powered lecture-hall projectors work better than home models.

REFERENCES

1. J. Tellinghuisen, "Laser-Induced Fluorescence in Gaseous I_2 Excited with a Green Laser Pointer," *J. Chem. Educ., 84*(2), 336–341 (2007).
2. R. M. Sutton, ed., "Resonance Radiation of Iodine Vapor," in *Demonstration Experiments in Physics,* pp. 475–476, McGraw-Hill: New York (1938).
3. H. F. Meiners, ed., "Resonance Radiation," in *Physics Demonstration Experiments,* p. 1213, Ronald Press: New York (1970).
4. G. P. Baxter and M. R. Grose, "The Vapor Pressure of Iodine between 50° and 95°," *J. Amer. Chem. Soc., 37*(5), 1061–1072 (1915).

Photochemistry

12.42

The Reversible Photochemical Bleaching of Thionine

Bleaching is the result of a chemical process that can be caused by absorption of light. Photoexcitation increases a molecule's electronic energy and can make it a better oxidizing (or reducing) agent. If the molecule is involved in reduction-oxidation (redox) reactions, the redox equilibria can be disturbed when light is being absorbed by the oxidizing agent. Oxidized and reduced forms generally have different absorption spectra, so color changes can signal redox changes.

When a deep violet solution is placed in a beam of bright white light, the solution initially becomes purple and then bleaches to nearly colorless where the light strikes it. When the light is extinguished, the violet color gradually reappears in the bleached region.

MATERIALS

0.03 g thionine (a salt of 7-amino-3-imino-3H-2-phenothiazine, $C_{12}H_9N_3S$)

600 mL distilled water

2.0 g iron(II) sulfate heptahydrate, $FeSO_4 \cdot 7H_2O$

10.0 mL 3 M sulfuric acid, H_2SO_4 (To prepare 100 mL of stock solution, carefully pour 17 mL of concentrated [18 M] H_2SO_4 into 75 mL of distilled water and dilute the mixture to 100 mL with distilled water.)

two 250-mL Erlenmeyer flasks

600-mL beaker

10-mL graduated cylinder

500-mL glass cylinder with an air-tight seal (A plastic cylinder may be used, as long as the plastic does not become stained by the thionine.)

slide projector

PROCEDURE [1]

Preparation

The following preparation should be carried out as quickly as possible to minimize the exposure of the solutions to air.

In a 250-mL Erlenmeyer flask, prepare 100 mL of a 0.001 M (nearly saturated) aqueous stock solution of thionine by dissolving 0.03 g of thionine in 100 mL of distilled water. In the other Erlenmeyer flask, dissolve 2.0 g of iron(II) sulfate heptahydrate in 50 mL of distilled water.

In a 600-mL beaker containing 250 mL of distilled water, combine all 50 mL of the iron(II) sulfate solution with 10 mL of the 0.001 M thionine solution and 10 mL of 3 M sulfuric acid solution. Dilute the mixture to 500 mL with distilled water. In the final solution, the concentration of iron(II) sulfate is 0.014 M, the concentration of thionine is 2×10^{-5} M, and that of sulfuric acid is 0.06 M.

Seal the solution into an air-tight colorless glass (or plastic) container no more than 10 cm in diameter.

Presentation

Irradiate the solution with a bright white light, such as the beam of a slide projector. Where the light strikes the solution, it will initially turn purple but bleach to colorless in about 15 seconds. When the light is extinguished, the solution will return to violet in 15–30 seconds. This process may be repeated indefinitely.

HAZARDS

The toxicological properties of thionine are unknown. Thionine is intensely colored and easily causes persistent stains to surfaces. Gloves should be worn when handling solid thionine to avoid staining the skin.

Since concentrated sulfuric acid is both a strong acid and a powerful dehydrating agent, it must be handled with great care. Spills should be neutralized with an appropriate agent, such as $NaHCO_3$, and then wiped up.

DISPOSAL

The sealed tube may be kept for repeated presentations of the demonstration. All waste solutions may be flushed down the drain with water.

DISCUSSION

In this demonstration, the central portion of a cylinder containing a solution of a violet dye is irradiated by a bright beam of white light. When the light first strikes the solution, the violet solution emits a red fluorescence, giving the irradiated solution a purple glow. As the irradiation continues, the violet color and the red fluorescence gradually fade, and the solution becomes clear and colorless. In the violet solution above the irradiated area, streams of lighter violet solution (Schlieren lines) move upward into the darker violet above. When the irradiating beam is extinguished, the colorless solution gradually returns to violet.

When a substance absorbs light, it absorbs energy. That energy can be emitted as light, can cause a chemical change, or can be converted to thermal energy that heats the substance. This demonstration shows the three fates of the energy from the absorbed light: 1. the red fluorescence visible when the violet solution is first irradiated is energy that is emitted; 2. the violet color fades because the dye undergoes a chemical transformation to a colorless form; and 3. the streaming that occurs at the border of the beam of light is a result of a decrease in the density of the solution as it becomes warmer when the absorbed light is converted to thermal energy.

That the solution is violet indicates that it absorbs visible light. (A solution that absorbs no visible light is colorless.) The color of a solution indicates the color of the light it does *not* absorb, so the solution in this demonstration is violet because the dye it contains does not absorb violet light, that is, it transmits violet light. In general, the color of the light absorbed is the complement of the color that is not absorbed. A color wheel, such as that in Figure 1, is a schematic representation of the relationship between complementary colors and shows that the complement of violet is yellow, so the violet solution absorbs yellow light. This is confirmed by the visible absorption spectrum of an aqueous thionine solution (Figure 2). The spectrum shows a maximum absorbance at 600 nm, which corresponds to yellow light.

The color of light is related to the energy it carries. The progression of colors in the visible spectrum from red and continuing sequentially through orange, yellow, green, blue, and violet corresponds to increasing energy of the light (Figure 3). (See the beginning of the introduction for a description of the relationship between light energy and color.) When absorbed light energy is emitted, it can have at most the same energy as that of the light absorbed. Most often, however, the emitted light is of lower energy than that absorbed, which is what is

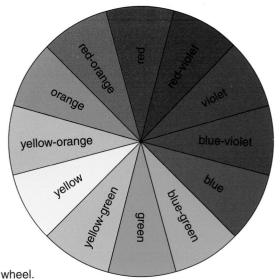

Figure 1.   Color wheel.

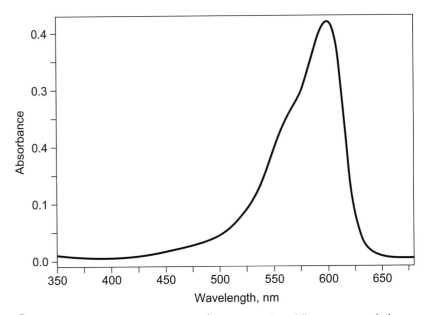

Figure 2.   Absorption spectrum of 5×10^{-4} M thionine in acidic aqueous solution.

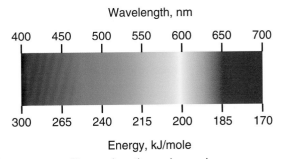

Figure 3.   Visible spectrum with wavelengths and energies.

observed in this demonstration. When the violet solution is irradiated with white light, it emits red light, which is lower in energy than the yellow light absorbed by the violet solution.

In addition to heating the solution and being emitted as red light, some of the energy absorbed by the violet solution causes a reaction in which the violet dye becomes colorless. The violet dye in the solution is thionine, which is a salt, usually the chloride or acetate, of the 7-amino-3-imino-3H-2-phenothiazine cation, whose molecular structure is

thionine

Thionine is related structurally to the common redox indicator, methylene blue, whose structure is

methylene blue

In both of these structures, the three rings are fully conjugated (the π electrons are delocalized over the entire molecule), which is why they absorb visible light and are colored. (This principle is described in the section *The Photochemistry of Vision* in the introduction, page 67, and in the discussion section of Demonstration 12.47, The Photobleaching of Carotene.)

Like methylene blue, thionine is also a redox indicator. Thionine molecules can undergo a two-electron reduction, which produces colorless leucothionine, as represented by the reduction half-reaction

thionine (T^+), violet   leucothionine (TH), colorless

In leucothionine, the three rings are no longer conjugated, the molecule does not absorb visible light, and the product is colorless. The standard reduction potential of this process is 0.38 V [2].

The solution also contains iron(II) ions from iron(II) sulfate. Iron(II) ions can be oxidized to iron(III) ions.

$$Fe^{2+} \longrightarrow Fe^{3+} + e^-$$

The oxidation potential for this reaction is −0.771 V [3]. The sum of the reduction potential of thionine and the oxidation potential of Fe^{2+} is negative, about −0.4 V. This result indicates that thionine cannot oxidize Fe^{2+} and is why the solution containing these two species remains violet. In the mixture, there is an equilibrium between the violet form of thionine and Fe^{2+} on one side and the colorless form and Fe^{3+} on the other.

violet   colorless

Under the conditions that this solution is prepared, the equilibrium position lies on the left side of the equation, and the solution is violet.

When the violet solution is irradiated with bright white light, the solution bleaches to colorless. This outcome indicates that the light has caused the redox reaction to shift from left to right, because the energy supplied by the absorbed light is sufficient to allow thionine to oxidize iron(II). The violet color of the thionine solution results from its absorption of some wavelengths of visible light, specifically wavelengths around 600 nm. The energy of a photon of light with a wavelength of 600 nm can be calculated using Planck's equation:

$$E = \frac{hc}{\lambda} = \frac{(6.626 \times 10^{-34}\ \text{J s})(3.00 \times 10^8\ \text{m s}^{-1})}{600 \times 10^{-9}\ \text{m}} = 3.3 \times 10^{-19}\ \text{J photon}^{-1}$$

This result indicates that the energy absorbed by thionine increases the energy of one of its electrons by 3.3×10^{-19} J.

The potential of an oxidation-reduction reaction is expressed in volts, and a volt is a joule per coulomb. A coulomb is an electrical charge, and the charge of one mole of electrons is 96,485 coulomb. Therefore, the energy absorbed by the violet thionine solution corresponds to a voltage as follows:

$$\left(3.3 \times 10^{-9}\ \frac{\text{joule}}{\text{photon}}\right)\left(1\ \frac{\text{photon}}{\text{electron}}\right)\left(6.023 \times 10^{23}\ \frac{\text{electron}}{\text{mole}}\right)\left(\frac{1\ \text{mole}}{96485\ \text{coulomb}}\right) = 2.0\ \frac{\text{joule}}{\text{coulomb}}$$

$$= 2.0\ \text{volt}$$

This result indicates that the oxidizing power of the excited molecule of thionine is about 2 volts greater than that of the ground-state molecule. Because 2 volts is much more than the 0.77 volts needed to oxidize Fe^{2+} to Fe^{3+}, the excited state thionine molecule can oxidize Fe^{2+}, and in the process it is converted to its colorless reduced form.

In general, an excited electronic state of a molecule can be a better oxidizing agent than its ground state. Why this is so can be visualized using a simple energy-level diagram for the molecule (Figure 4). The left side of the diagram shows an energy level for the highest-energy occupied molecular orbital (HOMO). Above it is the lowest-energy unoccupied molecular orbital (LUMO). At the right is an energy level for the highest-energy electron in the reducing agent, which in our case is Fe^{2+}. The two arrows in the HOMO show that it is full, and the LUMO is empty. The arrow in the Fe^{2+} represents the electron that is lost when it is oxidized to Fe^{3+}. All of the orbitals in thionine that are lower in energy than the Fe^{2+} orbital are full, so Fe^{2+} cannot transfer an electron to thionine. However, when thionine absorbs a photon of yellow light, one of the electrons is promoted from the HOMO to the LUMO (Figure 5). Now, there is an opening in a thionine orbital with lower energy than the electron in Fe^{2+}, and the electron from Fe^{2+} can transfer to thionine. This oxidizes Fe^{2+} to Fe^{3+} and starts the process of reduction of violet thionine to its colorless form.

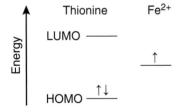

Figure 4.   Energy-level diagram for ground-state thionine and iron(II).

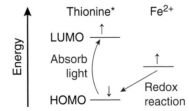

Figure 5.   Energy-level diagram for excited state of thionine and iron(II).

The energy supplied by the absorbed light drives the redox reaction to the right, to the colorless form. The colorless form will react spontaneously with any Fe^{3+} that it encounters and revert to the violet form and Fe^{2+}. The regenerated violet form will absorb another photon of light and go back to the colorless form, as long as there is an ample supply of the proper photons. As long as sufficient light is provided, the steady state will be maintained, and the redox reaction will stay toward the right, toward the colorless form. If the light is diminished, the reaction will shift back to the left, to the violet form.

The reaction of the colorless, high-energy form of thionine with Fe^{3+}, which is the reverse reaction of the equilibrium, is spontaneous and releases energy. Much of the energy it releases warms the reaction mixture. This thermal energy comes ultimately from the light that is absorbed by the violet solution. This absorbed energy causes a chemical reaction whose final products, leucothionine and Fe^{3+}, then react exothermically to regenerate the initial reactants, thionine and Fe^{2+}. The energy from this exothermic reaction is absorbed by the solution, which causes the solution to become warmer. Evidence for the warming of the solution is the mixing of colorless and colored solutions that occurs in the region above the illuminated solution (producing the Schlieren bands). As the illuminated solution becomes warmer, its density decreases, and the more dense violet solution above sinks down, displacing the less dense colorless solution.

The purple glow that appears when thionine is irradiated by bright light is produced by the emission of red light from the excited thionine molecules. The combination of reflected or transmitted violet light and emitted red light gives the solution a purple appearance. The red light emitted by the excited thionine molecules is fluorescence from the initially formed excited singlet state of the molecules. Fluorescence emission is at longer wavelengths than the corresponding excitation wavelengths; in this case red light is emitted from molecules that absorb yellow light. (The reason for this wavelength shift is illustrated in the Jablonski diagram, Figure 39, page 44 in the introduction to this book, and the accompanying discussion.) The difference in energy between the absorbed and emitted light is dissipated by increased molecular vibrations and by collisions with other molecules, which cause the solution to become warmer.

The thionine-Fe^{2+} photoredox system has been extensively studied for more than half a century to learn about how electric current can be produced from a photochemical process [4–10]. The ultimate goal is the conversion of the energy in sunlight to electricity. These studies have led to a simplified set of reactions that are responsible for the observations in this demonstration. (Much of the simplification is the neglect of the important but complicating role of hydronium ions and of different conjugate acid-base forms of the thionine and its derivatives.) In these reactions H^+ is hydronium ion, T^+ is thionine ground state, TH is leucothionine, TH^+ is semithionine (the one-electron reduction product—a radical cation—formed in the first thionine-Fe^{2+} reduction step discussed above), $_1T^{+*}$ is the thionine singlet excited state, and $_3T^{+*}$ is the thionine triplet.

$$T^+ + h\nu \text{ (yellow)} \longrightarrow {_1}T^{+*}$$

$${_1}T^{+*} \longrightarrow T^+ + h\nu \text{ (red)} \qquad \text{(fluorescence)}$$

$${_1}T^{+*} \longrightarrow {_3}T^{+*} \qquad \text{(rapid intersystem crossing)}$$

$${_3}T^{+*} + Fe^{2+} + H^+ \longrightarrow TH^+ + Fe^{3+}$$

$$2\,TH^+ \longrightarrow T^+ + TH + H^+ \qquad \text{(rapid disproportionation reaction)}$$

$$TH + Fe^{3+} \longrightarrow TH^+ + Fe^{2+}$$

$$TH^+ + Fe^{3+} \longrightarrow T^+ + Fe^{2+} + H^+$$

In this demonstration, the return to the initial violet state, thionine plus Fe^{2+}, takes several seconds after irradiation has been stopped. Therefore, the last two reactions must be relatively slow compared to the others. Thus, a bright light can drive the reaction almost completely to the colorless leucothionine, leaving so little thionine that its color and fluorescence are undetectable. At this photochemical steady state, the concentration of the semithionine (TH^+) is also very low because the disproportionation reaction occurs essentially every time two of these molecules meet. Under these conditions, the colorless solution contains Fe^{3+} at approximately twice the concentration of leucothionine (and a great excess of Fe^{2+}, which is present in about 700-fold excess over the thionine).

The photosensitive solution prepared in this demonstration will maintain its photosensitivity indefinitely, as long as it is sealed tightly against atmospheric oxygen, which will gradually oxidize Fe^{2+} to Fe^{3+}. When all of the Fe^{2+} has been depleted, the solution will no longer bleach upon exposure to light. As the chemical equations above indicate, increases in the concentration of Fe^{3+} will increase the rate at which the solution returns to violet after cessation of irradiation. Therefore, to achieve the slowest return of the violet color, the concentration of Fe^{3+} must be minimized. Because Fe^{2+} is oxidized when it is exposed to air, the solution used for this demonstration should be prepared and sealed into its container as quickly as possible.

REFERENCES

1. L. J. Heidt, "The Photochemical Reduction of Thionine," *J. Chem. Educ., 26,* 525–526 (1949).
2. E. Rabinowitch, "The Photogalvanic Effect I. The Photochemical Properties of the Thionine-Iron System," *J. Chem. Phys., 8,* 551–559 (1940).
3. D. R. Lide, ed., *CRC Handbook of Chemistry and Physics,* 90th ed., CRC Press: Boca Raton, Florida (2009).
4. R. Hardwick, "Kinetic Studies of the Thionine-Iron System," *J. Amer. Chem. Soc., 80,* 5667–5673 (1958); and a reinterpretation of the data, "Kinetic Studies of the Thionine-Iron System. II," *J. Phys. Chem., 66,* 349–350 (1962).
5. C. G. Hatchard and C. A. Parker, "The Photoreduction of Thionine by Ferrous Sulphate," *Trans. Faraday Soc., 54,* 1093–1106 (1961).
6. P. D. Wildes, N. N. Lichtin, and M. Z. Hoffman, "Solvent Effects on the Electron-Transfer Disproportionation Rate Constant of Semithionine Radical Cation," *J. Amer. Chem. Soc., 97,* 2288–2289 (1975).
7. P. D. Wildes and N. N. Lichtin, "Indirect Measurements of the Thionine-Leucothionine Synproportionation Rate Constant by a Photochemical Perturbation Technique," *J. Phys. Chem., 82*(9), 981–984 (1978).
8. P. D. Wildes, N. N. Lichtin, M. Z. Hoffman, L. Andrews, and H. Linschitz, "Anion and Solvent Effects on the Rate of Reduction of Triplet Excited Thiazine Dyes by Ferrous Ions," *Photochem. Photobiol., 25,* 21–25 (1977).
9. T. L. Osif, N. N. Lichtin, and M. Z. Hoffman, "Kinetics of Dark Back Reactions of Products of the Photoreduction of Triplet Thionine by Iron(II). Evidence for Association of Leucothionine and Semithionine with Iron(III)," *J. Phys. Chem., 82*(16), 1778–1784 (1978).
10. P. V. Kamat, M. D. Karkhanavala, and P. N. Moorthy, "Kinetics of Photobleaching Recovery in the Iron(III)-Thionine System," *J. Phys. Chem., 85,* 810–813 (1981).

12.43

Photochromic Methylene Blue Solution

A photochromic material is one that changes color reversibly when exposed to light. Photo-excitation increases molecular electronic energy and can make excited molecules better oxidizing (or reducing) agents. If these molecules are involved in reduction-oxidation (redox) reactions, their redox equilibria can be disturbed if the reducing agent absorbs light. Oxidized and reduced forms generally have different absorption spectra, so color changes can signal redox changes. When the excitation is removed, the equilibrium state and initial color are restored.

When a colorless solution is exposed to ultraviolet radiation, the solution turns blue. When the ultraviolet radiation is removed, the solution fades to colorless.

MATERIALS

0.05 g methylene blue, 3,7-bis(dimethylamino)phenothiazin-5-ium chloride, $C_{16}H_{18}N_3SCl$

540 mL 0.5 M hydrochloric acid, HCl (To prepare 1.0 L of stock solution, add 42 mL of 12 M HCl to 500 mL of distilled water and dilute the resulting solution to 1.0 L.)

0.2 g tin(II) chloride dihydrate, $SnCl_2 \cdot 2H_2O$

plastic or rubber gloves

1-L beaker

100-mL beaker

500-mL glass cylinder

stirring rod

dropper

long-wave (365 nm) ultraviolet lamp or black light

PROCEDURE

Preparation

Wearing gloves, dissolve 0.05 g of solid methylene blue in 500 mL of 0.5 M HCl in a 1-L beaker. This produces a dark blue solution that is about 3×10^{-4} M methylene blue.

In a 100-mL beaker, dissolve 0.2 g $SnCl_2 \cdot 2H_2O$ in 40 mL of 0.5 M HCl. This solution is about 0.02 M in $SnCl_2$.

While stirring the methylene blue solution in the beaker, very slowly add the $SnCl_2$ solution until the blue color just disappears. The reaction is quite slow and could take as long as 15 minutes and use about 18 mL of the $SnCl_2$ solution.

Pour the colorless solution into a 500-mL cylinder.

Presentation

Using a long-wave (365 nm) UV source or a black light, irradiate the solution at the middle of the cylinder, so that solution near the top and at the bottom are not irradiated. Over a span of about 30 seconds, where the solution is irradiated it will turn blue, whereas the solution that is not irradiated remains colorless.

Turn off the UV source. The solution will fade to colorless over the span of about 60 seconds. The coloring and fading process may be repeated several times.

HAZARDS

Methylene blue is intensely colored and easily causes persistent stains to surfaces. Gloves should be worn when handling solid methylene blue to avoid staining the skin.

Black light (long-wavelength ultraviolet light) can damage one's eyes. Shield the black-light source from the audience and never look directly at the source yourself.

DISPOSAL

All solutions may be flushed down the drain with water.

DISCUSSION

In this demonstration, the central portion of a cylinder containing a colorless solution is irradiated with a black light, a source of long-wave ultraviolet radiation. Where the light strikes the solution, the solution gradually turns blue. When the ultraviolet radiation is removed, the blue fades back to colorless.

The colorless solution contains methylene blue, which is commonly available as a salt in which the organic portion is a cation with an inorganic counter ion, such as chloride.

methylene blue

Methylene blue is a common oxidation-reduction indicator. The cation, MB^+, is blue, but will be colorless after undergoing a two-electron reduction to the leuco form, LB. The standard potential for this reduction is 0.53 volt at pH=0 [1].

MB$^+$ blue   LB colorless

The solution also contains tin(IV) ions, which were formed when tin(II) chloride was added to the methylene blue solution. Tin(II) undergoes oxidation to tin(IV), and the standard potential for the oxidation is −0.15 volt [2].

$$Sn^{2+} \longrightarrow Sn^{4+} + 2e^-$$

Under the acidic conditions used in preparing the solution for this demonstration, tin(II) reduces methylene blue to its colorless form. Enough tin(II) is added to shift the equilibrium between the blue and colorless forms of methylene blue substantially to the colorless form [3]. In this mixture the colorless form and tin(IV) are the lower-energy combination, while the blue form and tin(II) have the higher-energy.

$$MB^+ + Sn^{2+} \rightleftharpoons LB^- + Sn^{4+}$$

blue   colorless

The reduced, leuco, form of methylene blue is colorless. It does not absorb visible light, but it does absorb in the ultraviolet region. The absorption spectrum of the reduced form of methylene blue (Figure 1) shows that leucomethylene blue absorbs ultraviolet radiation, both in the region usually called long-wave, or black light (400 to 315 nm), as well as in the short-wave region (280 to 100 nm).

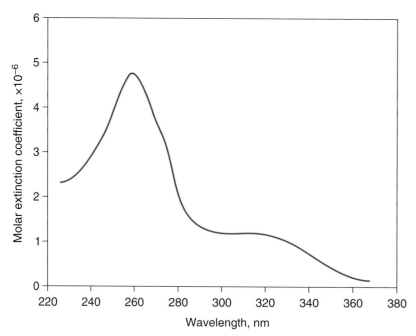

Figure 1.   Absorption spectrum of leucomethylene blue. [4]

In this demonstration, when the solution of leucomethylene blue is irradiated with black light, the ratio of the colorless to blue forms of methylene blue is shifted away from colorless toward the blue form. The ultraviolet energy absorbed by leucomethylene blue allows the colorless leucomethylene blue to reduce tin(IV) to tin(II) and reform the blue color. A steady state is set up in which there is a competition between the forward thermal rate of the redox reaction above and the new photochemical rate of the reverse reaction. In this steady state, there is enough of the blue form of methylene blue present to be visible to the audience. When the energy of the black light is removed, the equilibrium that favors colorless leucomethylene blue and tin(IV) is restored, and the solution becomes colorless. Compare the steady state here with the more complete treatment in Demonstration 12.52, Photochromism in Ultraviolet-Sensitive Beads. (Photochromic materials with reversible formation of color have been developed for filtering UV light, similar to eyeglass lenses that darken when exposed to UV from sunlight and fade when UV is not present. [5])

The equilibrium between tin(II) and methylene blue on one side and tin(IV) and leucomethylene blue on the other in this demonstration lies very much toward the colorless leucomethylene blue and tin(IV). Leucomethylene blue is not a strong enough reducing agent to reduce tin(IV) to tin(II). However, after it absorbs ultraviolet light, the leucomethylene blue does reduce tin(IV) and changes to methylene blue.

The explanation for this observation can be illustrated with simple orbital energy diagrams. Figure 2 shows an orbital energy diagram for the ground state of leucomethylene blue and tin(IV). The highest-energy occupied molecular orbital of leucomethylene blue is labeled HOMO, and the lowest-energy unoccupied orbital is labeled LUMO. The empty orbital of tin(IV) is at a higher energy than the HOMO of leucomethylene blue, so leucomethylene blue cannot transfer an electron to it. When leucomethylene blue absorbs

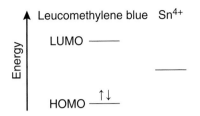

Figure 2. Ground-state energy diagram for leucomethylene blue and tin(IV).

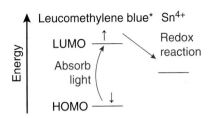

Figure 3. Excited-state energy diagram for leucomethylene blue and tin(IV).

ultraviolet energy, it enters an excited state in which one of its electrons is now in an orbital higher in energy than the empty orbital in tin(IV) (Figure 3). Now an electron can be transferred to tin(IV), starting the reduction of tin(IV) and the oxidation of leucomethylene blue. Therefore, in an excited state, leucomethylene blue is a stronger reducing agent than it is in the ground state.

The solution prepared for this demonstration undergoes degradation as it is used. The energy absorbed when the solution is irradiated by ultraviolet also causes irreversible changes in the leucomethylene blue, and these changes reduce the amount of leucomethylene blue in the solution. After many repeated colorless-to-blue-to-colorless cycles, the intensity of the blue color will diminish and eventually fail to appear. When this happens, the solution should be discarded.

REFERENCES

1. D. Harris, *Quantitative Chemical Analysis,* 6th ed, p. 303, W.H. Freeman: New York (2003).
2. D. R. Lide, ed., *CRC Handbook of Chemistry and Physics,* 90th ed., CRC Press: Boca Raton, Florida (2009).
3. G. Oster, N. Wotherspoon, "Photobleaching and Photorecovery of Dyes," *J. Chem. Phys., 84,* 157–158 (1954).
4. H. Obata, "Photoreduction of Methylene Blue by Visible Light in the Aqueous Solution Containing Certain Kinds of Inorganic Salts. II. Photobleached Product," *Bulletin of the Chemical Society of Japan, 34,* 1057–1063 (1961).
5. United Kingdom Patent 1,003,816, "Improvements in or Relating to Light Filters," American Optical Company (1965) uses 8.56×10^{-4} M methylene blue, 0.29 M HCl, and 2.95×10^{-3} M $SnCl_2$.

12.44

The Photochemical Reaction of Chlorine and Hydrogen

Gaseous atoms, with the notable exception of the noble gases, are very reactive species. Absorption of visible light by the colored diatomic halogens, chlorine, bromine, and iodine, produces reactive halogen atoms.

Shining light on a gas phase mixture of hydrogen and chlorine, for example, produces an explosive reaction, as in Demonstration 1.45, Photochemical Reaction of Hydrogen and Chlorine, in volume 1 of this series. A less violent example is Demonstration 12.53, The Photodissociation of Bromine and the Bromination of Hydrocarbons, in this volume.

12.45

The Effects of Solvents on Spiropyran Photochromism and Equilibria

Photochromism, the change of color produced by absorption of light and reversion to the original color when the light is removed, is a striking phenomenon with possible applications ranging from sunglasses to memory-storage devices. Photochromic molecules usually exist in two different-colored isomeric structures that are in equilibrium, with the more stable form predominating in the absence of light. Light absorption provides the energy required for isomerization and color change. Since the electronic and geometric structures of the isomers are different, their equilibria can be greatly influenced by their surroundings, liquid and solid solvent matrices.

A clear, colorless solution in a nonpolar solvent is irradiated with a photoflash unit—or is placed on the lighted stage of an overhead projector—and the solution turns dark blue. In the dark, the solution quickly (minutes) returns to its colorless state (Procedure A). A clear, almost colorless solution of the same solute in a polar solvent is flash irradiated with a photoflash unit and the solution turns pink. Illuminated on the stage of an overhead projector, the pink color is rapidly (seconds) bleached, or, left in the dark, the solution slowly (hours) returns to the original, almost colorless state (Procedure B). A clear, dark pink solution of the same solute in an even more polar solvent is placed on the lighted stage of an overhead projector and rapidly (seconds) bleaches to colorless. Left in the dark, the solution slowly (hours) returns to the original dark pink state (Procedure C). The solute in a light-yellow solid matrix on a microscope slide is warmed on a hot plate and the matrix turns deep orange-red. If kept in the dark, the orange-red color remains. If irradiated with a bright light, the color returns in a short time to the original yellow. These cycles can be repeated many times (Procedure D). The solute in a white solid polystyrene matrix is irradiated by black light, and the solid turns blue. When the blue solid is irradiated with white light and yellow light, the blue fades, more with the yellow than with the white (Procedure E).

MATERIALS FOR PROCEDURE A

10 mg spiropyran, 1′,3′-dihydro-1′,3′,3′-trimethyl-6-nitrospiro[2H-1-benzopyran-2, 2′-(2H)-indole] (abbreviated as 6-NO$_2$-BIPS), CAS No. 1498-88-0, Aldrich cat. no. 27,361-9

10.0 mL toluene

10.0-mL volumetric flask

small glass vial that can be sealed to prevent toluene leakage

photoflash unit (If the flash has a UV-absorbing filter, it should be removed to maximize the amount of UV radiation available from the flash.)

small glass Petri dish with cover (optional) (Note: plastic Petri dishes are likely to be etched by toluene.)

overhead projector

MATERIALS FOR PROCEDURE B

10 mg spiropyran, 1',3'-dihydro-1',3',3'-trimethyl-6-nitrospiro[2H-1-benzopyran-2, 2'-(2H)-indole] (abbreviated as 6-NO$_2$-BIPS), CAS No. 1498-88-0, Aldrich cat. no. 27,361-9

40 mL 2-propanol (isopropanol) (Either reagent-grade or 99% isopropyl alcohol [rubbing alcohol)] from a drugstore or supermarket are satisfactory.)

10.0-mL volumetric flask

25.0-mL volumetric flask

2.00-mL pipet

small glass or plastic container that can be sealed to prevent evaporation

photoflash unit (If the flash has a UV-absorbing filter, it should be removed to maximize the amount of UV radiation available from the flash.)

overhead projector

MATERIALS FOR PROCEDURE C

stock solution: 1 mg mL^{-1} spiropyran, 1',3'-dihydro-1',3',3'-trimethyl-6-nitrospiro [2H-1-benzopyran-2,2'-(2H)-indole] (abbreviated as 6-NO$_2$-BIPS), CAS No. 1498-88-0, Aldrich cat. no. 27,361-9, from the preparation for Procedure B

20 mL 2-propanol (isopropanol) (Either reagent-grade or 99% isopropyl alcohol [rubbing alcohol] from a drugstore or supermarket are satisfactory.)

25.0 mL volumetric flask

2.00-mL pipet

5.00-mL pipet

small glass or plastic container that can be sealed to prevent evaporation

photoflash unit (If the flash has a UV-absorbing filter, it should be removed to maximize the amount of UV radiation available from the flash.)

MATERIALS FOR PROCEDURE D

10 mg spiropyran, 1',3'-dihydro-1',3',3'-trimethyl-6-nitrospiro[2H-1-benzopyran-2, 2'-(2H)-indole] (abbreviated as 6-NO$_2$-BIPS), CAS No. 1498-88-0, Aldrich cat. no. 27,361-9

0.5 mL reagent-grade acetone

0.5 mL colorless nail polish

glass microscope slide

metal microspatula

5-mL vial

electric hot plate

device for measuring the surface temperature of the hot plate

bright white light such as a photoflood or overhead projector stage

MATERIALS FOR PROCEDURE E

10 mL toluene, C$_6$H$_5$CH$_3$

15 mg spiropyran, 1',3'-dihydro-1',3',3'-trimethyl-6-nitrospiro[2H-1-benzopyran-2,2'-(2H)-indole] (abbreviated as 6-NO$_2$-BIPS), CAS No. 1498-88-0, Aldrich cat. no. 27,361-9

clear, colorless polystyrene drinking cup

15-mL to 25-mL screw-capped vial

aluminum weighing dish, having a diameter of about 6 to 8 cm

hot plate (optional)

black light

75-watt or brighter halogen desk lamp, or overhead projector

index card

yellow transparency, about 10 cm square, such as a piece cut from a yellow transparent report cover

PROCEDURE A

Preparation

Dissolve 10 mg of spiropyran in 10.0 mL of toluene in a 10.0-mL volumetric flask (3.1×10^{-3} M). The solution will be essentially colorless. Put the solution in a clear, colorless glass capped container.

Presentation

Display the container so the audience can see that the contents are clear and colorless. Hold the vial over a photoflash pointed away from the audience. Trigger the flash. The irradiated solution will immediately turn blue. (Flashing more than once will produce a deeper blue solution.) The color will fade to its original colorless state in about 1 to 2 minutes. *Note*: Fluorescent room lights produce enough ultraviolet radiation to prevent the blue solution from becoming completely colorless, so it is necessary to put the sample in the dark to get to the completely colorless state.

Alternatively, put the solution in a small, covered Petri dish. Set the dish on the stage of an illuminated overhead projector. The projected image will show a colorless solution that turns blue over the course of a few seconds. Turn off the projector. The solution will return to colorless in about 1 to 2 minutes. Turn on the projector again to show that the solution has returned to its colorless condition. While illuminated, it will again turn blue.

A pedagogical disadvantage of this easier alternative presentation with the overhead projector is the requirement to explain that it is the near-ultraviolet light from the projector that is responsible for the color change. This is particularly important if this procedure is coupled with any of the following procedures where light (visible wavelengths) from an overhead projector *bleaches* the color formed by ultraviolet irradiation or a very polar environment.

PROCEDURE B

Preparation

Dissolve 10 mg of spiropyran in 10.0 mL of 2-propanol in a 10.0-mL volumetric flask. The dissolution is slow and will take at least overnight to complete. The solution will be dark pink. This is your stock solution, 1 mg mL^{-1} (3.1×10^{-3} M) of spiropyran. Store the solution in a sealed container to prevent solvent evaporation. The solution is light sensitive and will fade in a lighted room. After dissolution is complete and storage in the dark for a few hours, the solution should have an absorbance maximum of 0.40 ± 0.05 at its peak in the visible, about 547 nm.

Pipet 2.00 mL of stock (1 mg mL^{-1}) spiropyran solution into a 25.0-mL volumetric flask and make up to volume with 2-propanol. The solution, 0.08 mg mL^{-1} (0.25×10^{-3} M), will be very light pink. Store a sample of this solution in a clear glass or plastic container that is sealed to prevent solvent evaporation. The solution is light sensitive and will fade to colorless in a lighted room. This is not a problem—a faded solution will work well for the presentation.

Presentation

Display the container so the audience can see that the contents are clear and colorless or very light pink. Hold the vial over a photoflash pointed away from the audience. Trigger the flash. The irradiated solution will immediately turn pink. The color will gradually (hours) fade to a very light pink state.

PROCEDURE C

Preparation

Pipet 5.00 mL of distilled water into a 25-mL volumetric flask. Add 2-propanol to fill the flask about three-quarters full. Pipet 2.00 mL of the stock (1 mg mL^{-1}) spiropyran solution from Procedure B into the flask and bring it up to volume with 2-propanol. Upon standing in the dark for a few hours, this solution, 0.08 mg mL^{-1} (0.25×10^{-3} M), in a 20% water–80% 2-propanol solvent will turn deep pink and look similar to the more concentrated stock solution, although it will be a different pink. Store a sample of this solution in a clear glass or plastic container that is sealed to prevent solvent evaporation. The solution is light sensitive and will fade in a lighted room. Keep the solution in the dark for at least a few hours before doing the demonstration.

Presentation

Display the container so the audience can see that the contents are dark pink. Place the container on the lighted stage of an overhead projector. The irradiated solution will quickly (seconds) become colorless. In the dark, the color will gradually (hours) return to its dark pink state.

PROCEDURE D

Preparation

In a 5-mL vial dissolve 10 mg of spiropyran in 0.5 mL of acetone. The solution will be violet. Add 0.5 mL of colorless nail polish and swirl the mixture to blend it completely. The mixture will be yellow orange. Pour about one-third of the liquid in a line across a microscope slide and quickly spread the liquid over the surface using the tip of a metal spatula. This needs to be done quickly because the mixture will become too viscous to spread in 15 to 30 seconds. Avoid spreading the mixture to the edge of the slide because it may run under the slide if it reaches the edge. Repeat this with a second slide. Allow the slides to rest for a few hours to dry thoroughly. The coating will be pale yellow to orange. Once prepared, the slides may be reused for many presentations of the demonstration.

Shortly before presenting (within an hour) the demonstration, irradiate the coating on the slide with bright white light, for example, by putting it on the stage of an overhead projector, until the coating becomes pale yellow. The light source used should not heat the slide, because heating will prevent the coating from turning yellow.

Adjust the hot plate so its surface temperature is between 60 and 80°C.

Presentation

Display the slide with its yellow coating. It may be viewed directly, placed on the stage of an overhead projector, or displayed by video projection.

Place the slide, coated side up, on top of a warm hot plate. The coating will gradually (within minutes) turn dark red. Display the dark red coating.

Irradiate the coating with bright white light. The deep red coating will gradually (minutes) fade to pale yellow. A small opaque object, such as a key or paper clip can be placed between the coating and the light source, and its silhouette will remain deeply colored as its surroundings fade in the light. If an overhead projector is used, an index card may be placed on the stage of the projector, and the red-coated slide placed on top of the card with half the slide off its edge. The portion of the coating exposed to the light will fade to yellow, and the fading will be observable in the projected image. The index card may then be removed, and the projected image will show that the part of the coating that was on the card is still red, but will fade as it is exposed.

The heating-irradiating cycle may be repeated as often as desired. When the slide is stored in the dark, the coating slowly (days) regains the dark orange-red color and needs to be irradiated with bright light before another presentation of the effect of heating.

PROCEDURE E

Preparation

Put 10 mL of toluene into a vial having a capacity of at least 15 mL. Dissolve about 3 mg of spiropyran in the toluene. Close the vial with its cap.

Cut or break a portion of a colorless, transparent polystyrene cup into pieces small enough to fit into the vial. Put about 3 g of the polystyrene chips into the vial with the toluene-spiropyran solution. Seal the vial with its cap. Agitate the vial to mix its contents. Repeat the agitation periodically until the mixture becomes a uniform, viscous solution; this may take as long as a couple of days.

When the toluene-spiropyran-polystyrene solution is uniform, pour it into a mold, 6–8 cm in diameter, such as a disposable aluminum weighing dish. A glass container may also be used, but the bottom of the container must be narrower than the top for easier removal of the product. A beaker with flared rim will work, but a Petri dish will not. Place the dish in a hood or other well-ventilated area to allow the toluene to evaporate. When the toluene has evaporated, a stiff polystyrene disk will remain in the dish. At room temperature, it may take several days for all of the toluene to evaporate. Warming the dish to speed evaporation is not recommended, because warming can produce bubbles in the mixture.

When the mixture in the dish has hardened into a solid disk, remove it from the dish. If the dish is disposable aluminum, it may be easiest to peel the dish away from the disk. If a glass mold is used, it may be necessary to gently pry the disk away using a thin metal spatula around the edges of the disk. The disk will be hard, perhaps brittle, white or pale blue, and probably cloudy. Store the disk in the dark until the time of the presentation. When stored in the dark, the disk will become white or yellowish white.

Presentation

Display the white disk. Irradiate the disk with a black light (long-wave ultraviolet). The disk will rapidly turn blue.

If a halogen desk lamp is to be used as a source of white light, place the blue disk on the table top, cover about a third of it with an index card, another third with a yellow transparency, and leave a third uncovered. Place the halogen desk lamp next to the disk, turn it on, and adjust the lamp to brightly and uniformly illuminate the disk. Allow the lamp to shine on the disk for several minutes, until the uncovered part is a pale blue. Turn off the lamp and display the disk. The blue color will have faded, most where the disk was under the yellow transparency and least where it was under the card.

If an overhead projector is to be used as a source of white light, place an index card and a piece of yellow transparency on the stage of the projector. Set the disk on top of the card and transparency, so that about a third of the disk is on the card, a third on the yellow transparency and a third unobstructed. Turn on the projector and leave the disk in position for several minutes. Turn off the projector. The blue color of the disk will have faded most where it was over the yellow transparency and least where it was over the card.

Darken the room lights and place the disk under the black light. With the room lights dimmed, a red fluorescence will be visible from the disk, and the fluorescence will increase as the disk darkens.

HAZARDS

Toluene, 2-propanol, and acetone are flammable and must not be used in the vicinity of open flames.

Toluene is a possible cancer-causing compound and should be handled with protective latex gloves in a fume hood. Keep toluene solutions in closed containers to prevent exposure to the fumes.

Spiropyran, 1',3'-dihydro-1',3',3'-trimethyl-6-nitrospiro[2H-1-benzopyran-2,2'-(2H)-indole], is irritating to eyes, respiratory system, and skin, so it should be treated with care, and contact with the solid or breathing its dust should be avoided. The chemical, physical, and toxicological properties associated with exposure have not been thoroughly investigated. There is no indication that the spiropyran is a possible cancer-causing compound, but it is toxic with an LD_{50} of 56 mg kg^{-1} for intravenous administration in mice.

Black light (long-wavelength ultraviolet light) can damage one's eyes. Shield the black-light source from the audience and never look directly at the source yourself.

DISPOSAL

The solutions may be stored for several months without apparent degradation of the photochromic properties. For longer-term storage, the solvents may be allowed to evaporate in a hood and the residual solid sealed in the container to be reconstituted for a later demonstration. If the solutions are to be discarded, local ordinances for disposal of toluene should be followed. Solutions in 2-propanol may be disposed of down the drain with plenty of running water.

The glass slides with acrylic polymer coating and the polystyrene coating may be retained for repeated presentations. The glass slide may be discarded in a receptacle for broken and contaminated glass. The polystyrene disk may be discarded in a solid-waste receptacle.

DISCUSSION

Photochromism is the formation of color (Greek, *chroma*) in a system caused by shining light (Greek, *photos*) on it. Some of the most interesting and potentially important photochromic systems are those that are reversible, either thermally or photolytically (by shining a second, different wavelength of light on the colored system). Window glass and sunglasses that darken when bright sunlight containing a great deal of ultraviolet light shines on them and then revert to greater transparency when the light is less intense are one practical application of photochromism. The compounds are being studied as a means of photocontrolled motion in nanomechanical devices [1] and extensively explored for data storage in which information would be stored in the system by light of one wavelength and interrogated or erased by light of another wavelength [2].

The compound used in this demonstration and its derivatives have been extensively studied as models to help understand the molecular-level details of photochromism [3, 4, 5]. They have also been used in teaching-laboratory experiments [6, 7] that could be used to extend these demonstrations for interested high school or college students. The following is a brief overview that focuses on the structural properties of the species that are formed photolytically or thermally in this system and that are responsible for the differences observed in solvents of different polarity.

The structure of the spiropyran, 1′,3′-dihydro-1′,3′,3′-trimethyl-6-nitrospiro[2H-1-benzopyran-2,2′-(2H)-indole], is

(1)

This shows the three-dimensional structure of the molecule with the two double-ring systems held perpendicular to one another by the spiropyran linkage at the carbon through which the intersecting rings are bonded, the one at the center of the molecule. Because they are perpendicular to each other, the π-bonding systems at either end of the molecule cannot interact with one another to delocalize electrons over the whole molecule, so their energy levels are relatively high. Thus, the molecule absorbs light in the ultraviolet, but not the lower energy visible, and is colorless in solution.

The near-ultraviolet absorption of spiropyran is due largely to the benzopyran ring system (the right-hand rings in structure 1) and is responsible for the photochemical ring-opening, the photochromic reaction [8]. The bond between the ring oxygen and the carbon of the spiropyran linkage is relatively easily broken, and the most stable form of the open-chain product, a merocyanine, is the trans configuration,

(2)

In the planar structures shown, the electrons from both ring systems are delocalized over the entire molecule, which lowers the electronic energy levels. Thus, the merocyanine form of the compound absorbs visible light, and solutions containing this form are colored.

All the resonance structures for this open chain form of the molecule have either a zwitterionic structure (an overall neutral molecule with opposing charges localized on two different atoms) like that on the left or a quinoid structure (an overall neutral molecule with quinone-like bonding) like that on the right. In all the zwitterionic structures, the central bond joining the two halves of the molecule (circled in the drawings) is a double bond. In the quinoid structures, the central bond is a single bond flanked by double bonds. Thus, depending upon which resonance structures contribute most to the overall delocalized electronic structure, the central bond behaves almost like a single bond (quinoid structure), with relatively easy rotation of the ends of the molecule with respect to one another, or more like a double bond (zwitterionic structure) with restricted rotation.

The ring-opening reaction must initially form a cis structure, and such a structure must reform if the open-chain structure is to return to the spiro structure.

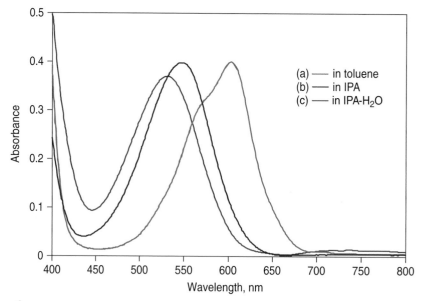

(3)

Both zwitterionic and quinoid structures of the cis form are also possible, as shown here. These molecules can either close the ring to form the spiro structure or the ends of the molecules can rotate with respect to one another to form the more stable trans conformations. The ease of transformation to the trans conformation is a function of the degree of double bonding at the central bond. The zwitterionic form with the central double bond will restrict this transformation more than the quinoid form with a central single bond.

The striking observations in these solvents are the *differences* in the color of the photoproduct in polar and nonpolar solvents, 2-propanol and toluene, and also the color of the solution in the very polar 2-propanol–water mixture. The absorption spectra of the colored species in these three solvents are shown in Figure 1. The absorbance maximum moves toward shorter wavelengths (a blue shift), that is, to higher energies, as the solvent polarity increases from toluene to 2-propanol to the 2-propanol–water mixture. This phenomenon is called *solvatochromism,* that is, solute color depends on the properties of the solvent. (The molar absorptivity of solvatochromic species decreases as the polarity increases with a range from about 50,000 to 10,000 L mol^{-1} cm^{-1} for various merocyanines. The absorption bands get broader with increasing polarity.)

Figure 1.   Absorption spectra of 6-NO$_2$-BIPS in (a) toluene, ~1 mg mL^{-1} (~3 × 10^{-3} M), a few seconds after flash irradiation by near-UV light, (b) 2-propanol (isopropyl alcohol, IPA) at ~1 mg mL^{-1} (~3 × 10^{-3} M), and (c) 80% 2-propanol–20% water at ~0.08 mg mL^{-1} (~0.3 × 10^{-3} M).

To help interpret these spectroscopic data, a measure of solvent polarity would be useful. One such empirical measure of the polarity of solvents has been developed by a German chemist, Christian Reichardt, and his coworkers [9, 10]. The method is based on the solvatochromism (the shift in the absorption maximum in the visible and near-infrared with change in solvent) of a particular dye, called Reichardt's dye, in different solvents [11].

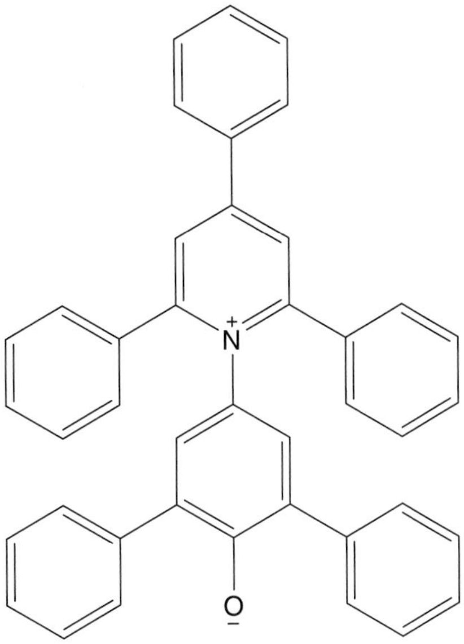

(4)

Reichardt's dye

2,6-diphenyl-4-(2,4,6-triphenyl-1-pyridinio)phenolate
(CAS No. 10081-39-7)

The electronic transition responsible for the absorption of light is the charge transfer that occurs as the electrons redistribute to reduce both the negative charge on the O-atom center and the positive charge on the N-atom center. The energy of the wavelength of maximum absorbance for this transition, E_T, increases as the solvent polarity increases and can be used as a relative measure of the polarity.

The reason for the shift in the energy of the absorbed light is that, as solvent polarity increases, the ground state of the absorbing molecule is more stabilized than the excited state. The ground state of Reichardt's dye, a zwitterion as depicted in the structure above, is more polar than the excited state, in which the charges are reduced, and is more stabilized by polar solvents. Thus, an increase in solvent polarity increases the energy of the charge-transfer transition. In the later development of this technique, a normalized, dimensionless scale, E_T^N, was devised that is based on assigning the polarity of water (very polar) as 1.00 and the polarity of tetramethylsilane (very nonpolar) as zero. Values of E_T^N for a large number of solvents, including mixtures [12] are available in the literature.

The merocyanine spectra in Figure 1 show that the energy difference between the ground and first excited state of the open-chain, trans merocyanine structure increases as the solvent polarity increases. This is similar to what happens for Reichardt's dye, which implies that the merocyanine ground state is more polar than the excited state and is more stabilized by polar solvents. Evidence for this interpretation comes from Figure 2, a plot of the energies of maximum absorbance, E_T (in kJ mol^{-1}), from Figure 1 as a function of Reichardt's solvent polarity parameter E_T^N.

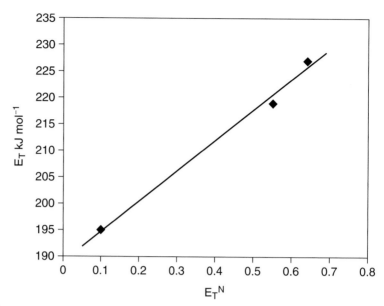

Figure 2.   Energies of maximum absorbance, E_T, from Figure 1 for the merocyanine in different solvents plotted against Reichardt's E_T^N polarity parameter.

The linearity of the plot in Figure 2 shows that the polarity of the solvents affects the merocyanine energy-level differences in the same way as for Reichardt's dye, so it is likely that the merocyanine energy transition is also the result of charge transfer. Another solution, the acetone solution of the spiropyran prepared in Procedure D is violet. The E_T^N for acetone is 0.355 [9], which puts its polarity directly between those for toluene and 2-propanol and, from Figure 2, predicts that its maximum absorbance should be at about 570 nm. This absorption is just at the green-yellow part of the visible spectrum, leaving the red and blue to be transmitted and produce the observed violet color.

Further evidence that the ground state of the open-chain, trans merocyanine becomes more stabilized as the solvent polarity increases comes from the appearance of the solutions after they have been in the dark and the spiropyran and merocyanine forms have come to equilibrium at room temperature. The amount of the colored, merocyanine form increases with polarity (from none detectable in toluene to greater amounts in the highly colored 2-propanol–water mixture), so we can conclude that the merocyanine form becomes more stable relative to the spiropyran form as solvent polarity increases. This conclusion makes sense because contributions from the highly polar zwitterionic structure of the merocyanine form will make its interactions with polar solvents more favorable than those for the relatively nonpolar spiropyran form.

If, in the excited state of the open-chain trans merocyanine, electron density is transferred from the more negative oxygen to the more positive nitrogen, the molecule will become less polar and not as stabilized by polar solvents as it is in the ground state. A schematic way to show this charge transfer is with a structure that is intermediate between the zwitterionic and quinoid structures,

(5)

In addition to this structure's lower polarity, the central double bond is weakened, is closer to a single bond, and is less of a restriction to the rotation that would convert this structure to a cis form, which could close to the spiro structure.

The most stable structure of the blue merocyanine formed by ultraviolet flash irradiation in nonpolar toluene is probably closer to the less polar quinoid structure than to the highly polar zwitterionic. Since rotation about the central bond is easier in the quinoid form, the activation energy for the trans-to-cis transformation should be relatively low. This low activation energy explains the rapid return to the colorless spiropyran that is represented schematically in Figure 3a.

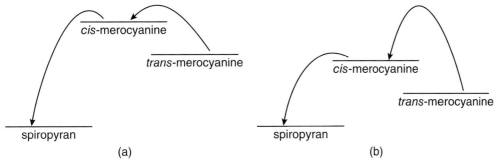

Figure 3.   Schematic diagrams illustrating the energy levels of the spiropyran-merocyanine system and activation energies for the merocyanine-spiropyran reaction in (a) nonpolar and (b) polar solvents.

Conversely, the most stable structure of the pink merocyanine formed by ultraviolet flash irradiation in the polar 2-propanol solvent is probably closer to the highly polar zwitterionic structure than to the less polar quinoid. Since rotation about the central bond is more difficult in the zwitterionic form, the activation energy for the trans-to-cis transformation should be relatively high. This high activation energy explains the slow return in the dark to the colorless spiropyran that is represented schematically in Figure 3b. The activation energy for the reverse cis-to-trans transformation is probably also high, as represented in the figure, which explains the slow reappearance in the dark of the pink merocyanine form after it has been bleached by irradiation.

Although the thermal merocyanine-to-spiropyran reaction in polar solvents is slow, it is greatly accelerated by irradiation with visible light. This simplified reaction scheme (omitting many intermediate steps) explains the effect of visible light.

$$\text{t-MC} \underset{\text{slow}}{\overset{h\nu\ (\text{visible})}{\rightleftharpoons}} \text{t-MC*} \longrightarrow \text{c-MC} \xrightarrow{\text{fast}} \text{SP} \qquad (\text{scheme 1})$$

In this reaction, t-MC is the ground-state trans merocyanine structure, t-MC* is the photoexcited state (structure 5 shown above), c-MC is the cis merocyanine structure, and SP is the spiropyran structure. The excited state, t-MC*, formed when t-MC absorbs visible light, can lose energy and return to the ground state, or it can transform to the cis structure and thence, by losing energy, to c-MC. As discussed above, the excited state of merocyanine is likely to have a central bond that does not hinder rotation as much as the ground state, so this transformation is relatively probable. The results of the dark reaction indicate that the thermal cis-to-trans reaction is slow compared to the ring-closing that yields SP, as noted in the reaction scheme. Visible light drives the reaction from left to right, and the overall result is depletion of t-MC to form SP. The brighter the visible light, the faster the bleaching.

If both near-ultraviolet and visible light are absorbed by the spiropyran-merocyanine system, as is the case with both photoflash and overhead projector irradiation, other pathways have to be added to reaction scheme 1.

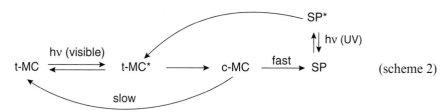

(scheme 2)

What is observed in this case depends on the relative amounts (intensities) of the ultraviolet and visible light at the wavelengths that are absorbed as well as on the relative concentrations of the spiropyran and merocyanine.

Under the conditions of these demonstrations, the reactions in polar solvents on the overhead projector, where the amount of ultraviolet is low, are dominated by the visible light absorption, and bleaching occurs. In the nonpolar solvent, toluene, on the overhead projector, a steady state is reached in which the visible and ultraviolet pathways are proceeding at the same rate and the intensity of the blue color remains constant. (If a container with a column of the solution is placed on the stage, there is a gradation in the blue color, which is darker nearer the light source. The absorbance in the ultraviolet region is high, so its intensity decreases substantially as the beam passes through the solution, thus decreasing the rate of the ultraviolet pathway that produces the colored merocyanine form.)

Compared to the overhead projector, the photoflash emission contains a higher proportion of ultraviolet light (although still not a large amount) and the amount of SP following the ultraviolet pathway in scheme 2 is increased. In the case of the nonpolar solvent, this results in a deeper blue photochromic solution (especially if the sample is flashed multiple times). In the polar solutions, the proportion of the spiropyran and merocyanine at thermal equilibrium (in the dark) is the factor that determines whether formation of colored merocyanine or colorless spiropyran predominates. The amount of the merocyanine formed in 2-propanol, ~0.3×10^{-3} M total concentration, is quite low—the solution is almost colorless. The photoflash produces more of the merocyanine via the ultraviolet pathway, and the solution becomes more highly colored. In the 2-propanol–water mixture, there is a substantial amount of the merocyanine form—the solution is dark pink—and the photoflash bleaches the solution as the visible light absorption dominates the photo process.

These explanations for the observed properties of this spiropyran-merocyanine system in solution are also applicable to the properties in the solid matrix, with the added factor of greatly hindered molecular motion in the solid. The deep red-orange color of the system at equilibrium in the matrix in the dark indicates that the nail-polish matrix is highly polar and stabilizes the open-chain merocyanine form with a large contribution from the zwitterionic structure. Irradiation with visible light converts the colored merocyanine to the colorless spiropyran, as in reaction scheme 1. The reaction is slower than in solution because the rigid matrix in which the molecules are embedded physically impedes the internal trans-to-cis rotation required for the transformation. The reaction is hastened by brighter light that increases the rate of formation of excited molecules.

The matrix also impedes the rate of return to the equilibrium state, which is much slower in the matrix than in solution. This thermal reaction is sped up by heating the matrix to a higher temperature to help overcome the high activation-energy barrier due to both the zwitterionic structure of the intermediates, as in Figure 3b, and the matrix effects. It is possible that heating may also soften the matrix and somewhat decrease its impeding effect.

Analysis of the combination of procedures in this demonstration provides a background in photochromism and solvatochromism that can be the basis for further explorations of this or other similar systems found in the references.

REFERENCES

1. A. Athanassiou, M. Kalyva, K. Lakiotaki, S. Georgiou, and C. Fotakis, "Laser Controlled Mechanical Actuation of Photochromic-Polymer Microsystems," *Rev. Adv. Mater. Sci.*, *5*, 245–251 (2003).

2. A. Tork, F. Boudreault, M. Roberge, A. M. Ritcey, R. A. Lessard, and T. V. Galstian, "Photochromic Behavior of Spiropyran in Polymer Matrices," *Applied Optics, 40*(8), 1180–1186 (2001).

3. N. P. Ernsting, B. Dick, and Th. Arthen-Engeland, "The Primary Photochemical Reaction Step of Unsubstituted Indolino-spiropyrans," *Pure Appl. Chem., 62*(8), 1483–1488 (1990).

4. S.-R. Keum, M.-S. Hur, P. M. Kazmaier, and E. Buncel, "Thermo- and Photochromic Dyes: Indolino-benzospiropyrans. Part I. UV-VIS Spectroscopic Studies of 1,3,3-Spiro(2H-I-benzopyran-2,2'-indolines) and the Open-chain Merocyanine Forms; Solvatochromism and Medium Effects on Spiro Ring Formation," *Can. J. Chem., 69*, 1940–1947 (1991).

5. J. T. C. Wojtyk, A. Wasey, P.M. Kazmaier, S. Hoz, and E. Buncel, "Thermal Reversion Mechanism of N-Functionalized Merocyanines to Spiropyrans: A Solvatochromic, Solvatokinetic, and Semiempirical Study," *J. Phys. Chem. A, 104*, 9046–9055 (2000).

6. H. E. Prypsztejn and R. M. Negri, "An Experiment on Photochromism and Kinetics for the Undergraduate Laboratory," *J. Chem. Educ., 78*(5), 645–648 (2001). The experiment is an exploration of the kinetics of the merocyanine-to-spiropyran reaction in ethanol and also data on the thermodynamics of the spiropyran-merocyanine system.

7. A physical chemistry laboratory experiment, "Photochromism of Spiropyrans," J. P. Hagen, California Polytechnic Institute San Luis Obispo, to determine the activation parameters for the kinetics of the decay of the colored species in toluene solution is available at http://chemweb.calpoly.edu/jhagen/Photochromism.pdf.

8. N. W. Tyer, Jr. and R. S. Becker, "Photochromic Spiropyrans. I. Absorption Spectra and Evaluation of the (-Electron Orthogonality of the Constituent Halves," *J. Am. Chem. Soc., 92*(5), 1289–1294 (1970).

9. C. Reichardt, "Solvatochromic Dyes as Solvent Polarity Indicators," *Chem. Rev., 94*, 2219–2368 (1994).

10. C. Reichardt, S. Löbbecke, A. M. Mehranpour, and G. Schäfer, "Pyridinium *N*-Phenoxide Betaines and Their Application to the Determination of Solvent Polarities, XXIV. Syntheses and UV-vis Spectroscopic Properties of New Lipophilic *tert*-Butyl- and 1-Adamantyl Substituted, Negatively Solvatochromic Pyridinium *N*-Phenolate Betaine Dyes," *Can. J. Chem., 76*, 686–694 (1998).

11. An easy demonstration of solvatochromism is to use an overhead projector to display the colors of Reichardt's dye in several different solvents, for example, 2-butanone (methyl ethyl ketone), 2-propanone (acetone), ethanol, and methanol or binary solvents such as water added a little bit at a time to a solution of the dye in acetone.

12. T. M. Xrygowski, P. K. Wrona, U. Zielkowska, and C. Reichardt, "Empirical Parameters of Lewis Acidity and Basicity for Aqueous Binary Solvent Mixtures," *Tetrahedron, 41*(20), 4519–4527 (1985).

12.46

A Copper Oxide Photocell

Harnessing the energy of sunlight to power our civilization is the ultimate solution for our energy needs. Among the ways this can be done is the direct conversion of light to electrical energy, producing a current flow in an external circuit.

A strip of copper metal is heated in the flame of a Bunsen burner until it darkens. A few drops of solution are placed on the darkened strip, and a copper wire inserted into the drops. An ammeter connected to the copper strip and the copper wire registers a change in current when a bright light shines on the copper strip.

MATERIALS

strip of copper metal, about 2 cm × 5 cm and 0.5 mm thick

about 10 cm of insulated 12-gauge copper wire

about 5 g solid sodium chloride, NaCl

several drops of 1 M acetic acid, $HC_2H_3O_2$, or vinegar

several drops of 1 M sodium nitrate, $NaNO_3$, solution

ring stand to support copper strip

stand with clamp to hold copper wire

Bunsen burner

250-mL beaker

ammeter with clip leads

light source, e.g., desk lamp or bright flashlight

PROCEDURE [1]

Preparation

Strip about 1 cm of insulation from each end of the copper wire. Clamp the wire vertically in a stand.

Clean the strip of copper metal with a paste made by combining some solid sodium chloride and a little 1 M acetic acid (or vinegar). Scrub both surfaces until they are a bright, shiny copper color. Rinse the strip with water, and dry it.

Presentation

Place the copper strip on the ring stand and heat the center with the flame of a Bunsen burner. The strip should be about 1 cm above the inner cone of the flame and in the upper cone. Heat until the upper surface just turns uniformly black, and the flame color on the sides of the strip has changed from green to orange. Turn off the burner and allow the copper strip to cool. The underside of the copper strip will have a circular dark, shiny spot where it was in the flame. This is the photoactive region.

Clip the black (negative) lead of the ammeter to the copper strip and place the copper on a flat, nonconducting surface, with the side that was down in the flame facing upward now. An inverted beaker works well for this. Place a couple drops of concentrated sodium nitrate solution into the center of the darkened circular region on the copper strip, forming a small pool. Adjust the copper wire in its stand so that the lower tip just touches the pool of solution but does not touch the copper strip. Clip the red (positive) lead of the ammeter to the other end of the copper wire.

The ammeter will register a small current, perhaps a few microamps. Shine a bright light on the dark area of the copper strip covered by the pool of solution. The current will increase. Remove the light, and the current will decrease.

HAZARDS

There are no particular hazards associated with this demonstration.

DISPOSAL

The piece of copper should be rinsed with water, and the oxide coating removed by buffing the copper with fine steel wool in preparation for repeated use in this demonstration.

DISCUSSION

A strip of copper that has been heated in air is used to prepare an electrode that is sensitive to light. The electrode is used to construct a simple electrochemical cell. The current generated by the cell increases when the electrode is illuminated and decreases when the light is removed. The current change is small but easily detected.

When a piece of copper metal is heated in a flame, a thin coating of copper oxide forms on the surface of the copper. Initially, the coating is red copper(I) oxide. As the copper is heated further with an excess of oxygen, a coating of black copper(II) oxide results. In this demonstration, the bottom surface of the copper strip is inside the flame, where there is less oxygen than in the surrounding air. Therefore, copper(I) oxide remains on the lower surface longer than on the upper surface, where there is more oxygen. Heating is terminated while the coating on the bottom of the copper is still mainly copper(I) oxide.

Copper(I) oxide was one of the first substances to be identified as a semiconductor, which is a substance that has an electrical conductivity intermediate between that of insulators and conductors. In an electrical conductor, current can be induced by applying even the smallest voltage across the conductor material, whereas in an electrical insulator, no current flows until an extremely high voltage is applied. In a semiconductor, current flows at an intermediate voltage (or higher), but no current flows at lower voltages.

The electrical behavior of semiconductors can be explained with band theory, which is an extension of molecular orbital theory [2]. In a semiconductor, the valence electrons occupy a group of molecular orbitals with closely spaced energies; these orbitals constitute the valence band. If the valence band is not filled with electrons, the electrons can move easily from one orbital to another. This is the situation in most metals, which are electrical conductors. However, if the valence band is full, the electrons cannot move easily, and the material is not a conductor.

Somewhat higher in energy than the valence band is another set of orbitals, which constitute the conduction band. In a semiconductor, the valence band is full, but the conduction band is empty. In insulators, the energy gap between the valence band and the conduction band is so large that the materials do not conduct electricity until an extremely high voltage is applied. In semiconductors, the energy gap between the valence band and conduction band is significant, but not so large as in insulators. For this reason, a semiconductor does not conduct electricity when the applied voltage is small. However, if a higher voltage is applied, then some of the electrons from the valence band can be elevated in energy to the conduction band, where these electrons can move, so the material conducts electricity at the higher applied voltage. See Demonstration 12.7, Light-Emitting Diodes: Voltage and Temperature Effects, for a more complete discussion of band theory applied to conductors, insulators, and semiconductors.

Electrons can also be elevated in energy from the valence band to the conduction band when a semiconductor material absorbs light energy. The electrons elevated to the conduction band can carry current (as can the vacancies, "holes," they leave behind in the valence band). In this way, semiconductors can convert light energy to electrical energy. The efficiency of energy converted by the copper(I) oxide produced in this demonstration is very low, so it is of little practical significance. Silicon-based semiconductors of much greater efficiency have been made, but they are rather expensive to produce, so solar-energy conversion using them is also costly. Much recent research has been devoted to reducing the cost of semiconductor solar-energy conversion, and inexpensive copper(I) oxide is the subject of renewed interest for this purpose.

The copper(I) oxide semiconductor in this demonstration converts light energy into electrical energy, the reverse of what happens in light-emitting diode (LED) semiconductors, in which electrons are driven by an applied voltage through the semiconductor. The electrons enter the conduction band, come to the band gap, and fall to lower energy. The energy lost by the electrons is emitted as light. (See Demonstration 12.7, Light-Emitting Diodes: Voltage and Temperature Effects, for a further discussion of LEDs.)

The photocell used in this demonstration is an example of a "wet" photocell, so called because it uses a liquid electrolyte, in this case an aqueous solution of sodium nitrate. There have been numerous studies of various wet photocells, but none of these wet cells has proven to be as efficient or stable as a totally dry system. Dry systems are not as easy to construct as wet systems, which is why the wet system was first to be discovered and why it is used in this demonstration. Wet photocells are also not as practical for solar-energy conversion as dry cells, so they have not been as extensively studied. As a consequence, the electrochemical processes that occur in the wet photocell used in this demonstration are not well understood [3].

REFERENCES

1. W. Noon, *How to Build a Solar Cell That Really Works,* Lindsay Publications: Bradley, Illinois (1990).
2. N. F. Mott, "Note on Copper-Cuprous Oxide Photocells," *Proceedings of the Royal Society of London. Series A, Mathematical and Physical Sciences, 171,* 281–285 (1939).
3. A. W. Copeland, O. D. Black, and A. B. Garrett, "The Photovoltaic Effect," *Chem. Rev., 31,* 177–226 (1942).

12.47

The Photobleaching of Carotene

Molecules of the yellow-orange compound carotene have an extended conjugated double-bond system that absorbs visible light. Molecules with such bonds can undergo chemical reactions initiated by the energy absorbed. Often, higher-energy (shorter wavelength) absorbed light results in more extensive reactions.

Paper stained yellow with carotene is exposed to white, blue, green, and red light from an overhead projector. After 30 to 60 minutes, the paper exposed to white light is bleached the most, while blue and green produce sequentially less bleaching, and red light does not bleach the yellow paper.

MATERIALS

2 ounces of baby-food carrots

25 mL heptane, C_7H_{16} (or similar hydrocarbon solvent, such as hexane or octane)

125-mL Erlenmeyer flask, with stopper

magnetic stirrer with stir bar

150-mm test tube

aluminum foil

test-tube rack or clamp and ring stand to hold test tube

glass transfer pipet (Pasteur pipet) and pipet bulb

piece of 11-cm diameter filter paper

120-mm Petri dish

forceps

stand with spring clips to suspend filter paper to dry

overhead projector

blue, green, and red transparent filters, about 2.5 cm × 8 cm

thin sheet of opaque cardboard (e.g., file folder), about 15 cm × 15 cm

thin sheet of opaque cardboard, about 30 cm × 30 cm, to cover overhead stage

small pieces of cellophane or masking tape

pencil

PROCEDURE

Preparation

Carotene is light sensitive, so the following steps should be carried out in dim light or in the dark when feasible.

Place about 2 ounces of baby-food carrots (pureed carrots and water) into an aluminum-foil-wrapped 125-mL Erlenmeyer flask and add 25 mL of heptane. Put a magnetic stir bar in the flask and stopper the flask. Stir the mixture vigorously for at least 30 minutes. Pour the mixture into an aluminum-foil-wrapped 150-mm test tube, cover the mouth with aluminum foil, and let the test tube stand for about 30 minutes to allow the layers to separate. A clear yellow layer will float on top of a cloudy orange layer. This clear layer is a solution of carotene in heptane.

Put a piece of 11-cm-diameter filter paper into a Petri dish. Use a transfer pipet to remove the clear yellow layer of hexane solution in the test tube and put it all on the paper in the Petri dish. Using forceps, remove the filter paper and suspend it to dry in the dark. The carotene on the paper will react slowly with oxygen in the air, even in the dark, so the dried paper should not be stored for more than a few days before use in the demonstration.

Presentation

On the stage of an overhead projector, arrange the transparent red, green, and blue filters and a small thin sheet of opaque cardboard as shown in Figure 1. Use small pieces of tape to hold the filters and opaque mask in place. The yellow circle in the figure indicates where to place the yellow filter paper impregnated with carotene on top of this arrangement, so that light from the projector will have to pass through the colored filters before hitting the paper. After putting the filter paper in place, turn on the projector, so the audience can see the placement of the colored filters, the filter paper, and the cardboard that prevents light from reaching part of the filter paper. Being careful not to disturb this set up, place a sheet of opaque cardboard over the overhead stage, so room light does not strike the filter paper.

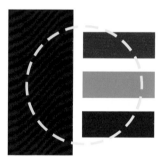

Figure 1. Arrangement of colored filters and opaque mask on the overhead stage with the yellow circle to show how the carotene-impregnated yellow filter paper should be placed over the array.

Allow irradiation by the projector to continue for 30–60 minutes and then remove the cardboard sheet to show the arrangement again. Turn off the projector and use a pencil to label the filter paper "red," "green," "blue," and "dark" where it is over the colored filters and opaque mask. Display the filter paper and/or pass it around the audience to observe the results. The area that was not irradiated serves as a control to account for possible color change that may occur thermally. The color will be completely bleached where unfiltered white light struck the paper and will be close to completely bleached by blue light transmitted by the blue filter. There will be no bleaching in the area where red light struck the paper. The area irradiated by the green light may show some bleaching, which will depend on the wavelengths transmitted by the particular green filter used.

HAZARDS

Heptane is flammable. Extinguish all flames in the vicinity when working with heptane.

DISPOSAL

Heptane solutions should be disposed of as organic waste in accord with local ordinances.

The filter paper may be discarded in a solid-waste recepticle.

The other materials may be saved for future presentations

DISCUSSION

Colored compounds absorb wavelengths of visible light that excite an electron to a higher energy state. In this demonstration, the bleaching of β-carotene (Figure 2) shows that such excitation can lead to chemical reaction by providing the molecules the energy and/or the electronic structure necessary to initiate reaction. The energy provided to the molecules depends on the energy of the light they absorb. As observed for carotene, absorption of the higher-energy blue light often produces more of a reaction than absorption of lower-energy light toward the red end of the spectrum.

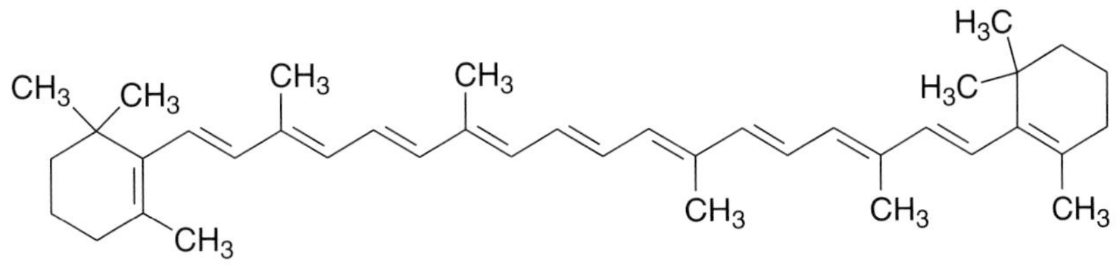

Figure 2.   Structure of all-*trans* β-carotene, the most abundant carotene in plants.

Carotene and/or similar compounds are often found together with chlorophyll in bacteria, archaea, and plants that use photosynthesis to produce carbohydrates from carbon dioxide and water. Many procedures for extracting chlorophyll from plant material extract carotenoids (and other colored compounds) as well. These compounds probably have at least a dual role in these organisms: harvesting light energy at wavelengths where chlorophyll does not absorb (Figure 3) and protecting chlorophyll from degradation by free radicals (especially singlet oxygen, a diradical [1]) that can be produced by reactions of excited-state chlorophyll itself [2, 3, 4, 5].

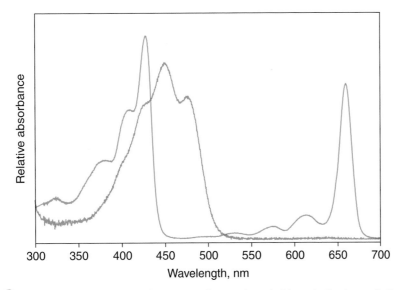

Figure 3.   Absorption spectra of β-carotene (orange) and chlorophyll *a* (green). Carotene absorbs green light that chlorophyll *a* does not.

Carotene and other carotenoids are also evident in the animal world, but no animal is able to synthesize them directly. (Very recently [6], two species of aphids have been found that have the genetic and molecular machinery to synthesize several carotenoids, giving them either a green or red color that provides camouflage from predators. The genes for this feat seem to have come from fungi by transfer into the aphid genome perhaps 30 to 80 million years ago.) Florida flamingoes and the shrimp they eat owe their pink color to the presence of these compounds in the algae at the bottom of their food chain [7]. In captivity in zoos, flamingoes must have carotenoids in their food in order to maintain the color and not return to their natural white plumage.

Of more importance to animals, including humans, is carotene's role as a precursor to retinol, the most common form of vitamin A, and, hence, to retinal, an essential component of rhodopsin in the visual system. (This explains why you are told to "eat your carrots" to improve your vision; see the introduction, page 67.) Cleaving carotene in two at the central double bond and adding oxygen (as –OH) to each new free-carbon end produces retinol (Figure 4). Most animals have enzymes that carry out this "cut-and-paste" transformation (plus the appropriate trans-cis isomerization and oxidation to make the required active form, *cis*-retinal, for rhodopsin). However, carnivorous animals, such as cats, do not have these enzymes, so they cannot make retinol from carotene and must get retinol by eating other animals that can make it.

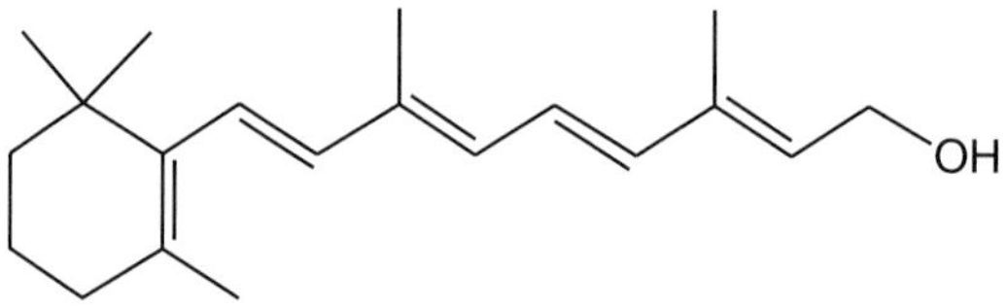

Figure 4.   Structure of *trans*-retinol.

The color of carotene is a result of the absorption of visible light with enough energy to excite an electron in its long, conjugated double-bond system from the highest occupied molecular orbital (HOMO) to the lowest unoccupied molecular orbital (LUMO). (See the introduction, page 39.) A simple model of the π–electron system in conjugated polyenes, like carotene, is the particle in a one-dimensional box. This model is a mainstay of the introduction to quantum theory because it embodies many of the features of actual quantum systems, such as quantized energy levels, in a readily calculable system. For the particle-in-a-box model, the HOMO-LUMO energy difference is proportional to $1/L^2$, where L is the length of the box. That is, the energy difference gets smaller as the box gets larger.

Application of the particle-in-a-box model to conjugated polyenes explains why the energy of light absorbed by the HOMO-LUMO transition gets smaller (the wavelength absorbed gets longer) as the polyene chain (the box) gets longer. For example, a polyene with six conjugated double bonds absorbs at about 360 nm in the ultraviolet, and a polyene with 11 double bonds (such as carotene) absorbs at about 450 nm in the visible. To calculate values that match these experimental values requires modifications of the simple particle-in-a-box model [8], but the principle is the same.

The photochemical bleaching of carotene involves processes that break or interrupt the conjugated system, so that the products have shorter conjugated polyene chains that do not absorb light in the visible. Thus, the color will disappear. Carotene bleaching also occurs slowly, over several hours, in the dark in the presence of air, whereas photolysis speeds the bleaching and likely occurs by different pathways that begin in carotene excited state(s). The products of these reactions are complex mixtures of oxygenated compounds, not all of which have been characterized. Many of these are secondary products that result from further reactions of the primary products.

One study that characterized a few products of β-carotene oxidation found, among many others, this series of shortened-chain compounds, which resulted from reactions at the double bonds in the carotene chain [9]:

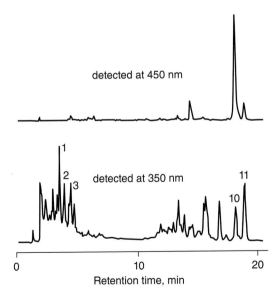

(1)

(2)

(3)

High-pressure liquid chromatography (HPLC) analysis of the reaction mixture after partial oxidation gave the results shown in Figure 5. The products emerging from the HPLC column were detected by monitoring the absorbance of light by the effluent at several wavelengths simultaneously. The results at a visible wavelength, 450 nm, and an ultraviolet wavelength, 350 nm, are shown. What you observe are many products that do not absorb visible light and that therefore account for the bleaching of the carotene color.

Figure 5.   HPLC analysis of the reaction mixture following partial oxidation of β-carotene. The detector output monitored at 450 nm (top) and 350 nm (bottom) are shown. The peaks labeled 1, 2, and 3, which are detected only in the ultraviolet, correspond to the compounds whose structures are shown in the text. The peaks labeled 10 and 11, which are detected by absorbance in both the visible and ultraviolet, are all-*trans*-β-carotene (Figure 1) and *cis*-β-carotene (cis at the central double bond in the β-carotene), respectively.

REFERENCES

1. B. Z. Shakhashiri, "Singlet Molecular Oxygen," in *Chemical Demonstrations,* vol. 1, pp. 133–145, University of Wisconsin Press: Madison (1983).
2. C. S. Foote and R. W. Denny, "Chemistry of Singlet Oxygen. VII. Quenching by β-Carotene," *J. Am. Chem. Soc., 90*(22), 6233–6235 (1968).
3. C. S. Foote, Y.C. Chang, and R. W. Denny, "Chemistry of Singlet Oxygen. X. Carotenoid Quenching Parallels Biological Protection," *J. Am. Chem. Soc., 92*(17), 5216–5218 (1970).
4. H. Zhang, D. Huang, and W. A. Cramer, "Stoichiometrically Bound β-Carotene in the Cytochrome b_6f Complex of Oxygenic Photosynthesis Protects against Oxygen Damage," *J. Biol. Chem., 274*(3), 1581–1587 (1999).
5. J. Glaeser and G. Klug, "Photo-oxidative Stress in Rhodobacter sphaeroides: Protective Role of Carotenoids and Expression of Selected Genes," *Microbiology, 151,* 1927–1938 (2005).
6. N.A. Moran and T. Jarvik, "Lateral Transfer of Genes from Fungi Underlies Carotenoid Production in Aphids," *Science, 328,* 624–627 (30 April 2010). See *Perspective,* T. Fukatsu, "A Fungal Past to Insect Color," *Science, 328,* 574–575 (30 April 2010).
7. L. Schenkman, ed., "Microbe Mascots," *Science, 328,* 553 (30 April 2010).
8. J. Autschbach, "Why the Particle-in-a-Box Model Works Well for Cyanine Dyes but Not for Conjugated Polyenes," *J. Chem. Educ., 84*(11), 1840–1845 (2007).
9. G. J. Handelman, J. G. M. van Kuuk, A. Chatteree, and N. I. Krinsky, "Characterization of Products Formed during the Autoxidation of β-Carotene," *Free Radical Biol. Med., 10,* 427–437 (1991).

12.48

Making a Cyanotype

A plan is often called a blueprint, as in "a blueprint for success." This terminology is derived from the use (beginning in the mid-nineteenth century) of blueprints, white lines on a blue background, for architectural plans (a process that has been largely displaced by modern copier technology). Artists still use the same blueprint process to produce cyanotypes.

A collection of opaque objects (Procedure A) or a photographic negative (Procedure B) is set onto a yellowish sheet of paper, and the paper is exposed to bright light. After a short time, the exposed portions of the paper turn blue. The paper is rinsed in water, and a blue negative image of the silhouettes of the objects or a positive image of the print on white paper is revealed.

MATERIALS FOR PROCEDURE A

2.0 g ammonium iron(III) citrate,* $(NH_4)Fe(C_6H_6O_7)_2$

0.8 g potassium hexacyanoferrate(III), $K_3[Fe(CN)_6]$, (potassium ferricyanide)

20 mL distilled water

tap water

8.5-in × 11-in sheet of white, smooth-surface Bristol board or other heavy paper of similar size and smooth finish

clean foam-sponge paintbrush, about 2 in wide

fan or hair dryer

opaque envelope or folder large enough to contain the paper

several opaque objects with recognizable silhouettes, e.g., fork, scissors

UV lamp, sun lamp, or bright fluorescent lamp (or sunlight for a sun print)

rectangular pan large enough to hold paper flat

MATERIALS FOR PROCEDURE B

2.0 g ammonium iron(III) citrate,* $(NH_4)Fe(C_6H_6O_7)_2$

0.8 g potassium hexacyanoferrate(III), $K_3[Fe(CN)_6]$, (potassium ferricyanide)

20 mL distilled water

tap water

8.5-in × 11-in sheet of white, smooth-surface Bristol board or other heavy paper of similar size and smooth finish

clean foam-sponge paintbrush, about 2 in wide

fan or hair dryer

opaque envelope or folder large enough to contain the paper

transparent negative, about 8 in × 10 in (See Preparation under Procedure B.)

---

* Ammonium iron(III) citrate is available in two forms: brown, 16.5–18.5% Fe, and green, 14.5–16.5% Fe. Both forms work, but green is more sensitive to light and reacts more quickly.

sheet of glass about the same size as the paper

UV lamp, sun lamp, or bright fluorescent lamp (or sunlight for a sun print)

rectangular pan large enough to hold paper flat

PROCEDURE A

Preparation

Dissolve 2.0 g of ammonium iron(III) citrate in 10 mL of distilled water. This solution is light sensitive and should be stored in an opaque bottle (or a bottle covered with aluminum foil) for no more than one week.

Dissolve 0.8 g of potassium hexacyanoferrate(III) in 10 mL of distilled water. This solution can be stored for up to one month in the light or in the dark.

Preparation of the paper should be carried out in a room with dim light. Sunlight and light from fluorescent lamps should be avoided, but moderate light from incandescent lamps is acceptable. For the remainder of the preparation, wear waterproof gloves to avoid skin contact with the solutions—they can stain the skin. Combine 2 mL of each of the two solutions and mix them thoroughly. Place the paper in the pan and use a foam-sponge paintbrush (or a sponge or cotton ball) to spread the solution uniformly over the surface of the paper. Dry the coating on the paper by placing it under a fan or by using a hair dryer. When the paper is dry, it should be protected from light by placing it in an opaque envelope or folder.

Presentation

Place the dry coated paper on a horizontal surface. Arrange several opaque, readily recognized objects on top of the paper. Scissors, keys, forks, spoons, hammers, and other tools work well. Expose the paper to bright light from a sun lamp or fluorescent lamp. Over a period of time that depends on the brightness of the light source, the yellow coating on the paper will turn blue where the light strikes it. When the fully exposed areas are dark blue, turn off the lamp and remove the objects from the paper. The paper will contain silhouettes of the objects on a blue background.

Immerse the paper in a pan of water and gently agitate it to wash off the remaining yellow coating. Empty the pan, refill with fresh water, and continue washing. (If this print is to be saved, repeat the fresh-water rinse one or two more times to be sure the unreacted coating has been completely removed.) When the yellow coating has been washed off, hang the paper to drip dry (or blow dry gently with a fan or hair dryer). The white silhouettes now show up in greater contrast on the blue background.

PROCEDURE B

Preparation

Dissolve 2.0 g of ammonium iron(III) citrate in 10 mL of distilled water. This solution is light sensitive and should be stored in an opaque bottle (or bottle covered with aluminum foil) for no more than one week.

Dissolve 0.8 g of potassium hexacyanoferrate(III) in 10 mL of distilled water. This solution can be stored for up to one month in the light or in the dark.

Preparation of the paper should be carried out in a room with dim light. Sunlight and light from fluorescent lamps should be avoided, but moderate light from incandescent lamps is acceptable. For the remainder of the preparation, wear waterproof gloves to avoid skin contact with the solutions—they can stain the skin. Combine 2 mL of each of the two solutions and mix them thoroughly. Place the paper in the pan and use a foam sponge paint brush (or a sponge or cotton ball) to spread the solution uniformly over the surface of the paper. Dry the coating on the paper by placing it under a fan or by using a hair dryer. When the paper is dry, it should be protected from light by placing it in an opaque envelope or folder.

Prepare a black-and-white negative approximately 8 in × 10 in. This may be done photographically. It may also be prepared from a color digital image by converting the color image to grayscale, inverting it to a negative image, and printing it on a transparency using a monochrome printer.

Presentation

Place the dry coated paper on a horizontal surface. Set the negative on the coated paper and cover it with a sheet of glass to hold the negative flat against the paper. Expose the paper to bright light from a sun lamp or fluorescent lamp. Over a period of time that depends on the brightness of the lamp, the yellow coating on the paper will turn blue where the light strikes it. When the fully exposed areas are dark blue, turn off the lamp and remove the glass and negative. The paper will contain a positive image in blue of the image from the negative.

Immerse the paper in a pan of water and gently agitate it to wash off the remaining yellow coating. Empty the pan, refill with fresh water, and continue washing. (If this print is to be saved, repeat the fresh-water rinse one or two more times to be sure the unreacted coating has been completely removed.) When the yellow coating has been washed off, hang the paper to drip dry (or blow dry gently with a fan or hair dryer). The positive image will now be visible in greater contrast as blue against white.

HAZARDS

Solutions of hexacyanoferrate should not be acidified because they can release poisonous hydrogen cyanide gas, HCN.

DISPOSAL

The solutions may be flushed down the drain with copious amounts of water.

DISCUSSION

In 1842 Sir John Herschel (British astronomer) invented the cyanotype process, so called because of the blue (cyan) color of the prints. Until the mid-twentieth century, it was used to make architectural blueprints (when other forms of reproduction displaced them) and it continues to be used by artists and photographers for making photographic art prints called *cyanotypes*. A large number of variations on the recipe for the photosensitive coating have been used over the years to produce different sensitivities and different effects in the final prints [1]. The recipe for this demonstration is simple, is easily made from materials such as those Herschel used, and produces prints in a reasonably short time.

citric acid

Herschel was interested in the reactions of light-sensitive compounds of metal ions, especially iron, with biologically occurring substances such as oxalic, tartaric, and citric acids. These are di- and tricarboxylic acids that are readily oxidized via decarboxylation (loss of carbon dioxide). For example, the citric acid cycle (tricarboxylic acid cycle, TCA), the central electron-harvesting pathway in aerobic respiration, involves two decarboxylation steps with electrons transferred to nicotinamide adenine dinucleotide, NAD^+. These acids form complexes with iron(III) ions that upon excitation by absorption of near-ultraviolet light transfer electrons to iron(III), reducing it to iron(II), with the acids undergoing oxidative decarboxylation.

In the cyanotype (blueprint) photosensitive coating, the iron(II) formed by this reduction quickly reacts with the hexacyanoferrate (ferricyanide) ion in a redox reaction to form Prussian blue (also known as Turnbull's blue)—iron(III) hexacyanoferrate(II), a mixed-valence compound containing both iron(II) and iron(III). The photochemistry of iron(III)-oxalate complexes, which have been the most extensively studied, as well as the formation and properties of Prussian blue, are discussed in more detail in Demonstration 12.49, An Iron(III)-Oxalate Actinometer. The oxalate system is more light sensitive than the citrate system and produces more rapid cyanotype printing, but the citrate is usually favored for artistic work because the slower color development can produce subtler color-gradation effects.

The cyanotype process in this demonstration can be used to examine the efficacy of sunscreens and sunblocks [2, 3]. For this purpose, the use of sunlight as the light source is preferable, since this is the radiation that the products are supposed to protect against. Color gradations in the resulting prints are useful for interpreting the relative protection of the products tested.

Ammonium iron citrate is an amorphous compound of varying composition that is commercially available in two forms, green or brown. The former contains a higher ammonium-to-iron ratio, is more sensitive to light, and reacts more quickly in this process. Although small amounts of discrete iron citrate compounds have been synthesized and characterized [4], iron ions and citrate usually form polymeric substances [5, 6]. In all cases, the structures depend on the charge of the citrate, usually negtive 3 or negative 4 (with loss of a hydrogen ion from the alcohol group at the central carbon). Thus, the amorphous state and variable composition of ammonium iron citrate is not surprising. Much of the research on iron citrate (and similar) complexes is aimed at learning more about their role in biological catalysis and sequestering or mobilizing iron for use by living organisms, which in some sense takes us full circle back to Herschel.

REFERENCES

1. M. Ware, *Cyanotype: The History, Science and Art of Photographic Printing in Prussian Blue,* National Museum of Science and Industry: London (1999).
2. G. D. Lawrence and S. Fishelson, "UV Catalysis, Cyanotype Photography, and Sunscreens," *J. Chem. Ed., 76*(9), 1199–1200 (1999).
3. G. D. Lawrence and S. Fishelson, "Blueprint Photography by the Cyanotype Process," *J. Chem. Ed., 76*(9), 1216a–1216b (1999).
4. I. Shweky, A. Bino, D. P. Goldberg, and S. J. Lippard, "Syntheses, Structures, and Magnetic Properties of Two Dinuclear Iron(III) Citrate Complexes," *Inorg. Chem., 33*(23), 5161–5162 (1994).
5. T. G. Spiro, G. Bates, and P. Saltman, "Hydrolytic Polymerization of Ferric Citrate. II. Influence of Excess Citrate," *J. Am. Chem. Soc., 89*(22), 5559–5562 (1967).
6. J. Strouse, S. W. Layten, and C. E. Strouse, "Structural Studies of Transition Metal Complexes of Triionized and Tetraionized Citrate. Models for the Coordination of the Citrate Ion to Transition Metal Ions in Solution and at the Active Site of Aconitase," *J. Am. Chem. Soc., 99*(2), 562–572 (1977).

12.49

An Iron(III)-Oxalate Actinometer

The study of photochemical reactions often requires measuring how much light (how many photons) is present in the beam from the light source. This measurement process is called *actinometry.*

Three dishes containing an almost colorless solution are placed on an overhead projector for varying lengths of time. An orange-red solution is added to each dish and to a fourth that has been kept in the dark. The solutions develop a blue-green color that is darker and bluer the longer the solution has been illuminated.

MATERIALS

0.30 g iron(III) nitrate nonahydrate, $Fe(NO_3)_3 \cdot 9H_2O$

0.20 g oxalic acid dihydrate, $H_2C_2O_4 \cdot 2H_2O$

0.50 g potassium hexacyanoferrate(III), $K_3[Fe(CN)_6]$ (potassium ferricyanide)

110 mL distilled water

250-mL Erlenmeyer flask

100-mL beaker

100-mL graduated cylinder

10-mL graduated cylinder

four 100-mm test tubes or 10-mL vials

four 60-mm Petri dishes

plastic dropper with 1-mL graduated stem

overhead projector

projection screen

PROCEDURE

Preparation

In a 250-mL Erlenmeyer flask, dissolve 0.30 g iron(III) nitrate nonahydrate in 100. mL distilled water. Add 0.20 g oxalic acid dihydrate to the solution. This results in a light yellow solution that contains iron(III) oxalate complexes. This solution must be kept in the dark, or the flask may be wrapped with foil to keep out light. In dim incandescent light (avoid fluorescent light or sunlight), prepare four 10-mL samples of this solution and keep them in the dark until used in the presentation.

In a 100-mL beaker, dissolve 0.50 g of potassium hexacyanoferrate(III) in 10. mL of water. This solution is light red-orange and may be stored for several weeks in the light or in the dark.

Presentation

In dim incandescent light (avoid fluorescent light or sunlight), put 10 mL of the iron(III) oxalate solution into each of four 60-mm Petri dishes. Keep one of the dishes in the dark. Set the other three on the stage of a lighted overhead projector. Remove one of the solutions and put it in the dark after 30 seconds, the second after 60 seconds, and the third after 120 seconds. Turn off the projector.

Place all four dishes on the unlit overhead projector stage. Add 1 mL of the red-orange potassium hexacyanoferrate(III) solution to each of the dishes and swirl them. Turn on the projector. The solutions will have developed a blue color, the intensity of which depends on the length of time the solution had been irradiated. The solution irradiated the longest will have the darkest blue color. The unirradiated solution will show almost no blue coloration but will be distinctly yellow due to the color of the dilute hexacyanoferrate(III) (ferricyanide) ion. The intermediate solutions will vary from green (yellow plus blue) to blue-green as the amount of blue color formed increases. Noting the initial variation in color should be done without delay. If the samples are left on the lighted overhead projector stage, the photochemical reaction of the iron(III) oxalate will continue and all samples will develop more of the blue color. (Compare with Demonstration 12.48, Making a Cyanotype.) To make clear that it is the iron(III)-oxalate solution that is photosensitive, it is irradiated prior to addition of the hexacyanoferrate(III) solution.

If a different size dish is used for the sample irradiation, adjust the volume of solution used so the depth of the liquid is 3–4 mm. In a greater depth of solution, it can be difficult to distinguish the colors of the irradiated samples. The intensity of overhead projector lamps is a variable in this demonstration. Check the projector that will be used to be sure that a 60-second irradiation gives a deep blue-green solution when the hexacyanoferrate(III) solution is added. If necessary, adjust the irradiation times.

HAZARDS

Do not acidify solutions containing the hexacyanoferrate(III) (ferricyanide) ion, because doing so can produce highly toxic hydrogen cyanide, HCN, gas.

DISPOSAL

The solutions may be flushed down the drain with copious amounts of water.

DISCUSSION

An actinometer is a device that measures the number of photons produced by a light source. A *chemical* actinometer, like the one demonstrated here, uses a photochemical reaction to determine the number of photons that enter the solution from the light source. The name "actinometer" comes from the word *actinic,* which refers to light that is able to bring about some change (for example, a photochemical reaction). The word is also often used in photography to denote light that will produce an image with the film and equipment being used.

A useful chemical actinometer requires a photochemical reaction that yields an easily analyzed product that is formed with high efficiency over a range of wavelengths. The wider the range of usable wavelengths over which the actinometer can be used, the more versatile it is, since it can be used to measure the intensities of many different light sources. One of the most widely used chemical actinometers is based on the photochemistry of iron(III) oxalate coordination complexes. The oxalate ion is a bidentate ligand, and the chemical actinometer used in photochemical research is prepared from the purified potassium trisoxalato iron(III) complex, whose anion is shown here.

For this semiquantitative demonstration, iron(III)-oxalate complexes are prepared in solution by mixing an iron(III) compound with oxalic acid. Under the concentration and acidic conditions of this demonstration, the mono-oxalato complex, $Fe(C_2O_4)^+$(aq), with water molecules as the other ligands predominates in the solutions.

When a solution of iron(III) oxalate is exposed to light, the solute is converted to iron(II) and carbon dioxide. An idealized representation of the reaction has this overall stoichiometry [1]:

$$2\ Fe(C_2O_4)^+(aq) \xrightarrow{h\nu} 2\ Fe^{2+}(aq) + 2\ CO_2(g) + C_2O_4^{2-}(aq) \qquad (1)$$

iron(III) oxalate   iron(II)

Iron(II) can also form oxalate complexes, but their formation constants are lower than those of iron(III), so it is shown here as the aqueous ion. Reaction 1 is a redox reaction, in which the coordinated oxalate ion is oxidized to carbon dioxide and iron(III) is reduced to iron(II). The stoichiometric half-reactions that describe in general terms the overall redox reaction 1 are

$$C_2O_4^{2-} \longrightarrow 2\ CO_2 + 2\ e^- \quad \text{oxidation} \qquad (2)$$

$$2\ Fe^{3+} + 2\ e^- \longrightarrow 2\ Fe^{2+} \quad \text{reduction} \qquad (3)$$

The pathway proposed for the steps following the primary photochemical event involves free radicals, notably $(OOCCOO\bullet)^-$, an oxalate ion that has lost one electron, and/or $(CO_2\bullet)^-$, the carbon dioxide radical anion. Here is a quite simplified version of this pathway:

$$Fe(C_2O_4)^+(aq) \xrightarrow{h\nu} {}^*Fe(C_2O_4)^+(aq) \qquad (4)$$

$${}^*Fe(C_2O_4)^+(aq) \longrightarrow Fe^{2+}(aq) + (OOCCOO\bullet)^-(aq) \qquad (5)$$

$$Fe(C_2O_4)^+(aq) + (OOCCOO\bullet)^-(aq) \longrightarrow Fe^{2+}(aq) + 2\ CO_2(g) + C_2O_4^{2-}(aq) \qquad (6)$$

Reaction 1 is the sum of these three reactions. If these three reactions proceeded to completion as written, two Fe^{2+}(aq) would be formed for each photon absorbed.

The quantum yield (the number of moles of a particular product formed—or a particular reactant destroyed—per mole of photons absorbed by the actinometer solution) is two for the formation of iron(II) in this idealized photochemical reaction. The quantum yield will be less than two if there are competing reactions that involve the reactive excited state and/or the free radicals; such reactions include disproportionation, combination, and reaction with other species. The observed quantum yield for the formation of iron(II) in this photochemical actinometer varies with wavelength, ranging from about 0.01 at 577 nm to 1.25 at 254 nm with quantum yields above unity at all wavelengths shorter than 435 nm [2]. Thus, more than one iron(II) is produced for each shorter wavelength photon absorbed by the iron(III) oxalate, but competing reactions keep the quantum yield below two, the ideal value for reaction 1.

There is uncertainty about the initial reaction of the photoexcited iron(III) oxalate, $*Fe(C_2O_4)^+$(aq). The essence of the question is whether reduction of iron(III) by intramolecular transfer of an electron from oxalate occurs before (or in concert with) the breaking of the iron oxalate bonds [3] (as implied in reaction 5) or whether the iron-oxalate bonds are broken before subsequent radical reduction of the iron(III) [4]. Under the conditions of the quoted literature studies, the predominant iron(III) oxalate complex was the tris-oxalate. There are differences in the observed photochemical-reaction intermediates at low pH (1–2), where the mono- and di-oxalato complexes predominate, and at higher pH (6–7), where the tris-oxalato complex predominates [5]. It is not clear whether the studies in [3, 4] apply directly to the mildly acidic conditions in this demonstration and the actinometric system on which it is based.

The blue color that appears when potassium hexacyanoferrate(III) (ferricyanide) solution is added to a solution containing iron(II) is caused by a redox reaction forming Turnbull's blue (or Prussian blue, see below), iron(III) hexacyanoferrate(II), a mixed valence compound containing both iron(II) and iron(III).

$$Fe^{2+}(aq) + K^+(aq) + Fe(CN)_6^{3-}(aq) \longrightarrow KFe^{III}[Fe^{II}(CN_6)](s) \qquad (7)$$

$$\text{red-orange} \qquad\qquad\qquad \text{deep blue}$$

The simplified product represented here is quite insoluble in water but forms so rapidly that the particles are colloidal and remain in suspension (which is why it is sometimes called *soluble Prussian blue*), giving the entire solution a blue color under the conditions of this demonstration. The absorption of red light that produces the intense blue color of the compound is the result of facile electron exchange between iron(II) and iron(III) through cyanide bridges, Fe(II)–CN–Fe(III), in the solid. The solid (*insoluble Prussian blue*) is difficult to characterize because it has a somewhat variable composition, $Fe^{III}_4[Fe^{II}(CN)_6]_3 \cdot xH_2O$, with $14 \leq x \geq 16$, but several methods have been used to show that the iron(II) is in an octahedral environment surrounded by carbon, and the iron(III) is in an environment of nitrogen and water. In the structure with 14 water molecules per unit cell, 6 are coordinated with iron(III) ions and 8 are in interstitial positions [6].

Adding a solution of hexacyanoferrate(II), ferrocyanide, to a solution containing iron(III) also produces Prussian blue. Prussian blue was also the first synthetic dye to find its way into the artist's palette. It was first synthesized around 1704 in Berlin, then the capital of Prussia. Turnbull's blue and Prussian blue are the same compound, the distinction arose because they were prepared by different methods, and they were not recognized at first as being the same [7].

REFERENCES

1. C. A. Parker and C. G. Hatchard, "Photodecomposition of Complex Oxalates—Some Preliminary Experiments by Flash Photolysis," *J. Phys. Chem., 63*(1), 22–26 (1959).

2. J. G. Calvert and J. N. Pitts, *Photochemistry,* p. 784, John Wiley and Sons: New York (1966).

3. I. P. Pozdnyakov, O. V. Kel, V. F. Plyusnin, V. P. Grivin, and N. M. Bazhin, "New Insight into Photochemistry of Ferrioxalate," *J. Phys. Chem. A, 112*(36), 8316–8322 (2008).

4. J. Chen, H. Zhang, I. V. Tomov, and P. M. Rentzepis, "Electron Transfer Mechanism and Photochemistry of Ferrioxalate Induced by Excitation in the Charge Transfer Band," *Inorg. Chem., 47*(6), 2024–2032 (2008).

5. B. A. DeGraff and G. D. Cooper, "On the Photochemistry of the Ferrioxalate System, " *J. Phys. Chem., 75*(19), 2897–2902 (1971).

6. F. Herren, P. Fischer, A. Ludi, and W. Haig, "Neutron Diffraction Study of Prussian Blue, $Fe_4[Fe(CN)_6]_3 \cdot xH_2O$. Location of Water Molecules and Long-range Magnetic Order," *Inorg Chem., 19,* 956–959 (1980).

7. B. Z. Shakhashiri, "Precipitates and Complexes of Iron(III)," in *Chemical Demonstrations,* vol. 1, pp. 338–343, University of Wisconsin Press: Madison (1983).

12.50

The Photoreduction of Silver Halide

The basis of the image-formation process in photographic film is the reduction of silver ion to silver metal when tiny silver halide crystals in the film gel absorb light.

A piece of white filter paper is treated with colorless solutions of silver nitrate and sodium chloride and then exposed to light. Where it is exposed to light, the paper darkens [1].

MATERIALS

about 5 mL 0.1 M silver nitrate solution, $AgNO_3$

about 5 mL 0.1 M sodium chloride solution, NaCl

three 120-mm diameter Petri dishes

round filter paper, about 11-cm in diameter

pair of sugical gloves, e.g., latex or nitile

forceps

small, flat opaque object that will fit in Petri dish, e.g., a key

halogen desk lamp or black light

video projection system (optional)*

PROCEDURE

Preparation

If the demonstration is to be presented to an audience too large to gather around a table to see the Petri dishes, prepare a video projection system, so that one Petri dish can be displayed.

Presentation

Place the piece of filter paper into one of the Petri dishes. Put on a pair of surgical gloves to protect your hands from silver nitrate stains. Pour enough 0.1 M $AgNO_3$ solution into the Petri dish to just cover the paper, about 5 mL. Using forceps, move the dampened filter paper to a second Petri dish. Pour enough 0.1 M NaCl solution into the dish to just cover the paper, about 5 mL. Using forceps, move the paper to the third Petri dish.

Place an opaque object, such as a key, onto the paper. Illuminate the paper from above with a bright light, such as a halogen desk lamp, or with a black light. Leave the light on until the paper has darkened perceptibly. Turn off the lamp.

Remove the opaque object from the paper to reveal its image in white on the paper.

HAZARDS

Silver nitrate solution can stain the skin.

---

* Various types of suitable video projection systems are described on page xxxiii.

Black light (long-wavelength ultraviolet light) can damage one's eyes. Shield the black-light source from the audience and never look directly at the source yourself.

DISPOSAL

The filter paper can be disposed of as solid waste. Residual silver nitrate solution can be mixed with residual sodium chloride solution and the precipitate of silver chloride collected and disposed of as heavy-metal waste in accord with local ordinances.

DISCUSSION

This demonstration shows how an image of an object can be produced by the action of light on silver chloride. A piece of filter paper is impregnated with solid silver chloride by first soaking the paper in a solution of silver nitrate and then adding sodium chloride to the dampened paper. Solid silver chloride forms in the paper by the reaction of aqueous silver nitrate with aqueous sodium chloride,

$$AgNO_3(aq) + NaCl(aq) \longrightarrow AgCl(s) + NaNO_3(aq)$$

in the net ionic reaction

$$Ag^+(aq) + Cl^-(aq) \longrightarrow AgCl(s)$$

When a crystal of solid silver chloride is exposed to light, the light energy causes an oxidation-reduction reaction to occur in the silver chloride crystals. An electron is transferred from chloride ions to silver ions, forming silver atoms and chlorine atoms.

$$\left(Ag^+ \; :\!\overset{\cdot\cdot}{\underset{\cdot\cdot}{Cl}}\!:^- \right)_{(s)} \xrightarrow{\text{light energy}} Ag^{\textbf{·}}_{(s)} + :\!\overset{\cdot\cdot}{\underset{\cdot\cdot}{Cl}}\!{\textbf{·}}_{(s)}$$
silver metal   chlorine atom

These atoms are trapped and separated in the crystal, and the reverse reaction cannot occur. The formation of tiny particles of silver metal causes the white silver chloride to darken. Where the silver chloride is protected from light by an opaque object, the silver chloride remains white.

This process forms the basis for black-and-white photography. In the photographic process, the darkened image is enhanced by a development process. In this process, the exposed silver chloride is treated with a reducing agent. The particles of silver formed by exposure to light catalyze the further reduction of silver chloride to silver, thus darkening and enhancing the image. A common reducing agent is hydroquinone.

$$HO-\!\!\!\left\langle\bigcirc\right\rangle\!\!\!-OH \longrightarrow O=\!\!\!\left\langle\bigcirc\right\rangle\!\!\!=O + 2\,H^+ + 2\,e^-$$

To prevent the image from darkening further on exposure to light, the remaining silver chloride is removed by dissolving it with a complexing agent, usually sodium thiosulfate, $Na_2S_2O_3$, solution and rinsing it away.

$$AgCl(s) + 2\,S_2O_3^{2-}(aq) \longrightarrow Ag(S_2O_3)_2^{3-}(aq) + Cl^-(aq)$$

This process is called *fixing.* Chemical-process photography has been largely replaced by electronic digital photography.

REFERENCES

1. J. J. Jacobsen and J. W. Moore, "Chemistry Comes Alive! Volume 3," *J. Chem. Educ.,* *76*(9), 1311–1312 (1999).

12.51

Photochemistry in Nitroprusside-Thiourea Solutions

The analysis of sulfur compounds, especially in biological systems, commonly uses nitroprusside in color-forming photoreactions. Photodissociation initiated by a charge-transfer electronic absorption and the reaction of a photoproduct cause the initial color change in this system. Thermal reaction of the small amount of intensely colored product returns the system to its initial color with only a small loss of reactants.

A light orange solution is exposed to bright light. After a short exposure, the solution turns deep blue. When the solution is placed in a boiling-water bath for a few minutes, the color of the solution returns to orange. This cycle can be repeated several times with the same solution.

MATERIALS

1.0 g sodium nitroprusside, $Na_2[Fe(NO)(CN)_5] \cdot 2H_2O$ (3.3×10^{-3} mol)

0.25 g thiourea $(H_2N)_2CS$ (3.3×10^{-3} mol)

8.0 g sodium bicarbonate, $NaHCO_3$ (9.5×10^{-2} mol)

300 mL distilled water

100-mL graduated cylinder

two 250-mL Erlenmeyer flasks

opaque bottle with capacity of at least 100 mL (a 250-mL Erlenmeyer flask covered in aluminum foil will work)

three 200-mm test tubes or three 100-mL beakers

three 50-mL graduated cylinders

150-mL beaker

slide projector or overhead projector

boiling-water bath

PROCEDURE

Preparation

Solution A: Pour 100 mL of distilled water into the opaque bottle. Add 1.0 g sodium nitroprusside and swirl the bottle to dissolve the solid. Store this solution in the dark (an opaque bottle or a foil-wrapped container) until it is used.

Solution B: Pour 100 mL of distilled water into a 250-mL Erlenmeyer flask. Add 0.25 g thiourea and swirl the flask to dissolve the solid.

Solution C: Pour 100 mL of distilled water into a 250-mL Erlenmeyer flask. Add 8.0 g sodium bicarbonate and swirl the flask to dissolve the solid. This solution is almost saturated and the solid may take some time to dissolve completely.

Very shortly before the presentation (or during the presentation), combine 30 mL of solution A, 30 mL of solution B, and 30 mL of solution C into a 150-mL beaker and stir to mix them. Divide this solution equally among three 200-mL test tubes or three 100-mL beakers. Keep these samples in the dark until ready for the presentation.

Presentation

Display the three test tubes or beakers. Reserve one in the dark to show the original color of the solution.

Expose the solutions in the other two test tubes to intense light by placing them in the beam of a slide projector, about 20 cm from the lens. The solutions will turn deep blue in about 2 minutes. Alternately, place the two beakers on the lighted stage of an overhead projector and watch the orange-to-blue color change on the projected image.

Place one of the tubes or beakers containing a blue solution into a boiling-water bath for several minutes. The heated solution will return to a light orange color. Remove it from the water bath when the color has returned. (The unheated solution will also return to a light orange, but takes several hours at room temperature.) After the heated solution cools to room temperature, it may again be exposed to intense light, and it will again turn blue. The sequence can be repeated several times.

HAZARDS

Thiourea is a suspect carcinogen, a mild skin irritant, and highly toxic. The oral LD_{50} in rats is 125 mg kg^{-1}.

Do not acidify solutions of sodium nitroprusside. Doing so may release poisonous hydrogen cyanide gas, HCN.

DISPOSAL

All solutions may be flushed down the drain with copious amounts of cold water.

DISCUSSION

In this photochemical reaction system, the sequence of color changes (orange → blue → orange) that can be repeated suggests that the reactions are reversible. Indeed, the demonstration was first introduced and has been used as an example of reversible photochemistry [1]. However, only a limited number of repetitions are possible, which means that a small amount of one or more of the original reactants is used up in each sequence, so the reactions are not truly reversible. However, the changes make an attractive example of a photodissociation reaction, and the fact that the reaction sequence is still tentative can make the point that mysteries remain and there is much to be discovered in chemistry.

The systematic name for sodium nitroprusside is sodium pentacyanonitrosylferrate(II). In the anion, the iron(II) cation is bonded to five cyanide, CN$^-$, ligands and one nitrosyl, NO$^+$, ligand. This bonding is inferred from the diamagnetism of the salt, which indicates that there are no unpaired electrons in the substance. This suggests that nitric oxide, NO, has donated its unpaired electron to iron to give NO$^+$, a d^6 electron configuration for iron, and a salt with all electrons paired [2]. The structure of the deep red (orange in dilute solution) anion is

The anion reacts both directly and photochemically with a variety of other species, including hydroxide, hydrogen sulfide, sulfite, ketones, thiourea, thiocyanates, and thiosulfates. Many of the products are intensely colored, and they have been used to identify specific compounds. An investigation of these color-forming reactions [3] that includes reaction with thiourea led to the original publication of this demonstration [1].

The reactions in nitroprusside anion photochemistry have been the subject of many studies [3–12]. Most of the references cited here involve photoinitiated reactions of nitroprusside with double-bonded sulfur compounds, especially the thiocyanate ion and thiourea. The reactions of these two species seem to parallel one another, with the exception that the deepblue product solution formed with thiourea is unstable and slowly decays while that from thiocyanate seems to be stable. Most of the quantitative studies have been done with thiocyanate and are an aid to understanding what likely happens in the thiourea reactions.

Photolysis of Nitroprusside Solutions

The nitroprusside absorption band at about 400 nm that is responsible for its deep red (orange in dilute solution) color arises from the excitation of an electron from an orbital largely localized on the iron to an orbital mostly associated with the NO [11]. This change, oxidation of the iron and reduction of NO^+, can be shown as

$$[Fe(II)(CN)_5NO^+]^{2-} \xrightarrow{h\nu} [Fe(III)(CN)_5NO]^{2-} \qquad (1)$$

light orange

The excitation also weakens the bonding between the iron and the NO, which can be lost to the solution.

$$[Fe(III)(CN)_5NO]^{2-} \longrightarrow [Fe(III)(CN)_5]^{2-} + NO \qquad (2)$$

The changes in pH following photolysis, along with isotopic labeling experiments, lead to the conclusion that the primary photochemical process for the nitroprusside ion is the loss of nitric oxide, NO, not the nitrosyl ion, NO^+ [11]. Although the pentacyano iron could rebind to the NO, the more likely process (in the absence of other possible ligands) is the reaction with water (aquation):

$$[Fe(III)(CN)_5]^{2-} + H_2O \longrightarrow Fe(III)(CN)_5(H_2O)]^{2-} \qquad (3)$$

The quantum yields for formation of the aqua complex, $[Fe(III)(CN)_5(H_2O)]^{2-}$, depend on the wavelength of the exciting light and are 0.18 and 0.35 for photolysis at 436 and 366 nm, respectively, in solutions buffered at pH 6 with phosphate [11].

Near-ultraviolet (black light) photolysis of nitroprusside in bicarbonate solution at the same concentration as in this demonstration produces the absorption spectrum shown in Figure 1. The formation of the peak at about 395 nm is consistent with formation of the

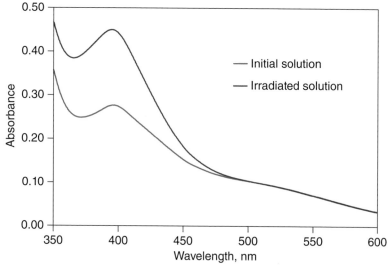

Figure 1.   Absorption spectra before and after photolysis of a 1.1×10^{-3} M sodium nitroprusside solution in 3.2×10^{-2} M bicarbonate solution.

aqua complex, $[Fe(III)(CN)_5(H_2O)]^{2-}$, which has an absorption maximum at 394 nm (molar absorptivity = 743 M^{-1} cm^{-1}) [8]. Its presence in the spectrum of the "unirradiated" initial solution shows that room light is sufficient to photoexcite the nitroprusside. Assuming that the peak is due to the aqua complex, the increase in absorbance shows that its concentration has increased by about 2×10^{-4} M, or about 2% of the nitroprusside has been photoconverted.

Addition of Thiourea to Photolyzed Nitroprusside Solutions

If thiocyanate or thiourea is added to a previously photolyzed solution of nitroprusside, the relatively rapid rate of formation of a blue solution can be followed by spectrophotometry at the wavelength of maximum absorption of the product, 580–590 nm (Figure 2) [7–10]. In these studies, the added sulfur compound was in excess over the amount of nitroprusside initially present and the formation of the blue product obeyed first-order kinetics. This result indicates first-order dependence on the concentration of the reactive iron complex in the photolyzed nitroprusside solution, which is consistent with this simple reaction for the nitroprusside-thiourea system:

$$[Fe(III)(CN)_5(H_2O)]^{2-} + S{=}C(NH_2)_2 \longrightarrow [Fe(III)(CN)_5(S{=}C(NH_2)_2)]^{2-} + H_2O \quad (4)$$

deep blue

The blue color in these products is likely the result of a charge-transfer electronic transition between the sulfur and iron centers. This transition has a large molar absorptivity (about 2700 M^{-1} cm^{-1} for the thiocyanate complex [8]), so a relatively small amount of product produces an intense color. If the molar absorptivity for the thiourea complex is similar, the data in Figure 1 show that the maximum concentration of this complex is about 3.1×10^{-4} M in a solution that is initially 1.1×10^{-2} M in both nitroprusside and thiourea (see next section). Only about 3% of the initially present nitroprusside and thiourea have reacted to give the dark blue color of the solution.

Photolysis of Nitroprusside-Thiourea Solutions

In this demonstration, a solution containing both unphotolyzed nitroprusside ion and thiourea is subsequently photolyzed to yield a blue solution. This reaction of the premixed reactants, as well as the similar one with thiocyanate, has been observed in several of the studies of nitroprusside photolysis [1, 3, 4, 7, 9, 10] but studied quantitatively in only one case [7]. The authors in this study found that in methanol solution, the formation of the blue color was first order under constant irradiation conditions and did not depend upon the concentration of thiourea. These results are consistent with photoexcitation of the nitroprusside as the primary photochemical process.

Although, it may seem reasonable to suppose that the reactions are the same as above, that simple mechanism leaves unexplained the observations that the initially formed blue color in the thiourea-nitroprusside system slowly fades (Figure 2) and that the rate of the initial reaction to form the blue product increases with increase in pH of the solution [8–10].

The data in Figure 1 show that the fading of the blue color following photoexcitation of the nitroprusside anion follows first-order kinetics (Figure 3). There is also evidence for a small growth in absorbance at about 395 nm as the blue product disappears and the possible appearance of an *isosbestic point* at about 440 nm in the spectra (Figure 2, inset). Isosbestic points (wavelengths) occur in a series of spectra when two species in the solution have the same molar absorptivity at a particular wavelength. If the total concentration of the two is constant, the absorbance at that wavelength is the same, no matter what the ratio of their individual concentrations. Thus, the substance that forms and absorbs at 395 nm and the blue product could be related, the first increasing as the latter decreases.

The product of this decay has not been isolated or characterized. One suggestion [10] is that the product is the aqua complex, $[Fe(III)(CN)_5(H_2O)]^{2-}$, which would be formed by the reverse of reaction 4. Although the observed increase in absorbance at 395 nm is consistent

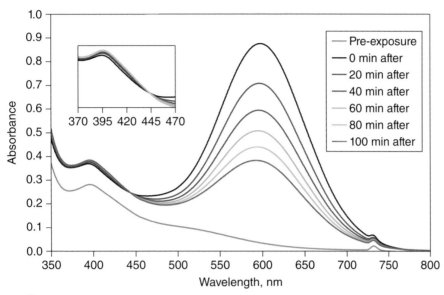

Figure 2. Absorption spectra of the blue photoproduct solution in the nitroprusside-thiourea system used in this demonstration. The spectra were taken as a function of time (in minutes) after the photoexcitation was stopped. The inset expands the portion of the spectrum that shows a small absorbance increase with time and a possible isosbestic point at about 440 nm. (The "blip" at about 735 nm is an instrumentation artifact.)

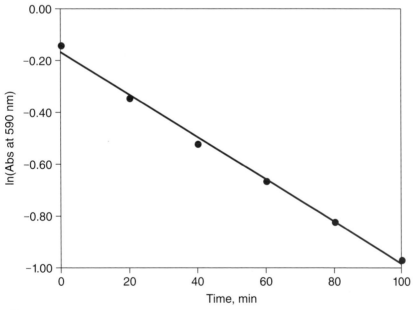

Figure 3. A first-order plot of the data in Figure 1: $\ln(A_{590})$ (corrected for absorbance by the original solution) as a function of time after photoreaction.

with formation of the aqua complex, the increase is small and inconsistent with the larger percentage decrease in absorbance of the blue product. Furthermore, if reaction 4 is the mechanism for formation of the blue product, its reverse cannot be the mechanism for its disappearance, since that would imply a reaction that "overshoots" its equilibrium state before recovering by the reverse reaction.

An alternative possibility is rearrangement of the blue complex so the thiourea ligand bonds to the metal center through one of its nitrogen atoms, thus losing the iron-sulfur charge-transfer transition responsible for the intense blue color.

$$[Fe(III)(CN)_5(S{=}C(NH_2)_2)]^{2-} \longrightarrow [Fe(III)(CN)_5((H_2N)_2C{=}S)]^{2-} \qquad (5)$$

deep blue   perhaps light orange or colorless

Such a rearrangement has been considered as a possibility in the thiocyanate system, where the authors conclude that the sulfur-bonded complex is favored by more than a factor of 10 [8]. For thiourea, steric effects are likely to limit the ability of the molecule to bind through a nitrogen and thus make the rearrangement, reaction 5, seem unlikely.

A more likely explanation is that the reactions are more complex than written above and, in particular, must take into account the acidity of the aqua complex.

$$[Fe(III)(CN)_5(H_2O)]^{2-} + H_2O \rightleftharpoons [Fe(III)(CN)_5(OH)]^{3-} + H_3O^+ \qquad (6)$$

The equilibrium constant for this reaction is about 4×10^{-9} ($pK_a = 8.3$) [8]. In their comprehensive study of the thiocyanate-nitroprusside system, Espenson and Wolenuk [8] found the rate of reaction of thiocyanate ion with the conjugate base (hydroxy complex) to be about 10^8 times as fast as with the conjugate acid (aqua complex). Thus, the observed increase in rate with increase in solution pH [8–10] is understandable. The pH in the bicarbonate solution used in this demonstration is about 8, which would mean that both the conjugate acid and base forms would be present in substantial amounts and reaction 7 could be the principal color-forming reaction.

$$[Fe(III)(CN)_5(OH)]^{3-} + S{=}C(NH_2)_2 \longrightarrow [Fe(III)(CN)_5(S{=}C(NH_2)_2)]^{2-} + OH^- \ (7)$$

Another complication in the nitroprusside photochemical reaction system is that there may be differences between the reactions when the sulfur-containing ligands are added after photolysis and when they are present during photolysis. West and his coworkers [9, 10] found that they did not get reproducible results when they added the sulfur-containing ligands immediately after nitroprusside photolysis. They obtained reproducible results after the photolyzed solutions had sat in the dark for 24 or more hours. This result suggests that some slowly attained equilibria control the concentration of the nitroprusside-derived species that reacts with the added ligand. No spectroscopic evidence for the possible changes in the photolyzed solution was presented.

A further experiment by these workers was designed to test for the presence of NO in the gas phase during photolysis of nitroprusside solutions. In the absence of the sulfur-containing ligands, no NO was detected, but it was detected when mixtures of the nitroprusside and ligands were photolyzed. Thus, it appears that the presence of the ligand may facilitate the loss of NO by the nitroprusside, so it forms faster than it can react with oxygen dissolved in the solution (see below) and some escapes into the gas phase.

The results outlined in the preceding paragraphs confirm the complexity of the nitroprusside photochemical reaction system, but do not offer enough data to resolve some of its puzzles. In particular, the slow fading of the blue color in the thiourea system remains unexplained and a challenge for further study.

Omitted in the preceding discussion is another complex series of acid-base reactions that begin with the NO that is lost from nitroprusside in the primary photochemical process. Some of the proposed mechanisms for the formation of the blue product involve various redox species formed in these acid-base reactions [5–7, 9, 10]. However, these mechanisms all begin with the loss of NO^+ in the primary photochemical process and so require a later step to oxidize the iron to form the blue complex. But, as discussed previously, the primary photochemical process probably produces NO and the required Fe(III) complex directly [11].

Nitric oxide does not react directly with water, but it does react with oxygen dissolved in the water to form dinitrogen tetroxide, N_2O_4, which disproportionates to form nitrous acid and nitrate ions.

$$2\,NO + O_2 \longrightarrow N_2O_4 \tag{9}$$

$$N_2O_4 + 2\,H_2O \longrightarrow HNO_2 + NO_3^- + H_3O^+ \tag{10}$$

These species, as well as the acidic $[Fe(III)(CN)_5(H_2O)]^{2-}$, are responsible for pH changes observed in photolysis of nitroprusside in unbuffered solutions containing dissolved air and added oxygen [11].

Although not implicated in the mechanism proposed for this demonstration, NO chemistry may be important as part of the chemistry of nitroprusside when used as a drug in treating hypertension, especially during possible heart attacks. Living things, including us, use many small molecules as neurotransmitters to send messages from one cell to another. Nitric oxide is one of these neurotransmitters and has a role in controlling the constriction and dilation of blood vessels. Many pharmaceuticals contain sulfur atoms doubly bonded to carbon and have the potential to react with nitroprusside, as in this demonstration. These kinds of possible drug interactions and interferences are the reason for studying nitroprusside reactions, photochemical and thermal, with these sulfur-containing compounds [12].

REFERENCES

1. I. W. Grote and J. H. Barnett, "A Simple Reversible Photochemical Experiment," *J. Chem. Educ., 10*(1), 43–44 (1933).
2. L. Pauling, "The Nature of the Chemical Bond. II. The One-Electron Bond and the Three-Electron Bond," *J. Am. Chem. Soc., 53,* 3225–3237 (1931).
3. I. W. Grote, "A New Color Reaction for Soluble Organic Sulfur Compounds." *J. Biol. Chem., 93,* 25–30 (1931).
4. O. Baudisch, "Radical Reactions with Certain Nitrogen Compounds: The Conversion of Benzene (Toluene, etc.) in Other Compounds at Low Temperature," *Science, 108,* 443–444 (1948).
5. R. P. Mitra, D. V. S. Jain, A. K. Banerjee, and K. V. R. Chari, "Photolysis of Sodium Nitroprusside and Nitroprussic Acid," *J. Inorg. Nucl. Chem., 25,* 1263–1266 (1963).
6. R. P. Mitra, B. K. Sharma, and S. P. Mittal, "Photolysis of Sodium Nitroprusside in the Presence and Absence of Air," *J. Inorg. Nucl. Chem., 34,* 3919–3020 (1963).
7. D. X. West and D. J. Hassemer, "Reactions of Sodium Pentacyanonitrosylferrate(II) with Some Thioureas in Methanol Solution," *J. Inorg. Nucl. Chem., 32,* 2717–2719 (1970).
8. J. H. Espenson and S. G. Wolenuk, Jr., "Kinetics and Mechanisms of Some Substitution Reactions of Pentacyanoferrate(III) Complexes," *Inorg. Chem., 11*(9), 2034–2041 (1972).
9. P. A. Stoeri and D. X. West, "Kinetics of the Reaction between Photolyzed Sodium Pentacyanonitrosylferrate(II) and Sodium Thiocyanate," *J. Inorg. Nucl. Chem., 36,* 2347–2350 (1974).
10. P. A. Stoeri and D. X. West, "Kinetics of the Reaction between Photolyzed Sodium Pentacyanonitrosylferrate(II) and Thiourea," *J. Inorg. Nucl. Chem., 36,* 3883–3884 (1974).
11. S. K. Wolfe and J. H. Swinehart, "Photochemistry of Pentacyanonitrosylferrate(2–), Nitroprusside," *Inorg. Chem., 14*(5), 1049–1053 (1975).
12. C. Glidewell and V. A. J. Musgrave, "Inorganic Drug Interactions: The Reaction between Thiobarbiturates and the Nitroprusside Ion," *Inorg. Chim. Acta, 167,* 253–356 (1989).

12.52

Photochromism in Ultraviolet-Sensitive Beads

Photochromic compounds isomerize and change color when they absorb light of appropriate energy and then slowly revert thermally to the original isomer and color. The temperature dependence of these processes helps to understand them.

Several ultraviolet-sensitive beads are irradiated under black light in an ice-water bath for a minute or two until all are highly colored. A few of the irradiated beads are simultaneously transferred to water baths at three different higher temperatures. The time required for the color to be lost decreases as the temperature increases (Procedure A). Beads in the four water baths are irradiated simultaneously for a few minutes. The intensity of the color produced decreases as the temperature increases (Procedure B).

MATERIALS FOR PROCEDURE A

12–15 ultraviolet-sensitive beads*

four Styrofoam coffee cups or other similar cups with a white interior

black light

temperature probe, preferably with a digital display readable by the audience
(A liquid-in-glass thermometer is also satisfactory.)

ice

cool tap water

warm tap water

color video camera interfaced to a computer for display to the audience

MATERIALS FOR PROCEDURE B

12 ultraviolet sensitive beads*

four Styrofoam coffee cups or other similar cups with a white interior

black light with a lamp long enough to simultaneously irradiate four cups

temperature probe, preferably with a digital display readable by the audience
(A liquid-in-glass thermometer is also satisfactory.)

ice

cool tap water

warm tap water

color video camera interfaced to a computer for display to the audience

---

* Several kinds of ultraviolet-sensitive beads are available from Educational Innovations, www.teachersource.com. Beads that change from white to purple on irradiation seem to be best for this demonstration, but others might be used in an inquiry study to see if and how they differ in their temperature dependencies. Beads vary in the intensity of the color produced, so for this demonstration, place several in an ice-water bath, irradiate with black light for 2–3 minutes, and select those that produce the darkest color.

PROCEDURE A

Preparation

Fill one of the cups with an ice-water mixture. Use the other three cups—cool and warm tap water and ice—to make three water baths at about 10–15°C, 20–25°C, and 30–35°C. Set up the video camera so it is looking down on the bench where the cups will be placed. Zoom and focus on a field that is wide enough for the audience to view all three cups simultaneously (about 20 cm).

Presentation

Place 6–9 beads in the ice-water bath, turn on the black light, and begin irradiating the beads floating in the bath. Place 2 more beads in each of the temperature baths and place them under the video camera, so the audience can see the white, unirradiated beads floating in each bath. Use the temperature probe to measure the temperature of each bath and record them. Designate three audience members as timekeepers, one for each temperature bath.

After the beads in the ice-water bath have been irradiated for at least 2–3 minutes, very quickly transfer 2–3 irradiated beads to each temperature bath. The timekeepers should note the time, to the nearest second, when the beads are added to their bath. As the color of the beads fades, the audience should indicate when the first and last bead in each bath has returned to the color of the unirradiated beads in the bath and the timekeepers should note these times. The color will probably disappear in less than 30 seconds in the warmest bath, in 1–2 minutes at the intermediate temperature, and in 4–6 minutes at the lowest temperature.

PROCEDURE B [1]

Preparation

Fill one of the cups with an ice-water mixture. Use the other three cups—cool and warm tap water and ice—to make three water baths at about 10–15°C, 20–25°C, and 30–35°C. Set up the video camera so it is looking down on the bench where the cups will be placed. Zoom and focus on a field that is wide enough for the audience to view all four cups simultaneously (about 25 cm). Position the black light so it can irradiate the surface of the water in the cups without interfering with the camera's view of them.

Presentation

Place 3 beads in each of the temperature baths and place the baths under the video camera, so the audience can see the unirradiated beads floating in each. Turn on the black light so all four baths are irradiated. Use the temperature probe to measure the temperature of each bath and record them. Wait at least 2–3 minutes after irradiation has begun to convince the audience that no further change is going to occur, and then note the relative intensities of the colors of the beads at the different temperatures. The color will be most intense in the coolest bath (ice-water) and least intense in the warmest with intermediate intensities at the intermediate temperatures.

HAZARDS

Black light (long-wavelength ultraviolet light) can damage one's eyes. Shield the black light source from the audience and never look directly at the source yourself.

DISPOSAL

Save the beads to be used again and pour the water down the drain.

DISCUSSION

Photochromic compounds (see Demonstration 12.45, The Effects of Solvents on Spiropyran Photochromism and Equilibria) sensitive to long-wavelength ultraviolet radiation (UV-B) are present in many consumer products, including decorative items like the beads in this demonstration, nail polish, hair barrettes and clips, and clothing decorated with pictures imprinted with dyes that change color in sunlight. A more practical use for these compounds is in photosensitive, self-darkening eyeglass lenses, in which the plastic lens is coated with layers of photochromic dyes [2]. One of the core building blocks of the naphthopyran photochromic dyes used in many of these lenses is shown here in its pyran (colorless) and open-chain (colored) form.

pyran

open-chain

Absorption of ultraviolet light by the pyran form leads to breaking of the oxygen-carbon bond shown in red in the pyran ring, which is responsible for the color change because the more highly conjugated open-chain form absorbs light in the visible region as well as in the ultraviolet. Derivatives of this structure that can undergo many transitions between the pyran and open-chain forms without too much degradation are used for eyeglass lenses. The search for new derivatives and other photochromic compounds is an active area of research and development as manufacturers seek to improve the performance of these lenses.

The proprietary structures of the photochromic dyes incorporated in the beads used in this demonstration are very likely to be similar to the pyrans used in eyeglass lenses or the spiropyran used in Demonstration 12.45. Although we do not know the exact structure of the dye, the results from this activity can be used to determine some of its properties. Another possible use for these beads is to test the efficacy of different sunscreens [1].

In Procedure A we found that the rate of loss of the color of the irradiated beads is a function of temperature and is faster at higher temperatures. The color-loss reaction is likely to be first order because individual molecules rearrange from the open-chain form to the cyclic (pyran, or spiropyran) form. The rate law for the loss of the Colored form, C, is

$$-\text{rate of loss} = k[C] = Ae^{-E_a/RT}[C] \tag{1}$$

The temperature dependence of the reaction is captured in the Arrhenius expression for the rate constant, which introduces the activation energy for the color-loss reaction. For two different temperatures,

$$\frac{-\text{rate}_2}{-\text{rate}_1} = \frac{Ae^{-E_a/RT_2}[C]}{Ae^{-E_a/RT_1}[C]} = \frac{e^{-E_a/RT_2}}{e^{-E_a/RT_1}} = e^{-\frac{E_a}{R}\left(\frac{T_1-T_2}{T_1 T_2}\right)} \tag{2}$$

$$\ln\left(\frac{-\text{rate}_2}{-\text{rate}_1}\right) = -\frac{E_a}{R}\left(\frac{T_1 - T_2}{T_1 T_2}\right) \qquad (3)$$

The concentration factors cancel because we began with the same amount of the colored form at each temperature and the Arrhenius frequency factor, A, is independent of temperature (at least over a short range of temperatures).

For the white-to-purple photochromic beads, we generally find that the color-loss reaction is about 3- to 4-fold faster for every 10°C increase in temperature. That is, for a temperature that is 10°C higher, the decay time gets shorter and the loss of color takes a third to a quarter as long as at the lower temperature. If we assume that T_1 = 293 K (20°C), T_2 = 303 K (30°C), and the rate increases by about 3.5-fold, we have

$$\ln(3.5) = -\frac{E_a}{8.314 \text{ J K}^{-1} \text{mol}^{-1}}\left(\frac{293 - 303}{293 \cdot 303}\right)$$

$$E_a = 92 \text{ kJ mol}^{-1}$$

This value probably has an uncertainty of at least 10%, but as a ballpark estimate, it shows that the activation energy is quite high, which is, of course, consistent with the observed large dependence on temperature from which it is derived.

The results in Procedure B show that, under constant irradiation with black light, the beads come to an unchanging intensity of color that is dependent on the temperature. The intensity of the color varies inversely with temperature—less of the colored form of the photochromic dye is formed at higher temperatures. In this case, a steady state is set up in which the rate of formation of the colored form is balanced by its rate of loss in the thermal reaction examined in Procedure A. [3]

The rate of formation of the colored form is

$$\text{rate of formation} = (aI_0\phi)[L] \qquad (4)$$

Here I_0 is the flux of photons, a is the fraction of photons absorbed by a unit concentration of the dye in the bead, L is the color*Less* form of the dye, and ϕ is the quantum yield for formation of the colored form of the photoexcited dye molecule. The factors inside the parentheses are essentially independent of temperature over the short range of interest for these beads.

At the steady state, the formation and decay rates are equal, and equations 1 and 4 can be equated and rearranged to give the ratio of the colored to colorless forms of the dye in the bead,

$$\frac{[C]}{[L]} = \frac{aI_0\phi}{k} \qquad (5)$$

Since k increases with temperature, the steady-state ratio decreases with temperature and the intensity of color developed is lower at higher temperatures, which provides a visual demonstration of the steady state kinetics. A further path of inquiry would be to change the flux of photons (brightness of the light source) to test whether the steady-state color is a function of this variable, as equation 5 predicts. The use of these beads to test the UV protection provided by sunscreens [1] depends on the validity of this prediction.

REFERENCES

1. T. Trupp, "Putting UV-Sensitive Beads to the Test" (JCE Classroom Activity 36), *J. Chem. Educ., 78,* 468A–468B (2001).
2. B. Erickson, "Self-darkening Eyeglasses," *C & E News, 87*(15), 54 (13 April 2009).
3. J. A. Bell, "Visualizing the Photochemical Steady State with UV-Sensitive Beads," *J. Chem. Educ., 78,* 1594 (2001).

12.53

The Photodissociation of Bromine and the Bromination of Hydrocarbons

Some compounds that are not very reactive in the dark will dissociate to yield highly reactive fragments when they absorb light. These compounds include chlorine (see Demonstration 12.44, The Photochemical Reaction of Chlorine and Hydrogen) and bromine.

A bright light bleaches the clear brownish-orange color of a solution of bromine in a hydrocarbon solvent and the bleached solution becomes slightly cloudy. Solution not exposed to the light retains its clear brownish-orange color (Procedure A). Testing the vapor phase above the irradiated solution reveals the presence of an acid in the vapor. No acid is present over the unirradiated solution (Procedure B).

MATERIALS FOR PROCEDURE A

200 mL of a saturated hydrocarbon such as hexane, cyclohexane, heptane, etc. (Do not use branched hydrocarbons that contain tertiary hydrogens because these may be reactive enough to undergo a relatively rapid thermal reaction with bromine.)

about 0.5 mL of bromine, Br_2 (See Hazards section before handling.)

5-mL glass graduated cylinder

250-mL sealable glass cylindrical container, such as a glass-stoppered graduated cylinder (For short-term storage, a cork or rubber stopper may be used.)

slide projector or other source of a directed beam of bright white light

MATERIALS FOR PROCEDURE B

200 mL of a saturated hydrocarbon such as hexane, cyclohexane, heptane, etc. (Do not use branched hydrocarbons that contain tertiary hydrogens, because these may be reactive enough to undergo relatively rapid thermal reaction with bromine.)

about 0.5 mL of bromine, Br_2 (See Hazards section before handling.)

5-mL glass graduated cylinder

sealable glass container to store the bromine/hydrocarbon solution

two 250-mL beakers without spouts

two flat plates of glass to cover the beakers to prevent loss of volatiles

two pieces of blue litmus paper or pH indicator paper

water in a small beaker, to dampen the indicator papers

small pair of tongs or tweezers

overhead projector

PROCEDURE A

Preparation

Fill the sealable cylindrical container about 80% full with the hydrocarbon. In a fume hood, use the 5-mL graduated cylinder to measure the bromine and add it to the hydrocarbon.

(Using a dropper to measure and transfer liquid bromine is not recommended because the high vapor pressure of the liquid is enough to eject the corrosive liquid before it can be transferred.) Swirl or stir the mixture to get a uniformly colored solution. Seal the cylinder and keep it in the dark or wrap it with aluminum foil until the presentation. Do not store this solution for more than a day or so.

Presentation

Place the cylinder containing the bromine-hydrocarbon solution in front of the lens of an overhead projector (or slide projector) so that the beam from the lens passes through only a portion of the solution. After about 30 to 60 seconds, the brownish-orange color of the mixture in the beam will bleach, and the bleached region will become slightly turbid.

PROCEDURE B [1]

Preparation

Put the hydrocarbon in the sealable container. In a fume hood, use the 5-mL graduated cylinder to measure the bromine and add it to the hydrocarbon. (Using a dropper to measure and transfer liquid bromine is not recommended because the high vapor pressure of the liquid is enough to eject the corrosive liquid before it can be transferred.) Swirl or stir the mixture to get a uniformly colored solution. Seal the container and keep in the dark until the presentation. Do not store this solution for more than a day or so.

Presentation

Add 100 mL of the bromine-hydrocarbon solution to each beaker and cover with the glass plates. Place one of the beakers in the dark (out of bright light) and the other on the lighted overhead projector stage. After about 30 to 60 seconds, the brownish-orange color of the solution on the projector will bleach and become slightly turbid, darkening the projected image.

Dampen a piece of blue litmus paper or pH indicator paper. Slightly lift the cover on the irradiated beaker and, using tongs or tweezers to hold the paper, insert it into the vapor space above the solution. The paper will indicate the presence of a strong acid, and the vapor may become foggy. Repeat the test with the indicator in the vapor over the unirradiated beaker still containing a clear brownish-orange solution. There will be no indication of the presence of acid in the vapor.

HAZARDS

Hydrocarbons are flammable. There should be no flames in the area where they are being handled and a fire extinguisher suitable for hydrocarbon fires should be available.

Bromine is a volatile corrosive liquid that is harmful to skin and mucous membranes and attacks many organic compounds. Handle the pure liquid in a fume hood and wear protective gloves.

The vapor over an irradiated solution of bromine in a hydrocarbon contains hydrogen bromide, HBr, that escapes from the solution. Avoid breathing this vapor, which is harmful to mucous membranes. Spraying a mist of water or solution of sodium hydrogen carbonate ($NaHCO_3$, baking soda) into the vapor will dissolve and/or neutralize the acid.

DISPOSAL

Exposing solutions containing residual bromine to bright light, including sunlight, will destroy the bromine (by reaction with the hydrocarbon). The hydrogen bromide formed in the reaction can be removed from the hydrocarbon, if necessary, by mixing the irradiated hydrocarbon solution with a saturated aqueous solution of sodium hydrogen carbonate (baking soda) in a separatory funnel. The resulting aqueous solution can be drawn off and disposed of down the drain. The remaining hydrocarbon may be reused for this demonstration or placed in the appropriate organic-waste container for disposal according to local ordinances.

DISCUSSION

This demonstration has been used for a long time as a classroom demonstration of the light-induced reaction between bromine and a hydrocarbon [2]. The brownish-orange color of bromine solutions in organic solvents shows that bromine molecules absorb visible-light wavelengths in the blue-violet region of the spectrum. As shown by the spectrum in Figure 1, the maximum absorbance is at about 415 nm. The action of visible light in this demonstration illustrates the photodissociation of bromine molecules that have absorbed these wavelengths and then dissociated to give reactive bromine atoms. The atoms initiate a free-radical chain reaction resulting in the loss of bromine and the production of brominated hydrocarbons and hydrogen bromide, HBr (detected in Procedure B), which is not soluble in the nonpolar solvent.

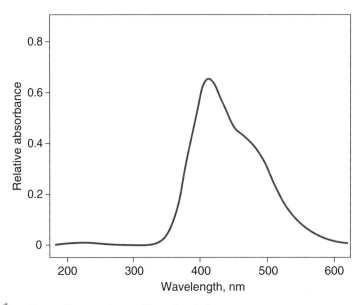

Figure 1.   Absorption spectrum of bromine in hexane.

In the dark at room temperature, a tiny fraction of bromine molecules dissociate thermally and the atoms undergo these same reactions. Although the dark reaction is much slower overall, solutions of bromine in hydrocarbon will eventually react completely to yield these same products and decolorize the solutions, which is why the solutions for this demonstration cannot be stored for long periods of time.

The photobromination of hydrocarbons begins with the absorption of visible (blue-violet) light by a bromine molecule, Br—Br, which can result in the dissociation of the molecule to give two bromine atoms, Br$^\bullet$:

$$Br-Br \xrightarrow{h\nu} 2\,Br^\bullet \tag{1}$$

The shared pair of electrons in the molecular bond and the unpaired electron on the atoms are shown explicitly as a reminder that the atoms are free radicals and to help keep track of the origin and fate of these electrons. Reaction 1 is the initiating step in a free-radical chain reaction that is responsible for the observations in this demonstration.

The first step in the propagation of the chain is abstraction (removal) of a hydrogen atom from the alkane, R—H, by a bromine atom.

$$R-H + Br^\bullet \longrightarrow R^\bullet + H-Br \tag{2}$$

The alkyl free radical, R$^{\bullet}$ produced in reaction 2 continues the chain by reacting with a bromine molecule to produce another bromine atom to keep the chain going.

$$R^{\bullet} + Br—Br \longrightarrow R—Br + Br^{\bullet} \tag{3}$$

Repetition of reactions 2 and 3 rapidly depletes the concentration of bromine molecules and forms a brominated alkane product, R—Br, and hydrogen bromide, H—Br.

In competition with these chain-propagating steps are chain-terminating steps in which the free radical species, Br$^{\bullet}$ and R$^{\bullet}$, combine to form nonradical products.

$$Br^{\bullet} + Br^{\bullet} \longrightarrow Br—Br \tag{4}$$

$$R^{\bullet} + Br^{\bullet} \longrightarrow R—Br \tag{5}$$

$$R^{\bullet} + R^{\bullet} \longrightarrow R—R \tag{6}$$

These termination reactions are going on all the time during the reaction, but, since the concentrations of the radical species are low, they are relatively infrequent and there may be thousands or even millions of propagation cycles before one or both of its free radicals undergoes one of these chain termination reactions. When the light is turned off, reaction 1 ceases and there is no longer a fresh source of free radicals, so those that are left quickly get used up in these termination reactions and the reaction stops.

The final reaction products in the alkane solvent in this demonstration are brominated alkanes, R—Br; hydrogen bromide, H—Br; and a very small amount of alkane dimers, R—R; but almost no bromine molecules. Since a bromine atom can abstract a hydrogen atom from several different positions on an alkane molecule, several different alkyl radicals, R$^{\bullet}$, can be formed, so several different brominated alkanes and alkane dimers are produced. The brominated alkanes and alkane dimers are nonpolar and soluble in the alkane solvent. But hydrogen bromide is quite polar and is not very soluble in the alkane, so it begins to come out of solution, creating a cloudy (turbid) solution. The H—Br can escape into the vapor phase over the solution, where it can be detected by acid-base indicator paper when it dissolves in water on the dampened paper and reacts to form hydronium ions.

Brominated hydrocarbons are useful starting materials in many synthetic pathways because they provide a reactive site on otherwise relatively unreactive molecules. Although photobromination can sometimes be used to prepare such compounds, it has only limited use, because often more than one site on the hydrocarbon reacts with the bromine atoms. Bromine atoms are not, however, indiscriminate reactants and have been used in experiments where the products of reaction are used to determine which hydrogen atoms are most readily abstracted. That method is based on the fact that the reaction to form H—Br and R$^{\bullet}$ from R—H and Br$^{\bullet}$ is endothermic and on the assumption (the Hammond postulate) that the transition state for endothermic reactions lies far toward the products. In this case, one of the products of the reaction, H—Br, is identical for all possible reaction sites, so the overall reactivity will be largely determined by the strength of the R—H bond. For simple hydrocarbons, the order of increasing reactivity of the hydrogen atoms noted in textbooks is generally primary < secondary < tertiary.

REFERENCES

The topic of photobromination is covered in most introductory organic chemistry textbooks because it is easily observed and its mechanism is a paradigm for free-radical reactions as well as an introduction to the relative stability of hydrocarbon free radicals.

1. I. Perina and B. Mihanovic, "A Classroom Demonstration of Aliphatic Substitution," *J. Chem. Educ., 66*(3), 257 (1989).
2. F. von Konek and A. Loczka, "Vorlesungsversuch zur Demonstrierung der chemischen Lichtwirkung," *Berichte der Deutschen Chemischen Gesellschaft, 57*(4), 679–680 (1924).

12.54

The Photochemical Formation and Reaction of Ozone

Life on the surface of the Earth is protected from short-wavelength ultraviolet radiation from the sun (UV-C) because oxygen molecules, O_2, in the upper atmosphere absorb this high-energy light. The absorbed energy causes the oxygen molecule to dissociate to form atomic oxygen, O, which reacts with oxygen molecules to form ozone, O_3. At the surface of the Earth, ozone formed by other photochemical processes reacts with organic molecules in the air to form haze and smog, which have adverse effects on life.

A short-wavelength UV lamp is used to irradiate the air in a clear, colorless glass bottle. A twist of lemon peel dropped into the bottle produces swirls of white smoke. No smoke is produced by lemon peel in an untreated bottle.

MATERIALS

two clear, colorless glass bottles, about 500- to 1000-mL capacity, with lids

a small quartz, low-pressure mercury lamp (that can be suspended in one of the bottles) and power supply*

piece of heavy corrugated cardboard, about 10-cm square

knife to cut the cardboard and prepare lemon peel

opaque cardboard box or other means to cover the bottle and protect the eyes of the audience when the lamp is on

two strips of lemon peel about 1 cm × 3 cm

sheet of flat black cardboard or other black background for viewing the results

few drops of turpentine (an alternative to the lemon peel)

source of oxygen, O_2, gas (optional)

video projection system to show the demonstration in large lecture halls**

PROCEDURE

Preparation

Make a slit in the piece of heavy cardboard and slip it around the power cord to the lamp at a position that will suspend the lamp in about the middle of one of the bottles when the cardboard is resting on the opening. (Different lamps may require a different setup for suspending the lamp inside the bottle. For safety purposes, there must be a way to cover the setup so the eyes of the audience and presenter can be protected when the lamp is on.) Shortly before the demonstration (or during its presentation, if desired), cut two 1-cm × 3-cm strips of peel from a lemon.

---

* A suitable source of UV irradiation is the Pen-Ray low-pressure mercury lamp produced by UVP (Upland, California) and distributed by many scientific supply companies. Be sure the lamp is a source of 253.7 nm radiation, because some lamps have envelopes made of materials (such as Pyrex or Vycor) that are not transparent at this wavelength.

** Various types of suitable video projection systems are described on page xxxiii.

Presentation

Suspend the lamp in one of the bottles, cover the setup with the box that will protect the audience from any UV radiation that might escape from the setup, and turn the lamp on for about a minute (or longer for a larger bottle). Turn off the lamp, remove the protective covering, take the lamp out of the bottle and put on the lid to contain the ozone that has been produced. Set the treated and untreated bottles in front of a black background to help the audience more easily see what will (or will not) happen inside the bottles. Twist one of the strips of lemon peel, remove the lid from the bottle containing ozone, drop the peel into the bottle and replace the lid. Twist the second strip of lemon peel, drop it into the second, unirradiated, bottle, and put its lid on. Wisps of white smoke will form in the bottle containing ozone and, if the lemon is quite fresh, will fill the bottle. No smoke will form in the second bottle. (Alternatively, a few drops of turpentine may be substituted for the lemon peel, and the reaction will be with the volatile organic compounds produced by pine trees.)

Optionally, if a source of oxygen gas is available, the bottles can be filled with oxygen by flowing oxygen in for several seconds to displace the air in the bottles before the above steps are carried out. The amount of ozone produced, and hence the amount of smoke from the reaction with the oils from the lemon, will be larger. Using only air does, however, reinforce the fact that the oxygen content of the atmosphere can form enough ozone to cause a significant amount of reaction with organic vapors.

HAZARDS

Short-wavelength UV radiation (UV-C) is used to kill bacteria in germicidal applications and is harmful as well to human skin and eyes. Do not expose anyone to the lamp when it is turned on, even when it is inside a glass bottle that transmits very little UV radiation.

Ozone attacks the respiratory system and is particularly harmful to people with asthma and other allergy problems. Long exposure can prove fatal. Do not open the bottle and breathe its contents after the ozone has been photosynthesized. The amount of ozone produced in this procedure is small and is used up by reaction with the oil from the lemon peel.

DISPOSAL

Lemon peel can be disposed of as trash and the rest of the materials saved for future presentations.

DISCUSSION

This demonstration incorporates two aspects of ozone photochemistry and reactivity, one in Earth's upper atmosphere and one at its surface. Life on Earth is protected by the photoreactions of oxygen and ozone in the upper atmosphere (see the introduction, page 55). These reactions remove essentially all the UV-C and most of the UV-B radiation from the sun. In this demonstration, the intense emission at 253.7 nm from a low-pressure mercury lamp represents the UV-C from the sun that is absorbed mainly by oxygen molecules and causes their dissociation to oxygen atoms. The atoms combine with oxygen molecules to form ozone, which then absorbs UV-B to reform oxygen atoms and molecules. This cycle of photolytic and thermal reactions produces a steady-state concentration of ozone that is the Earth's UV-B shield.

At the surface of the Earth, thermal reactions of nitrogen and oxygen (in internal combustion engines and high-temperature industrial processes) produce nitrogen oxides that can photodecompose to produce oxygen atoms. These atoms combine with the abundant oxygen molecules in the air to form ozone, which, in contact with plant and animal life, is harmful and toxic (see the introduction, page 56). Toxicity is further increased by the reaction of ozone with volatile organic compounds (VOCs) in the air that come from many sources, including incompletely burned fuels, organic solvents used in cleaning and industrial processes, and plants like pine trees. In this demonstration, the role of the VOCs is played by D-limonene, the compound that gives lemons their characteristic smell (Figure 1).

Figure 1. Structure of D-limonene.

The reaction of ozone with limonene (and other VOCs) produces oxygenated products that are less volatile than the reactants and that readily form aerosols of tiny particles suspended in the air. Since these particles scatter light, much like the tiny water droplets that make up clouds and fog, they can be seen as a mist or smoke, as in this demonstration. One of the primary products of the reaction of limonene with ozone is 4-acetyl-1-methylcyclohexene (Figure 2) [1].

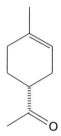

Figure 2. Structure of 4-acetyl-1-methylcyclohexene.

Further reaction of ozone with these primary products leads to secondary products that contain more oxygen atoms and are even less volatile. This mixture of compounds produces tiny particles that can have adverse health effects, especially for people with respiratory problems. The Air Quality Index includes particulates among the contaminants measured and provides a warning when the outdoor atmospheric concentrations in a locality become high enough to be dangerous. There is no general warning of this sort for indoor air quality, but research on indoor air indicates that effects, much like those in this demonstration, occur when ozone (from the outdoors or indoor sources) reacts with VOCs from household products. Among these VOCs are limonene and pinene from products like air fresheners and cleaners that have an added "lemony" or "piney" odor [2, 3].

REFERENCES

1. D. Grosjean, E. L. Williams, E. Grosjean, J. M. Andino, and J. H. Seinfeld, "Atmospheric Oxidation of Biogenic Hydrocarbons: Reaction of Ozone with (β-Pinene, D-Limonene and trans-Caryophyllene," *Environ. Sci. Technol.*, *27*, 2754–2758 (1993).
2. T. Wainman, J. Zhang, C. J. Weschler, and P. J. Lioy, "Ozone and Limonene in Indoor Air: A Source of Submicron Particle Exposure," *Environmental Health Perspectives*, *108*(12), 1139–1145 (2000).
3. H. Destaillats, M. M. Lunden, B. C. Singer, B. K. Coleman, A. T. Hodgson, C. J. Weschler, and W. W. Nazaroff, "Indoor Secondary Pollutants from Household Product Emissions in the Presence of Ozone: A Bench-Scale Chamber Study," *Environ. Sci. Technol.*, *40*, 4421–4428 (2006).

Index to Volumes 1–5

Numbers before the colon are volume numbers; numbers after the colon are page numbers.

Absolute zero,
  determination of, 2:28, 2:33
Absorption spectra, 5:6
  atomic, 5:35, 5:41
  of honeybee photopigments, 5:69
  of indicators, 5:43
  of retinal cells, 5:68
  of transition metal complexes, 1:262
Acetic acid,
  conductivity of, 3:326
  ionization of, 3:155
  strength of, 3:136, 3:155
Acetone,
  evaporation of, 1:5
  oxidation of, 2:216
  reaction with aldehydes, 4:65
  vapor pressure of, 2:85
Acetylene,
  reaction of chlorine with, 2:224
Acid rain, 2:184, 3:116, 3:125
Acid strength, 3:136, 3:155, 3:158
Acid-base indicators, 2:114
  preparation of, 3:27
  theory of, 3:18
Acid-base titration, 3:152
Acids,
  dissociation of, 3:140
  electrical conductivity of, 3:140
  preparation of stock solutions of, 3:30
  reactions of carbonates with, 3:96
Acids and bases,
  classical properties of, 3:58
Acrylic polymer, 1:235, 1:237, 5:144
Actinometer, 5:287
Activity series, 4:101
Addition polymer,
  poly(methyl acrylate), 1:235
  poly(methyl methacrylate), 1:237
  polybutadiene, 1:231
  polystyrene, 1:241
  polysulfur, 1:243
Afterimage, 5:208
Air,
  solubility of, 3:280
Air pollution, 3:341
Air pressure, 1:5, 2:6, 2:9, 2:12

Alcohol,
  as monomer in condensation polymer,
    1:216
  combustion of, 1:13
  photoreaction, 5:57
  solvent effect, 5:261
Alcohol gel, 3:360
Aldol condensation, 4:65
Alfin catalyst, 1:231
Alka-Seltzer,
  buffering action of, 3:186
Alternating current,
  electrolysis with, 4:198
  with gas discharge tubes, 5:105, 5:213
Aluminum,
  amphoteric properties of, 3:128
  anodization of, 4:253
  foil electroscope, 5:156
  reaction of bromine with, 1:68
  reaction of iron oxide with, 1:85
Aluminum chloride,
  as Lewis acid, 3:192
Aluminum electrode, 4:198
Aluminum soap, 3:337
3-Aminophthalhydrazide. *See* Luminol
Ammonia,
  as ligand, 1:299
  diffusion of gaseous, 2:59
  electrical conductivity of aqueous, 3:326
  oxidation of, 2:214
  preparation and properties of, 2:202
  reaction of hydrogen chloride with, 2:211
  reaction of iodine with, 1:96
  solubility of, 2:202
Ammonia fountain, 2:205
  multiple, 3:92
Ammonium acetate,
  electrical conductivity of aqueous, 3:326
Ammonium acrylate, 5:144
Ammonium chloride,
  as source of ammonia gas, 2:202
  as source of nitrogen gas, 2:154
Ammonium chloride ring, 2:59
Ammonium dichromate,
  decomposition of, 1:81
Ammonium iron(III) citrate, 5:283

Ammonium nitrate,
    heat of solution of, 1:8
    reaction of zinc with, 1:51
Amphoteric properties, 3:128
Aniline,
    basic properties of, 3:158
    polymerized with formaldehyde, 1:225
Anodized aluminum, 4:253
    dyeing of, 4:253
Antacids,
    neutralizing capacity of, 3:162
Antimony,
    reaction of chlorine with, 1:64
Antimony(III)
    complexes of chloride ion with, 1:324
Aqua regia, 3:83
Arrhenius acid-base theory, 3:58
Arrhenius equation, 5:302
Arsenic(III) sulfide,
    precipitation of, 4:80
Aspirator, 2:12, 2:81, 5:243
Atmospheric photochemistry, 5:55
Atomic spectra, 5:34
Avogadro's hypothesis, 2:44
Azeotropy, 3:263

Bakelite, 1:219
Balloon,
    effect of pressure on size of, 2:12
    effect of temperature on size of, 2:24
    explosive, 1:106
    flame-resistant, 3:239
    methane-filled, 2:196
Barium chloride, 5:108
Barium emission spectrum, 5:108
Barium hydroxide,
    base strength of, 3:136
    conductivity titration of, 3:152
    endothermic reaction of, 1:10
Barking dog, 1:74
Barometer,
    construction of, 2:9
Base strength, 3:158
Bases,
    preparation of stock solutions of, 3:30
Bearnaise sauce, 3:351
Beating heart,
    mercury, 4:260
Beer, 3:345
Beer-Lambert law, 5:42
Belousov-Zhabotinsky oscillating reaction, 2:257,
    2:262, 2:266, 2:270, 2:273, 2:276, 2:280
Benzaldehyde,
    reaction of acetone with, 4:65
Bioluminescence, 5:64
Bipyridine,
    as ligand, 1:194, 5:118

Bird,
    drinking, 3:249
Birefringence, 5:167
Bis(2-chloroethyl)ether,
    immiscibility of, 3:229
Bisulfite,
    reaction of iodate with, 4:16, 4:26
Black-body radiation, 5:23, 5:91, 5:96
Bleach,
    electrolytic production of, 4:181
    reaction of hydrogen peroxide with, 1:133
Blind spot, 5:82
Blood,
    catalyst from, 2:122
Blood vessels, 5:72, 5:299
Blue bottle experiment, 2:142
Blueprint, 5:283
Boiling at reduced pressure, 1:5
Boiling point,
    effect of pressure on, 2:81
Boiling-point elevation, 3:297
Borax,
    formation of gel with, 3:362
Boyle's law, 2:14, 2:20
Brass,
    reaction of acid with, 3:83
Briggs-Rauscher oscillating reaction, 2:248
Bromate,
    reaction of manganese(II) with, 4:83
Bromide ion, 5:228
    copper complexes with, 1:314
Bromination, 5:304
Bromine,
    diffusion of aqueous, 3:317
    diffusion of gaseous, 2:63
    liquid-vapor equilibrium of, 2:75
    photochemical reaction with hydrocarbons,
        5:304
    reaction of aluminum with, 1:68
    reaction of phosphorus with, 1:72
Bromothymol blue,
    as pH indicator, 3:33
Bronsted-Lowry acid-base concept, 3:10
Buckminsterfullerene, 5:62, 5:90
Buffer solution,
    action and capacity of, 3:173
    Alka-Seltzer as, 3:186
    preparation of, 3:33
    theory of, 3:20
Bunsen burner,
    flame structure and temperature of,
        1:116, 5:62
Buoyancy of air, 3:100
Butane,
    determination of molar mass of, 2:48
t-Butyl chloride,
    hydrolysis of, 4:56

Calcite,
  birefringence of, 5:167
Calcium acetate,
  formation of alcohol gel with, 3:360
Calcium carbonate,
  birefringence of, 5:167
  in calcite, 5:19
  reaction of acetic acid with, 3:155
Calcium chloride, 5:108
  boiling-point elevation by, 3:297
  emission spectrum of, 5:108
  heat of solution of, 1:36
Calcium fluoride,
  etching glass with, 3:80
Calcium hydroxide,
  in preparation of ammonia, 2:202
Calcium oxide,
  in glass, 5:139
  reaction of water with, 1:19
Camouflage, 5:81
Can,
  collapsing, 2:6
Candle,
  combustion of, 2:158
  flame emission, 5:87
Canned heat, 3:360
Cannon,
  carbon dioxide, 3:96
  hydrogen, 1:109
  hydrogen-oxygen, 1:106
Carbon dioxide,
  as product of combustion, 2:158
  catalyzed aqueous reactions of, 2:122
  from carbonated beverage, 2:109
  in aqueous solution, 1:330, 2:114
  preparation and properties of, 2:106
  reaction of limewater with, 1:329
  reaction of magnesium with, 1:90
  reaction of sodium hydroxide with, 3:100
  reactions in aqueous solutions of, 2:114
  reactions with limewater of, 1:329
  released by acids from carbonates, 3:96
  solubility in water (table), 1:333
  transport across soap film, 2:228
Carbon disulfide,
  as solvent for white phosphorus, 1:74
  reaction of nitrogen(I) oxide with, 5:124
  reaction of nitrogen(II) oxide with, 1:117
  volume of mixing of, 3:225
Carbonated beverage,
  carbon dioxide from, 2:109
Carbonates,
  reactions with acids of, 3:96
Carbonic anhydrase, 2:122
Carotene, 5:277
Casein,
  effect of pH on solubility of, 3:188

Catalysis,
  in ammonium nitrate decomposition, 1:51
  in oxidation of acetone, 2:216
  in oxidation of ammonia, 2:214
  in preparation of oxygen, 2:137
  of dextrose oxidation, 2:142
Catsup,
  thixotropy of, 3:364
Cellulose,
  in formation of rayon, 1:247
Cellulose nitrate, 1:43
Charles's law, 2:28
Chemical garden, 3:379
Chloride ion,
  complexes of antimony(III) and, 1:324
  complexes of cobalt(II) and, 1:280
  complexes of lead(II) and, 1:286
  in photoreduction of silver, 5.291
  quenching of fluorescence, 5:227
Chlorine,
  electrolytic generation of, 4:166, 4:181
  photoinduced reaction of, 1:121
  preparation and properties of, 2:220
  reaction of acetylene with, 2:224
  reaction of antimony with, 1:64
  reaction of hydrogen iodide with, 2:224
  reaction of hydrogen peroxide with, 1:133
  reaction of hydrogen with, 1:121
  reaction of iron with, 1:66
  reaction of phosphorus with, 1:70
  reaction of sodium with, 1:61
2-Chloro-2-methylpropane,
  hydrolysis of, 4:56
Chromate ion,
  movement in conducting solution, 4:150
Chromium,
  amphoteric properties of, 3:128
Chromium(III) oxide,
  from ammonium dichromate, 1:81
Chromium(VI),
  electrolysis of, 4:232
Cinnamaldehyde,
  reaction of acetone with, 4:65
Cobalt(II),
  complexes of, 1:280
Coin-operated reaction of copper with nitric
    acid, 3:83
Coke bottle,
  explosion in, 1:106
Cold pack, 1:8
Collapsing can, 2:6
Colligative properties,
  boiling-point elevation, 3:297
  freezing-point depression, 3:290
  osmosis, 3:283, 3:286, 3:390
  vapor pressure, 3:242
Colloids, 3:353

Color,
  of pH indicators, 3:33
  theory of, 1:260
Combining volumes of gases, 2:167, 2:190
Combustion,
  in flames, 5:61
  of metals, 5:98
  under water, 1:40
Common ion effect, 3:155
Concentration cell, 4:140
Condensation polymer,
  aniline-formaldehyde, 1:225
  nylon, 1:213
  phenol-formaldehyde, 1:219
  phenolphthalein-terephthalate, 1:229
  polyurethane, 1:216
  rayon, 1:247
  resorcinol-formaldehyde, 1:222
  thiokol rubber, 1:245
  urea-formaldehyde, 1:227
Conductivity,
  effect of ion-exchange resin on, 3:146
  end-point determination by, 3:152
  of acid solutions, 3:140
  of gases, 2:90
  of solids, 5:114
  of solutions, 3:326
Contact angle, 3:307
Cool light, 1:125
Copper,
  as catalyst, 2:216
  electroplating, 4:212
  half cell, 4:101
  photocell, 5:274
  reaction of nitric acid with, 2:163, 3:83
  reactions with mineral acids of, 3:72
Copper(II),
  complexes of, 1:314, 1:318
  complexes of bromide ions and, 1:314
  emission spectrum of, 5:108
  precipitation by sulfide of, 4:140
Copper(II) acetate,
  reaction of hydrazine with, 4:224
Copper(II) bromide,
  electrolysis of aqueous, 4:209
Copper(II) chloride, 5:108
Copper(II) ion movement in conducting
    solution, 4:150
Copper(II) sulfate,
  diffusion in water of, 3:317
  electrolysis of, 4:212
  heat of hydration of, 1:23
  in formation of osmosis membrane, 3:390
  in rayon formation, 1:247
Copper hexacyanoferrate,
  osmosis through, 3:390
Copper mirror, 4:224
Copper oxide, 5:274

Copper-magnesium cell, 4:137
Copper-zinc cell, 4:119
Cornstarch,
  dilatancy of mixture of water and, 3:364
Cottrell precipitator, 3:341
Coulometer, 4:189
Crystallization,
  in gels, 3:372
  of sodium acetate, 1:27
  of sodium thiosulfate, 1:31
  of thymol, 1:34
Cyalume light sticks, 1:146
Cyanide ion,
  complexes of nickel(II) and, 1:299
Cyanotype, 5:283
Cyclohexane,
  boiling point elevation of, 3:297
  bromination of, 5:304

Dalton's law, 2:41
Dangerous demonstrations, 1:124, 1:201
Daniel cell, 4:119
Dehydration of sugar by sulfuric acid, 1:77
Deionizer,
  effect on solution conductivity of, 3:146
Density,
  of liquids, 3:229, 3:323
Dextrose,
  catalyzed oxidation of, 2:142
  reaction of silver nitrate with, 4:240
Dialysis, 3:286
1,6-Diaminohexane,
  polymerization of, 1:213
Diapers, 3:368
Dichloroethane,
  polymerized with sulfur, 1:245
Dichroism, 5:151, 5:154
Dielectric, 5:33
Dielectric properties of liquids, 3:329
Diffraction, 5:14, 5:87, 5:93, 5:103, 5:131
Diffusion,
  Graham's law of, 2:69
  in liquids, 3:317
  of bromine vapor, 2:63
  relative rates of, 2:59
  through fritted glass, 2:69
  through porous cup, 2:55
1,3-Dihydroxybenzene,
  polymerization of, 1:222
Dilatancy, 3:364
Dilution,
  heat of, 1:17
Dimethylglyoxime
  complex of nickel(II) and, 1:299
Dimethylsulfoxide, 1:156
Dissociation,
  of weak acid, 3:140

Distillation, 3:263, 5:134
    fractional, 3:258
Drinking bird, 3:249
Drops,
    shape of, 3:307
Dry cell, 4:111
Dry ice,
    as source of carbon dioxide gas, 1:329,
        2:114, 2:228
    combustion of magnesium in, 1:90
    volume expansion on sublimation of, 2:24
Dumas method,
    determination of molar mass by, 2:51
Dust explosion, 1:103

EDTA. *See* Ethylenediaminetetraacetic acid
Effusion,
    Graham's law of, 2:72
Egg,
    osmosis through outer membrane of, 3:283
Egg in the flask, 2:24
Egg white,
    action of nitric acid on, 3:70
    whipped foam from, 3:345
Electric conductivity,
    of acids, 3:140
    of aqueous solutions, 3:140
    of gases, 2:90
    of liquids, 3:326
    titration endpoint by, 3:152
Electroless plating, 4:224, 4:240
Electrolysis,
    of water, 4:156, 4:170
Electromagnetic spectrum, 5:2
Electron diffraction, 5:36
Electroscope, 5:157
Emission,
    atomic, 5:103, 5:108
    electrogenerated, 5:118
    from diode, 5:111
Emulsion, 3:351
Endothermic process, 1:5
Endothermic reaction, 1:8, 1:10
Enzyme,
    from blood, 2:122
Equilibrium,
    effect of pressure on gaseous, 2:180
    effect of temperature on gaseous, 2:180
    in gaseous nitrogen(II) oxide, 2:180
    of antimony(III) complexes, 1:324
    of carbon dioxide in solution, 1:329, 2:122
    of cobalt(II) complexes, 1:280
    of lead(II) complexes, 1:286
    of mercury(II) complexes, 1:271
    of silver(I) complexes, 1:293
    redox, 5:249, 5:256
    solvent effect on, 5:261

Equilibrium constant,
    notation of, 1:251
Etching glass,
    with hydrogen fluoride, 3:80
Ethanol,
    as Lewis base, 3:192
    gel formation, 3:360
    oxidation of, 2:216
    thermal expansion of, 3:234
    volume of solution of water and, 3:225
Ethyl acetate,
    volume of carbon disulfide and, 3:225
Ethylenediamine,
    complexes of nickel(II) and, 1:299
Ethylenediaminetetraacetic acid (EDTA),
        5:118
Evaporation,
    endothermic, 1:5
Explosion,
    of powder, 1:103
Explosive gas,
    electrolytic production of, 4:156, 4:166

Faraday, Michael, 2:158, 5:22, 5:90
Flame, 5:61, 5:87
Flashbulb,
    photographic, 1:38
Flour,
    explosion of, 1:103
Fluorescence, 5:45
Foam, 3:345
    polyurethane, 1:216
    rigid, 1:216, 3:348
Formaldehyde,
    polymerized with aniline, 1:225
    polymerized with phenol, 1:219
    polymerized with resorcinol, 1:222
    polymerized with urea, 1:227
    reaction of sulfite with, 4:70
Fractional distillation, 3:258
Free-radical, 5:57
    bromination, 5:304
    polymerization, 1:235, 1:237, 1:241
Freezing-point depression, 3:290
Fuel cell, 4:123
Fumed silica, 3:364

Galvanizing, 4:244
Gas,
    handling of, 2:3
    physical properties of (table), 2:5
Gas evolution oscillator, 2:301
Gas laws,
    Avogadro's hypothesis, 2:44
    Boyle's law, 2:14, 2:20
    Charles's law, 2:28
    Dalton's law, 2:41

Gas solubility, 3:92
    effect of temperature on, 3:280
Gel,
    combustible, 3:360
    disappearing, 5:141
    growing crystals in, 2:305
    of poly(acrylamide), 3:368
    of sodium silicate, 3:372
Genie in bottle, 2:137
Glass,
    electrolysis through, 4:247
    hydrogen fluoride etching of, 3:80
    refractive index of, 5:12, 5:135
Glycerine,
    potassium permanganate oxidation of, 1:83
Gold,
    reaction of aqua regia with, 3:83
Graham's law, 2:59, 2:69, 2:72
Granite, 3:125
Gravity,
    effect on drop shape of, 3:307
Gravity cell, 4:119
Guncotton, 1:43

Half cells, 4:101
Halogen lamp, 5:97
Heat capacity,
    of water, 3:239
Heat of crystallization,
    of sodium thiosulfate, 1:36
Heat of solution,
    of calcium chloride, 1:36
Hemoglobin, 1:156, 5:52
Henry's law, 3:280
Hexamethylenediamine,
    as monomer in nylon, 1:213
Hot pack, 1:36
Household products,
    acid-base properties of, 3:50, 3:65, 3:162
    carbon dioxide from, 3:96
    foams from, 3:345
    surface tension of, 3:301
    thixotropic properties of, 3:364
    water absorption by, 3:368
Hydrate,
    of copper(II) sulfate, 1:23
    of manganese(II) sulfate, 3:269
Hydrazine,
    reaction of copper(II) sulfate with, 4:224
Hydrochloric acid,
    azeotrope of, 3:263
    chlorine from, 2:220
    electrolysis of, 4:166
    heat of neutralization of, 1:15
    hydrogen from, 2:128
    reaction of antimony(III) with, 1:324
    reaction of metals with, 1:25

    reaction of sodium hydroxide with, 3:276
    strength of, 3:136
Hydrogen,
    atomic spectrum of, 5:34, 5:103
    explosion of, 1:106, 2:131
    preparation of, 2:128
    reaction of chlorine with, 1:121
Hydrogen bromide, 5:305
Hydrogen chloride,
    diffusion of gaseous, 2:59
    fountain, 2:205
    infrared spectrum of, 5:39
    Lewis-acid properties of, 3:192
    preparation of, 2:198
    reaction of ammonia with, 2:211
Hydrogen fluoride,
    etching glass, 3:80
Hydrogen iodide,
    reaction of chlorine with, 2:224
Hydrogen peroxide,
    as source of oxygen gas, 2:137
    decomposition to oxygen, 1:133
    reaction of chlorine with, 1:133
    reaction of iodide with, 4:37
    reaction of luciginen with, 1:180
Hydrogen-oxygen cell, 4:123
Hydrolysis, 3:103, 5:64

Ice bomb, 3:310
Ideal solution,
    and boiling-point elevation, 3:297
    and freezing-point depression, 3:290
Immiscible liquids, 3:229
Incandescence, 5:5, 5:23, 5:61, 5:91, 5:96, 5:98
India ink,
    as colloid, 3:358
    as light-scattering agent, 5:131, 5:163, 5:175
Indicator,
    acid-base, 2:114, 3:18, 3:27, 5:43, 5:307
    oxidation-reduction, 2:142, 5:257
Interfacial tension, 3:305, 3:307
Interhalogen compounds, 2:224
Invisible painting, 3:47
Iodate,
    reaction of bisulfite with, 4:16, 4:26
Iodide,
    complexes of mercury(II) and, 1:271
    complexes of silver(I) and, 1:293
    fluorescence quenching by, 5:227
    reaction of hydrogen peroxide with, 4:37
    reaction of iron(III) with, 4:51
Iodine,
    diffusion in solution, 3:317
    diffusion in water, 3:313
    fluorescence of gaseous, 5:243
    reaction of ammonia with, 1:96
    reaction of zinc with, 1:49, 4:134

Iodine clock reaction, 4:16, 4:26, 4:29,
    4:37, 4:44
Ion migration,
    in conducting solution, 4:150
Ion-exchange resins, 3:146
Ions,
    electric conductivity of, 3:326
Iron,
    combustion of, 5:98
    formation of molten, 1:85
    reaction of chlorine with, 1:66
    reaction of hydrochloric acid with, 1:25
    reaction of sulfur with, 1:55
Iron(III),
    complexes of, 1:338
    oxidation of iodide by, 4:51
Iron(III) citrate,
    in cyanotype formation, 5:283
Iron(III) oxalate,
    in actinometer, 5:287
Iron(III) oxide,
    as combustion product, 5:100
    reaction of aluminum with, 1:85
Isocyanate,
    as monomer in polyurethane, 1:216
Isopropanol,
    combustion of, 1:13
    solvent effect of, 5:262

Jablonski diagram, 5:44
Jumping rubber, 1:231

Kinetic energy,
    from chemical reaction, 2:216
Kinetic molecular theory,
    and gas diffusion, 2:54, 2:59, 2:63
    simulator, 2:96

Landolt reaction, 4:16, 4:29
    color variations of, 4:26
Lasers, 5:48
    polarization of light from, 5:175
Lauric acid,
    in preparation of aluminum soap, 3:337
Lead,
    amphoteric properties of, 3:128
    combustion of, 1:93
    electrolysis of, 4:189
Lead(II),
    complexes of, 1:286
Lead nitrate,
    electrolysis of aqueous, 4:205
Lead storage battery, 4:115
Leaf,
    chlorophyll in, 5:60
    copper plating of, 4:212

Lewis acids and bases,
    definition of, 3:12
    properties of, 3:192
Liesegang rings, 2:305
Light,
    initiation of reaction by, 1:121
    polarized, 3:386
    scattering of, 3:353
Light waves, 5:12
Light-box,
    construction of, 3:33
Light-emitting diodes (LEDs), 5:49, 5:111
Lightsticks, 1:146, 5:63
    additive color mixing with, 5:184
Lime, 1:19
Limestone, 1:329, 3:125
Limewater,
    reactions of carbon dioxide and, 1:329
Limiting reagent, 4:33
Liquid,
    density of, 3:323
    dielectric properties of, 3:329
    diffusion in, 3:317
    elastic, 3:337
    electric conductivity of, 3:326
    miscibility of, 3:323
    temperature dependence of volume of, 3:234
    viscosity of, 3:313
Liquid crystal,
    display, 5:5, 5:21, 5:171
Liquid nitrogen, 2:24
    effect on LEDs, 5:111
Liquid oxygen,
    preparation and properties of, 2:147
Liquid-vapor equilibrium, 2:75, 2:85
Lithium chloride,
    emission spectrum of, 5:108
    heat of solution of, 1:21
Luciginen,
    chemiluminescence from, 1:180
Lucite, 1:237
Luminescence sensitizers, 1:202
Luminol,
    chemiluminescence from, 1:156, 1:168, 1:175
Lycopodium,
    explosion of, 1:103

Magnesium,
    burning of, 1:38
    combustion in carbon dioxide of, 1:90
    reaction of acid with, 4:137
    reaction of hydrochloric acid with, 1:25
Magnesium sulfate,
    heat of hydration of, 1:36
Magnetic field,
    from electric current, 4:97
    visualized with iron filings, 5:22

Magnetic properties,
    of liquid oxygen, 2:147
Manganese(II),
    oxidation by bromate of, 4:83
Manganese(II) sulfate,
    temperature effect on solubility of, 3:269
Manganese dioxide,
    as catalyst, 2:137, 2:220
Marble,
    acid-neutralizing properties of, 3:125
Mass of book,
    using Boyle's law to determine, 2:20
Mayonnaise, 3:351
Mean free path, 2:63
Mercury,
    emission spectrum of, 5:103
Mercury(II),
    complexes of iodide ions and, 1:271
Mercury(II) iodide, 4:29
Mercury barometer, 2:9, 2:14, 3:234
Mercury beating heart, 4:260
Mercury drops, 3:307
Metal oxides,
    basic properties of, 3:109
Methane,
    combustion of, 1:113, 2:196
    filling balloon with, 2:196
    preparation and properties of, 2:193
Methanol,
    as solvent, 3:266
    density of, 3:323
    distillation of, 3:258
Methyl acrylate,
    polymerization of, 1:235
Methyl methacrylate,
    polymerization of, 1:237
Methyl violet, 3:192
Methylene blue, 5:252
    catalyst in oxidation of dextrose, 2:142
    photochromic, 5:256
Mineral oil,
    density of, 3:229
Miscibility of liquids, 3:229
Molar mass of gas,
    and musical pitch, 2:88
    determination of, 2:48, 2:51
Molten salt,
    electrolysis of, 4:247
Multiple ammonia fountain, 3:92
Musical pitch,
    and molar mass of gas, 2:88

Naphthalene,
    boiling-point elevation by, 3:297
Natural gas,
    blue flame from, 5:90
    combustion of, 1:113

Neutralization,
    heat of, 1:15
Nickel(II),
    complexes of, 1:299
    electrolysis of aqueous, 4:228
Nitric acid,
    reaction of copper with, 2:163, 3:83
    reaction of protein with, 3:74
Nitrocellulose, 1:43
Nitrogen,
    emission from, 5:62, 5:65
    liquid, 2:147, 5:111
    oscillating generation of, 2:301
    preparation and properties of, 2:154
Nitrogen(II) oxide,
    preparation and properties of, 2:163
    reaction of carbon disulfide with, 1:117
    reaction of oxygen with, 2:167
    solubility of, 2:163
Nitrogen(IV) oxide,
    acidic properties of, 3:116
    equilibrium in, 2:180
    formation of, 3:83
Nitrogen oxides in smog, 5:56, 5:309
Nitrogen triiodide, 1:96
Nitroprusside, 5:293
Nonideal solutions, 3:225
Nonmetal oxides,
    acidic properties of, 3:109
Nucleophilic substitution, 4:63
Nylon,
    formation of, 1:213
    in triboelectric series, 5:159

Old Nassau clock reaction, 4:29
Olive oil, 3:301
Optical activity,
    and polarized light, 5:163
    in calcite, 5:167
    in lasers, 5:175
    in liquid crystals, 5:171
Orange and black clock reaction, 4:29
Orange electrode, 4:107
Orange tornado,
    from mercury(II) iodide precipitate, 1:271
Osmosis, 3:283, 3:286, 3:390
Ostwald process, 2:214
Overhead projector, 2:147, 2:297, 3:33, 3:41,
    3:50, 3:58, 3:65, 3:103, 3:109, 3:125,
    3:128, 3:173, 3:254, 3:329, 3:353, 3:386,
    3:390, 4:97, 5:141, 5:146, 5:163, 5:167,
    5:181, 5:189, 5:204, 5:208, 5:215, 5:223,
    5:243, 5:261, 5:277, 5:287, 5:293, 5:304
Oxalyl chloride,
    chemiluminescence from, 1:153
Oxides,
    acidic and basic properties of, 3:109

Oxygen,
  bonding in molecular, 1:141
  chemiluminescence from singlet, 1:133
  formation of ozone from, 5:308
  liquid, 2:147
  phosphorescence quenching by, 5:52, 5:233
  preparation and properties of, 2:137
  reaction of dextrose with, 2:142
  reaction of hydrogen with, 1:106, 4:123
  reaction of nitrogen(II) oxide with 2:167
  reaction of sulfur with, 2:190
Oxygen-hydrogen cell, 4:123
Ozone, 5:308

Painting with indicators, 3:47
Paper,
  reaction of sulfuric acid with, 3:70
Paper cup,
  boiling water in, 3:239
Paramagnetism,
  of liquid oxygen, 2:147
Partial pressures, 2:41
Penny,
  as catalyst, 2:216
  coated with zinc or tin, 4:263
Periodate,
  reaction of thiosulfate with, 4:86
Peroxyacetone, 1:46
Peroxydisulfate,
  reaction of iodide with, 4:44
pH,
  effect on protein solubility of, 3:188
pH changes,
  upon electrolysis of water, 4:156, 4:170
pH indicator,
  from plants, 3:50
  universal, 3:41, 3:103, 3:109
pH indicator solutions,
  preparation of, 3:27
Phase separation,
  of liquids, 3:266
Phenol,
  acid properties of, 3:158
  polymerized with formaldehyde, 1:219
Phenolphthalein,
  polymerized with terephthalate, 1:229
Phenyl oxalate,
  chemiluminescence from, 1:146
Phosphorescence, 5:46
Phosphorus,
  chemiluminescence from, 1:186
  combustion of, 1:74
  combustion under water of, 1:40
  reaction of bromine with, 1:72
  reaction of chlorine with, 1:70
  reaction of potassium chlorate with, 1:99

Phosphorus pentoxide,
  reaction of water with, 1:43
Photobromination, 5:56
  of hydrocarbons, 5:304
Photocell, 5:274
Photochemical reaction,
  of chlorine and hydrogen, 1:121
  color change from, 5:249, 5:246, 5:261,
    5:277, 5:283, 5:287, 5:291, 5:293,
    5:300, 5:304
Photoelectric effect, 5:28
Photofluorescence, 2:284
Photoinitiation, 1:121
Photoisomerization, 5:53
Photons, 5:31
Photoreceptor cells, 5:68
Photosynthesis, 5:59
Photovoltaic cells, 5:58
Planck hypothesis, 5:26
Plants,
  acid-base indicators from, 3:50
Plastic sulfur, 1:243
Platinum black,
  formation of, 4:123
Plexiglas, 1:237
Polarized light, 5:19
  effect of sugar solution on, 3:386
  from a laser, 5:175
  in calcite, 5:167
  in LCD, 5:171
  variation with wavelength, 5:163
Pollution control, 3:341
Poly(vinyl alcohol),
  formation of gel with, 3:362
Polyacrylamide,
  refraction with, 5:141
  superabsorbency of, 3:368
Polyethylene oxide,
  properties of solution of, 3:333
Polymer,
  acrylic, 1:235, 1:237
  iridescence of, 5:154
  refraction with, 5:141
  photochromic, 5:264, 5:300
Polymer solution,
  properties of, 3:333, 3:335
Polystyrene, 1:241, 5:154, 5:155, 5:261, 5:263,
    5:265, 5:266
  as photochromic matrix, 5:264
Polysulfide, 1:245
Polyurethane, 1:216
Polyvinyl alcohol, 3:362
Polyvinyl chloride,
  acid combustion products from, 3:121
  generating static charge with, 5:156
Pop bottle,
  explosive reaction of hydrogen in, 1:106

Porous cup,
    flow of gas through, 2:54
Potassium, 5:110
Potassium carbonate,
    salting out with, 3:266
Potassium chlorate,
    reaction of phosphorus with, 1:99
    reaction of sugar with, 1:79
Potassium chloride, 5:108
Potassium hexacyanoferrate(III), 5:283, 5:287
Potassium iodide,
    electrolysis of aqueous, 4:174, 4:189
    phosphorescence quenching by, 5:227
Potassium permanganate,
    reaction of glycerine with, 1:83
Powder explosions, 1:103
Priestley, Joseph, 2:167
Primary colors, 5:181, 5:189
Protein,
    effect of pH on solubility of, 3:188
    reaction of nitric acid with, 3:74
PVC (polyvinyl chloride),
    acid from combustion of, 3:116
    triboelectric properties of, 5:156
Pyrogallol, 1:175
    oscillating oxidation by bromate of, 2:289
Pyrophoric lead, 1:93

Raoult's law, 3:242
Rayon, 1:247
Receptive fields, 5:78
Rectifier,
    aluminum, 4:198
Red cabbage,
    acid-base indicator from, 3:52
Reference electrode, 4:107
Reflection, 5:7
Refraction, 5:10, 5:135, 5:145, 5:167
Resorcinol,
    polymerized with formaldehyde, 1:222
Retina, 5:72
Retinal, 5:67
Rhodopsin, 5:67
Rubber, 1:245
Ruthenium,
    chemiluminescence from, 1:194
    electrogenerated chemiluminescence of, 5:118

Saponification, 3:58
Sealed-bag reactions, 3:100
Sebacoyl chloride,
    as monomer of nylon, 1:213
Seltzer, 3:280
Semiconductors, 5:49
    conductivity of, 5:114, 5:275
Sensitizers of emission, 1:133, 1:153, 1:180,
    1:202

Silica gel, 3:372
Silicate garden, 3:379
Silver, 5:30, 5:152, 5:153, 5:291, 5:292
    half cell, 4:101
Silver(I),
    complexes of, 1:293, 1:307
    concentration cell, 4:144
    electrolysis of aqueous, 4:236
    photoreduction of, 5:291
Silver chloride, 5:292
Silver halide,
    photoreduction of, 5:291
Silver mirror, 4:240
Silver nitrate,
    electrolysis of aqueous, 4:205
    reaction of dextrose with, 4:240
Silver series, 1:307
    potentiometric, 4:144
Simethicone,
    as antifoaming agent, 3:345
Singlet state,
    definition of, 1:127, 5:44
    in lightstick, 5:63
    phosphorescence quenching by, 5:52
Siphon,
    tubeless, 3:333
Slaking of lime, 1:19
Slime, 3:362
Smoke rings, 2:211
Soap,
    as liquid crystal, 5:174
    making, 3:61
Soap bubbles,
    colors of, 3;381, 5:133
    exploding, 1:108
    floating, 2:228
    methane-filled, 1:113
    shapes of, 3:381
Soap films,
    shape and color of, 3:381
Soda water, 3:280
Sodium,
    electrolytic production of, 4:247
    emission spectrum of, 5:103, 5:108
    reaction of chlorine with, 1:61
Sodium acetate,
    buffer, 3:155
    crystallization of, 1:27
Sodium chloride,
    electrolysis of aqueous, 4:181
    emission from, 5:108
    fluorescence quenching by, 5:227
Sodium nitrate,
    electrolysis of molten, 4:247
Sodium nitroprusside, 5:293
Sodium peroxide,
    reaction of aluminum with, 1:59
    reaction of sulfur with, 1:57

Sodium thiosulfate,
    crystallization of, 1:31, 1:36
    disproportionation of, 3:353
    heat of crystallization of, 1:36
Solids,
    reaction between, 1:10
Solubility of a salt
    effect of temperature on, 3:269
Solubility of gases,
    effect of temperature on, 3:280
Solvent effect, 5:255
    on color of solution, 5:261
    on rate of reaction, 4:56
Spiropyran, 5:261
Spontaneous combustion, 1:74, 1:93
Static electricity,
    effect on stream of liquid of, 3:329
Steam,
    superheated, 2:93
Sterno, 3:360
Styrene,
    polymerization of, 1:241
Styrofoam, 3:348
Sublimation, 2:78
    of dry ice, 2:26
    of iodine, 1:49
Suds, 3:345
Sugar,
    dehydration by sulfuric acid, 1:77
    reaction of potassium chlorate with, 1:79
Sugar solution,
    effect of polarized light on, 3:386, 5:163, 5:175
    osmotic pressure of, 3:286
Sulfide,
    precipitation of copper(II) with, 4:140
Sulfite,
    aqueous chemistry of, 4:9
    reaction of formaldehyde with, 4:70
Sulfur,
    combustion of, 2:184
    dissolving in sodium hydroxide, 3:61
    plastic, 1:243
    polymerized with dichloroethane, 1:245
    precipitation of, 4:77
    reaction of iron with, 1:55
    reaction of oxygen with, 2:190
    reaction of sodium peroxide with, 1:57
    reaction of zinc with, 1:53
Sulfur dioxide,
    acidic properties of, 2:184, 3:120
    preparation of, 2:184
    volume produced by burning sulfur, 2:190
Sulfuric acid,
    dehydration of sugar by, 1:77
    heat of dilution of, 1:17
    reaction of paper with, 3:74
    reaction of water with, 3:75
    titration with barium hydroxide, 3:152

Sunset,
    sulfur, 3:353, 5:18, 5:19, 5:160, 5:162
Superabsorbent polymer, 3:368
    refraction in, 5:141
Supercooling,
    of sodium acetate, 1:27
    of sodium thiosulfate, 1:31
    of thymol, 1:34
Surface spreading,
    of oil on water, 3:301
Surface tension, 3:301
    and absorption of water, 3:305
Synapse, 5:73

Tannic acid,
    oscillating oxidation by bromate of, 2:294
Temperature,
    dependence of emission on, 5:93, 5:98
    effect on color of gas, 2:180
    effect on color of solution, 1:282
    effect on LEDs, 5:111
    effect on phosphorescence, 5:241
    effect on photochromism, 5:300
    effect on volume of gas, 2:24
Terephthaloyl chloride,
    polymerized with phenolphthalein, 1:229
Tesla coil,
    ignition with, 2:216
Thermal expansion,
    of liquids, 3:234
Thermite reaction, 1:85
Thermometers, 3:234
Thiocyanate ions,
    complexes of cobalt(II) and, 1:280
Thionine, 5:249
Thiosulfate,
    aqueous chemistry of, 4:10
    disproportionation, 4:77
    reaction of periodate with, 4:86
Thiourea, 5:293
Thixotropy, 3:364
Thymol, 1:34
Tin(II) chloride,
    electrolysis of aqueous, 4:205
    in photochromic solution, 5:256
Tin-coated coin, 4:263
Tissue paper,
    absorbency of, 3:305
Titration,
    acid-base, 3:152, 3:162
Titration curve,
    recording of a, 3:167
Tornado,
    from mercury(II) iodide precipitate,
        1:271
Traveling waves, 2:297
Triboluminescence, 5:65

Triplet state,
  definition of, 1:127, 5:44
  phosphorescence from, 5:52, 5:232
Tyndall effect, 3:353
  cause of blue sky, 5:18
  polarization of light from, 5:160

Universal indicator, 3:41
Urea,
  polymerized with formaldehyde, 1:227

Vapor pressure, 1:5, 2:85
  and boiling point, 2:81
  as function of temperature, 2:75, 2:78
  of liquids, 3:242
  of solution, 3:254
Vaporization, 2:75
Vegetable oil,
  refractive index of, 5:135
Viscosity,
  of liquids, 3:313
Visible radiation,
  table of wavelength and energy of, 1:126, 5:2
Volcano,
  ammonium dichromate, 1:81
Volume change upon mixing solutions, 3:276
Volume of gas,
  temperature effect on, 2:24
Volume of mixing, 3:225

Water,
  electrolysis of, 4:156, 4:189
  expansion upon freezing of, 3:310
  hard, 3:345
  heat capacity of, 3:239
Water glass, 3:372, 3:379

Zinc,
  amphoteric properties of, 3:128
  half cell, 4:101
  photoelectric effect on, 5:156
  reaction of acid with, 4:130
  reaction of ammonium nitrate with, 1:51
  reaction of iodine with, 1:49, 4:134
  reaction of sulfur with, 1:53
Zinc(II) sulfate,
  electrolysis of aqueous, 4:244
Zinc sulfide,
  as dopant, 5:47
  in glow-in-the-dark material, 5:226, 5:236,
    5:240, 5:242
Zinc-acid cell, 4:130
Zinc-coated coin, 4:263
Zinc-copper cell, 4:119
Zinc-iodine cell, 4:134

Illustration Credits

American Chemical Society, from Jerry A. Bell, *Chemistry: A General Chemistry Project of the American Chemical Society*

Pages 6, 16 (figures 17 and 18), 17, 22, 25, 29, 36, 37

Robert A. Bloodgood, University of Virginia School of Medicine

Page 73 (figure 64b)

Dr. Christine Curcio, Department of Opthalmology, University of Alabama at Birmingham School of Medicine, from C. A. Curcio, K. R. Sloan, R. E. Kalina, A. E. Hendrickson, "Human Photoreceptor Topography," *Journal of Comparative Neurology, 292,* 497–523 (1990).

Page 75 (figure 67)

© Art Wolfe / Artwolfe.Com

Pages 81 (figure 72a), 82 (figure 73b)

© Brian Kenney

Page 82 (figure 73a)

American Chemical Society, from Wujian Miao and Allen J. Bard, "Electrogenerated Chemiluminescence. 77. DNA Hybridization Detection at High Amplification with [Ru(bpy)$_3$]$^{2+}$-Containing Microspheres," *Analytical Chemistry, 76*(18), 5379–5386 (2004).

Page 121

Coatings by Sandberg, Inc., manufacturer of dichroic glass

Page 152

© Copyright the Trustees of The British Museum

Page 153 (figure 2)